RELATIVE ROLE OF EUSTASY, CLIMATE, AND TECTONISM IN CONTINENTAL ROCKS

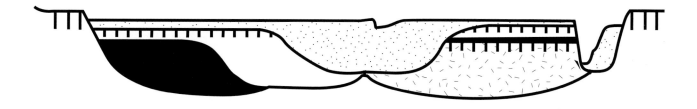

Edited by
Keith W. Shanley
Amoco Production Company, Box 800, Denver, CO 80202

and

Peter J. McCabe
U.S. Geological Survey, MS 939, Box 25046, Federal Center, Denver, CO 80225

Tulsa, Oklahoma, U.S.A.

July, 1998

SEPM thanks the following for their generous contributions to Relative Role of Eustacy, Climate, and Tectonism in Continental Rocks:

Gary Kocurek
The University of Texas
Austin, Texas

The U.S. Geological Survey
Denver, Colorado

RELATIVE ROLE OF EUSTASY, CLIMATE, AND TECTONISM IN CONTINENTAL ROCKS:
AN INTRODUCTION

The renaissance in stratigraphy over the last two decades has been largely driven by the belief that stratigraphic packaging is determined by allocyclic controls. An understanding of the controls on stratigraphy allows us to make better predictions about the nature and geometry of strata within areas of basins where data is more limited. From an economic standpoint, one can better estimate the potential for oil, gas, coal, or mineral accumulations. Ten years ago discussions concerning the importance of driving mechanisms was polarized with many stratigraphers tending to emphasize the importance of a single allocyclic control. Cyclostratigraphers emphasized the importance of climatic cycles. Event-stratigraphers studied the role of major events such as meteorite or comet impacts, volcanic episodes or widespread oceanic anoxia. Tectonostratigraphers examined the role of tectonism. Sequence stratigraphers, who divided strata on the basis of time-significant surfaces such as unconformities and marine-flooding surfaces, demonstrated the importance of eustasy. Although some zealots remain, most stratigraphers now recognize the importance of taking a more holistic approach to understanding allocyclic controls. A special session was convened at the 1994 Denver annual meeting of the AAPG/SEPM entitled "Allocyclic controls on nonmarine stratigraphy". This session was enthusiastically attended and featured papers that demonstrated a wide range of approaches to developing an understanding of alluvial architecture. This volume represents a collection of these papers and as such brings together the results of research where authors have examined the relative importance of eustasy, climate, and sediment supply in determining the nature of lithologies and the style of packaging of continental strata. We hope that readers will appreciate the complexities of nonmarine deposits and will begin to understand that the relative importance of allocyclic controls on architecture vary both in time and space.

Accommodation is the space available for sediment accumulation. The magnitude and rates of change of accommodation are critical in determining the packaging of strata. In the shallow-marine realm changes in accommodation are linked directly to changes in relative sea-level, which is in turn a function of eustasy and local rates of subsidence or uplift. In the nonmarine realm accommodation is somewhat harder to define and can be a function of (1) climate and tectonism in the source area, (2) the climate and tectonism at the site of deposition, (3) the climate and tectonism downstream of the depositional site, and (4) relative sea-level changes. The dynamic surface which defines the upper limit of accommodation is encompassed by such concepts as lake level, the graded profile of a river, and the water table in a wet eolian or organic system. Various viewpoints on the controls on accommodation and the application of sequence stratigraphic concepts to continental strata are reviewed in Shanley and McCabe (1994).

One major consequence of the new understanding of the role of allocyclic factors in sedimentation has been a reexamination of long-held beliefs. For example, twenty years ago, in the age of facies models, the interconnectedness of fluvial channel sandstones was generally attributed to the type of river system. Channel sandstones of braided rivers were thought to be more interconnected than those of meandering rivers because of the greater amount of overbank sedimentation associated with lower gradient and more sinuous rivers. Today, we consider the rate of increase in accommodation as an equally important factor. Slower rates of rise in accommodation will result in more extensive reworking of fluvial strata and a higher preservation potential and connectivity of channel deposits, no matter what the fluvial system.

Several papers in this volume address the controls on sedimentation in Quaternary and Recent fluvial systems. Observations and conclusions presented in these papers should be of particular interest to stratigraphers and sedimentologists studying more ancient alluvial strata and their correlation to marine strata. Hovius draws data from many rivers throughout the world to assess the factors which control the rates of sediment supply by rivers. Ethridge, Wood and Schumm critique current sequence stratigraphic models of alluvial strata. They discuss geomorphological studies which demonstrate that it is difficult to distinguish between the effects of various controls on fluvial systems. Their conclusions are reminiscent of the work of Kendall and Lerche (1988) who found that the effects of tectonics and eustasy could not be separated in marine strata. An examination of the coastal plain of Texas by Blum and Price allows a comparison of the relative role of eustasy and climate in controlling alluvial plain strata during the Quaternary. Their work documents the formation of widespread erosion surfaces within alluvial strata that reflect changes in sediment supply caused by climate changes that are unrelated to changes in relative sea level.

Several papers in this volume also discuss the role of allocyclic controls on the deposition of Mesozoic and Tertiary fluvial strata. Through a study of the Mannville Group (Cretaceous) of Alberta, Cant shows how the character of alluvial facies are determined by their sequence stratigraphic settings and discusses the importance of varying rates of subsidence. Dalrymple, Prosser, and Williams discuss the regional evolution of different architectural styles in fluvial strata using the concept of the stream equilibrium profile with examples drawn from the Statfjord Formation (Triassic-Jurassic) in the North Sea. Legaretta and Uliana discuss the formation of sequences of fluvial strata in the Upper Cretaceous of the Neuquén basin of Argentina: a succession characterized by internal drainage. In a study of the Morrison Formation (Jurassic) of southern Utah, Robinson and McCabe discuss the allocyclic controls on evolution of a thick braided river succession and the resulting alluvial architecture. The evolution of an incised valley system in a tectonically-active area is described by Dam and Sønderholm from the Paleocene of the Nuussuaq Basin in Greenland. Papers by Holbrook and White, on the Dakota Group (Cretaceous) of New Mexico, and Way and others, on the Lakota and Cloverly

Formations (Cretaceous) of Wyoming, Montana, and South Dakota, demonstrate how variations in fluvial architecture may be controlled by tectonic movements during deposition.

Distinguishing the relative importance of tectonics, climate, and relative sealevel in ancient fluvial systems remains an extremely difficult problem. The nature of Quaternary and Recent deposits are such that we can, with some measure of sophisitication determine the relative importance of these controls. The ability to acquire radiometric age dating, combined with an independent understanding of both climate history as evidenced by changes in flora and fauna as well as sealevel as evidenced by isotope and paleotologic studies offer researchers the opportunity to place nonmarine stratigraphy in a broad, relatively well-constrained perspective. In more ancient alluvial strata, however, these problems are not so easily addressed. In these strata, careful attention should be given to (1) the development of paleosols within interfluve deposits, and (2) the stratal correlation of nonmarine and shallow marine deposits. This may allow one to distinguish between widespread erosion surfaces that are controlled by changes in sediment flux (which is often associated with climatic change) from surfaces that reflect base-level changes driven by changes in relative sealevel. Trying to resolve the relative importance of these allocyclic controls is of more than passing academic interest. An understanding of these issues will contribute to better predictive models that can be used in the projection of reservoir trends in the subsurface.

Facies models used to show coals accumulating as peats in low-lying mires on floodplains, on coastal plains, and on delta-tops. These environments, however, are generally not suitable for the accumulation of thick low-ash peat that is a prerequisite for the formation of coal. Recognition of the importance of raised mires resolves some problems, but sequence stratigraphic concepts offer a whole new perspective on the relationship between coal and associated clastic strata. For example, there may be a significant temporal difference between many coals and underlying clastic strata, and, in some cases, the two may be separated by an unconformity. It is no longer realistic to assume that coal (as peat) accumulated in close proximity to environments indicated by associated clastic rocks. The timing of clastic versus organic accumulation in an area is determined by the interplay of allocyclic controls. Wet/dry climatic cycles and water-table fluctuations associated with changes in relative sea-level can determine where and when low-ash peat will accumulate. The paper in this volume by Diessel demonstrates how sequence stratigraphic concepts can be used to explain the distribution of various coal facies in the Permian of the Sydney Basin in Australia and the Miocene of Germany.

Changes in lake-level are strongly influenced by local climatic fluctuations and tectonism. Consequently, fluctuations are often of a higher temporal frequency than those of sea-level and they can be of greater magnitude. The resulting strata, may be considered in a sequence stratigraphic framework of highstand, lowstand and transgressive systems tracts. However, lakes may have a very different character at different lake levels. For example, a lake may be a playa with evaporite deposition at lowstand, a saline lake with mixed carbonate and clastic deposition at an intermediate stand, and a freshwater lake dominated by deltaic and turbidite sedimentation at highstand. In this volume Milligan and Chan discuss variations in the style of coarse-grained deltas in highstand and lowstand systems tracts of the Pleistocene of Lake Bonneville, Utah—the location of G.K. Gilbert's classic study. The paper by de Wet, Yocum and Mora discusses the relative role of climate and tectonism in determining the lithologies and stratigraphic packaging of the Triassic in the Gettysburg basin of Pennsylvania.

The interpretation of eolian facies has also been influenced by the realization of the importance of allocyclic controls. Physical and temporal changes in the elevation of groundwater tables significantly impact the sedimentology of eolian systems as well as the nature and degree of preservation of these deposits. The interpretation of super surfaces, for example, can be related to climatic and tectonic controls in both wet and dry eolian systems (Kocurek and Havholm, 1993). The paper in this volume by Carr-Crabaugh and Kocurek shows how this type of approach can be used to apply sequence stratigraphic concepts to a widespread eolian deposit—the San Rafael Group (Jurassic) of Utah.

Major advances have been made in the last few years in understanding the role of allocyclic controls in determining the nature and stratigraphic style of continental strata. There is the potential, however, for significantly expanded research in this field. In particular, the development of sequence stratigraphic models for nonmarine strata are still in their infancy compared to models developed for the marine realm. Such models could be particularly useful in (a) predicting the location and degree of interconnectedness of oil and gas reservoirs in stratigraphic traps associated with fluvial and lacustrine settings, (b) predicting variations in coal bed thickness and quality, (c) understanding and predicting flow patterns in aquifers and hydrocarbon reservoirs in nonmarine strata, and (d) predicting the location of mineralization associated with channel sandstones. We hope that the papers in this volume not only shed new light on how allocyclic factors control deposition of continental strata but also point the way for future research in this exciting field.

Keith W. Shanley
Amoco Production Company, Box 800,
Denver, Colorado 80202

Peter J. McCabe
U.S. Geological Survey, MS 939, Box 25046, Federal Center,
Denver, Colorado 80225

REFERENCES

KENDALL, C. G. ST. C. AND LERCHE, I., 1988, The rise and fall of eustasy, *in* Wilgus, C. K., Posamentier, H., Ross, C. A., and Kendall, C. G. ST. C., eds., Sea-level Changes: An Integrated Approach: Tulsa, Society of Economic Paleontologists and Mineralogists Special Publication 42, p. 3–17.

KOCUREK, G. AND HAVHOLM, K. G., 1993, Eolian sequence stratigraphy—a conceptual framework, *in* Weimer, P., and Posamentier, H., eds., Siliciclastic Sequence Stratigraphy—Recent Developments and Applications: Tulsa, American Association of Petroleum Geologists Memoir 58, p. 393–409.

SHANLEY, K. W. AND MCCABE, P. J., 1994, Perspectives on the sequence stratigraphy of continental strata: American Association of Petroleum Geologists Bulletin, v. 78, p. 544–568.

CONTENTS

PART I
ALLUVIAL DEPOSITS

CONTROLS ON SEDIMENT SUPPLY BY LARGE RIVERS

NIELS HOVIUS*

Department of Earth Sciences, University of Oxford, Parks Road, Oxford, OX1 3PR, England

ABSTRACT: Sediment supply to continental basins is determined at first order by the rate at which source areas are eroded. Factors govering the rate of denudation are investigated using data from 97 intermediate and large drainage basins around the world. By multiple regression, five variables are selected from a larger set of climatic, topographic and hydrological estimators as being most effective at explaining global variance in specific sediment yield. The five are drainage area, maximum height of catchment, specific runoff, mean annual temperature and temperature range. These factors may be most adequate at describing the influence of weathering, hillslope erosion and fluvial transport on the rate of denudation, but they only explain half of the observed variance of that rate. It is demonstrated that rate of uplift of rock is a control on regional denudation rates of an importance similar to, or greater than that of climatic and topographic factors. Sediment supply reflects the way in which surface processes have interacted with the tectonic input of material to the regolith. Estimates of sediment supply can be based on knowledge of precipitation, relief and runoff but should be refined using information on the rate of uplift of rocks. Alternatively, the volume of sediment supplied may be estimated from the tectonic setting and size of the source area of the material.

INTRODUCTION

The development of fluvial stratigraphy in sedimentary basins in a continental setting is controlled at first order by three factors: accommodation space, sediment supply and the hydraulic characteristics of the fluvial system. Over the past few decades, the interpretation of internal structure and arrangement of bed forms and the geometry of strata in continental facies, in terms of the interaction between fluid flow and sediment particles, has been one of the central issues in the sedimentological literature. Although the nature and geometry of these architectural elements are mostly determined by autocyclic factors, their large-scale arrangement seems to be a function of the interplay between different allocyclic controls. Sequence stratigraphic and numeric modelling studies have eagerly employed allocyclic processes that affect accommodation space to explain continental stratigraphic relations, ignoring, at large, other factors that may play an equally critical role (Posamentier et al., 1988; Posamentier and Vail, 1988; Aubrey, 1989; Van Wagoner et al., 1990; Flint, 1993; Kvale and Vondra, 1993). This approach may be partly inherited from work in the marine realm, which is still biased towards the importance of relative sea-level changes. To a large extent though, it is due to the fact that inclusion of other controlling factors in conceptual models of sedimentary basin fills would involve enormous complications.

Recently, a number of workers have stressed the role of sediment supply in the development of sedimentary sequences. Several general discussions of continental sequence stratigraphy and its conceptual basis (Cloetingh et al., 1993; Puigdefabregas, 1993; Shanley and McCabe, 1994) have noted that over the past few years the role of sediment supply, classically considered as one of the predominant controls on stratigraphy, has been ignored. Schlager (1993) demonstrated that changes in sediment supply may have as large an effect on sequence geometry as changes in accommodation space. Furthermore, the importance of the interplay between the rate of change in accommodation space and the rate of sediment supply as the main control on stratigraphy has been illustrated by a number of field and modelling studies of sedimentary basin infill (e.g., Jervey, 1988; Galloway, 1989; Aigner et al., 1991; Swift and Thorne, 1991; Sinclair and Allen, 1992; Blum, 1994; Burgess and Allen, 1996). Having re-established the significance of sediment supply to the development of sedimentary sequences, quantification of this factor has now become of fundamental importance to progress in sequence stratigraphical work and numerical modelling of various sedimentary environments.

To make valid assumptions about the role of variations in sediment supply in the development of the basin fill, it is necessary to consider factors that govern the flux of clastic material from source areas into depositional basins. At first order, sediment flux is determined by the rate of denudation of the source area. This paper investigates the primary factors that govern denudation of the world's major drainage basins. Data is presented on the sediment yield, sediment transport and source area characteristics of 97 intermediate and large river catchments ($>2.5 \times 10^4$ km^2), including most of the largest rivers of the world. The drainage basins of these rivers are spread across a range of climatic, tectonic, topographic, and geomorphic zones, so that insight can be gained in the relative importance of each of these factors to the functioning of sediment routing systems. Numerous workers have tried to explain the global pattern of sediment yield both in terms of relief of drainage basins (Ahnert, 1970, 1984; Pinet and Souriau, 1988; Einsele, 1992; Milliman and Syvitski, 1992; Summerfield and Hulton, 1994), and in terms of amount of precipitation (Langbein and Schumm, 1958; Fournier, 1960; Douglas, 1967; Wilson, 1973; Jansson, 1982, 1988; Ohmori, 1983). Not all of these attempts have been equally effective. Here, I will use multivariate statistical techniques to investigate the major controls on sediment yield on a regional to subcontinental scale.

SEDIMENT YIELD

The distinction between sediment production, sediment load, sediment deposited and sediment yield is an important one in the use of any dataset on the flux of sediment through fluvial systems. The total amount of sediment eroded from a source area is termed the sediment production. Depending on the distribution and the magnitude of the accommodation space in the sediment routing system, part of the eroded material will be deposited; the remainder is transported through the river system as a sediment load. The sediment load at the river mouth is termed the sediment yield, representing the net-denudation of a catchment. The sediment leaving a drainage basin may eventually become part of the depositional component in a delta or deeper marine environment. For stratigraphic modelling purposes, another factor, sediment supply, is most relevant. Due to differences in the rate of sediment production and transport,

*Present address: Department of Geosciences, Pennsylvania State University, 540 Delke Bldg., University Park, PA 16802
Relative Role of Eustasy, Climate, and Tectonism in Continental Rocks, SEPM Special Publication No. 59

sediment supply varies both spatially and temporally within a drainage basin and is very hard to measure directly. Using figures for sediment supply based on measurements of yield, load, or deposition rates, without being aware of the fundamental differences between these variables, may lead to important misinterpretations.

Material can be transported into sedimentary basins by a number of processes, of which fluvial processes are, at present, by far the most important. Over the past two decades, a number of data sets have been published that contain information on sediment transport by major rivers (Holeman, 1968; Meybeck, 1976, 1979; Milliman and Meade, 1983; Walling and Webb, 1983; Berner and Berner, 1987; Pinet and Souriau, 1988; Milliman and Sivitsky, 1992). These data sets, however, focus on the sediment load of rivers, at the interface of the continental and the marine realm and contain little or no detail on the source area or the transport history of the material. Only in a limited number of cases is it known where exactly sediment was derived from, how it was liberated, what processes were responsible for its initial erosion and in what manner it was provided to the fluvial system. Each of these steps is of prime importance to the quality and quantity of material being put into the fluvial system. Most geomorphic processes involved in these steps act on a scale of 10^{-2} to 10^3 m and are only quantifiable for small drainage basins. The input of sediment into continental basins by intermediate and large river systems remains a poorly understood quantity.

As few attempts have been made to document sediment production in a systematic way, it is, at present, hard to put realistic brackets on sediment supply to subaerial deposition areas. Instead, to infer values for sediment input into the fluvial system, we have to resort to the little we know about the sediment load of rivers in their downstream reaches. The sediment load at the mouth of a river reflects the sum total of all erosional and depositional processes that occur within the drainage basin. With an increase in catchment area, there is an increase of the relative importance of depositional processes. This is illustrated by the inverse relationship between specific sediment yield (calculated as the amount of erosion from a unit surface area per unit time averaged over the entire drainage basin) and drainage basin area, noted by a number of workers (e.g., Schumm and Hadley, 1961; Milliman and Meade, 1983; Milliman and Syvitski, 1992). The filtering effect of the depositional part of a river system should be kept in mind when analysing sediment load data, and it is clear that any figure for upstream erosion and sediment supply calculated from these data is in fact a minimum estimate. This estimate may deviate from the real value by up to an order of magnitude.

Erosion in the upstream part of a catchment has two components. Material is removed from hillslopes by a range of denudational processes that provide sediment to the fluvial system. Rivers then carry (part of) the sediment out of the source area. The total amount of material leaving the upstream part of a drainage basin will therefore be a function of both the rate of hillslope denudation and erosive power and transport capacity of the fluvial system. The following provides an overview of the components of the sediment supply system

Denudation is generally considered to be a product of the interaction between erosivity (i.e. the potential of denudational process systems to remove material from a certain locality) and the erodibility of the material in that locality. Erodibility depends on the characteristics of the regolith, such as its physical and chemical composition, infiltration capacity, cohesion and angle of friction. Due to this wide range of controlling factors, it is difficult to measure erodibility directly, and no universal method of assessment has been developed (Selby, 1993). However, many regolith characteristics relevant to erodibility depend upon the dominant weathering processes, which are in turn affected by climatic influences and the lithology of the underlying bedrock. Climatic conditions most important to weathering are temperature, temperature range and availability of water. Assuming that the influence of local differences in lithology is averaged out on a regional to subcontinental scale, further consideration of factors affecting the rate and character of weathering will be limited to aspects of temperature and precipitation.

Processes involved in the removal of material from slopes are controlled by a number of climatic and topographic factors. Many denudational processes, such as rainsplash, sheetwash, rill erosion, stream flow, mud- and debris flow and land slide, involve a certain amount of water, while their occurrence and effectiveness is also controlled by the length and angle of a slope. In terms of precipitation, both mean annual precipitation and maximum precipitation rate may be of importance, as some denudational processes depend on the long term availability of water whereas others reflect peak rates of rainfall. Slope aspects, although highly relevant, are more problematic. The topographic information required to assess the importance of local slope to the global variation of denudation rate should be gathered on a scale of several hundred meters to a few kilometers. Detailed digital topographic models are not yet available on a global scale, so it is not possible to assess local slope for the drainage basins under consideration. More general topographic variables such as mean height and maximum height of the drainage basin, and the overall angle of the catchment are used instead.

Transport capacity of a fluvial system can be viewed in terms of slope angle of the river bed and depth of flow. For many rivers, relevant information on flow depth is not readily available, but total annual runoff may be used as a substitute, as long as one does not want to calculate the actual shear stress exerted by the flow.

THE DATA

Information on the amount of sediment transported by major rivers has been compiled by a number of workers. Milliman and Meade (1983) and Milliman and Syvitski (1992) have made frequently cited compilations of data on the transport of solids, while much larger datasets have been generated by Jansson (1988) and Dedkov and Moszherin (1992). Concentrations of dissolved matter in the world's major rivers have been described by Meybeck (1976, 1979, 1988) and Walling and Webb (1983). These, and other (Holeman, 1968; Lisitzin, 1972; Pinet and Souriau, 1988; Meade, 1992; Enos, 1991; Leeder, 1991; Summerfield and Hulton, 1994) compilations, make use of data from reports published by national and international organisations, and from scientific papers in fields ranging from hydrological engineering to soil science to oceanography. The measurement techniques and observation times used to derive these data are not uniform, however, and hence there is great variation in the quality of data available. Often, data are taken from compilations without any notion of their quality, thus recycling potentially inaccurate information.

Modern river sediment loads seldom represent natural quantities. Human activities have caused both increases and decreases in sediment yield. Deforestation and poor soil conservation have enhanced soil erosion over the past several thousand years on a global scale. More recently, urbanization has locally induced subdued erosion rates. Large-scale river training, dredging and mining, irrigation, and hydro-electric works may also have a spectacular influence on sediment yield. Training of natural rivers changes the flow regime, affecting sediment transport capacity. Artificial levees prevent flooding, thus reducing storage of sediment in overbank deposits. Sediment load may be lowered considerably by dredging of river beds, or increased by disposal of mining debris. Irrigation works reduce water discharge in main channels, generally reducing sediment transport capacity. Finally, dams have a profound impact on both water discharge and sediment load of a river. Discharge characteristics may be changed, causing downstream alterations in the sediment transport capacity of a river. Dams also interrupt the sediment flux in a river, trapping material in artificial reservoirs.

Although many workers have documented changes in sediment flux caused by specific human activities, it is often impossible to calculate their combined effect on the sediment yield from a drainage basin. Even when, for instance, pre-dam data are available, the recorded sediment loads will still reflect changes in land use that occurred over a much longer period. Depending on the character and relative importance of human activities, their net effect may be to increase or decrease sediment load with respect to natural quantities. First-order estimates of natural sediment yields exist for some of the major drainage basins of the world. In this paper, I cite estimated natural yields, where available. However, in the majority of cases, anthropogenic components of recorded sediment yields are unknown.

Table 1 lists estimates of the discharge and solid and solute load of 97 major rivers around the world, as measured at or near the outlet of the drainage basin. These data have largely been derived from the previously mentioned compilations, though a number of additions and corrections have been made. Where possible, unrepresentative data have been avoided and, when available, figures are used that have been corrected for human impact. It should be noted that neither discharge, nor sediment load data are based on continuous assessment, such that peak rates of transport and discharge are often not reflected in the average annual totals. Bed load has not been included in the figures for mechanical sediment transport. It is commonly assumed that bed load amounts to at least 10% of the suspended load. There is, however, little scientific evidence supporting this generalization. As bed load is essentially unknown, the figures presented here are for suspended load only.

Table 1 also lists a number of general characteristics of the drainage systems and estimates for a total of 11 topographic, climatologic and hydrologic variables that may influence water discharge and sediment output. Several other variables that may influence sediment yield can be calculated by combining two or more of these parameters. A definition of all the variables considered in the analysis of controls on specific sediment yield is given in Table 2.

General Characteristics Drainage Area

The general drainage area characteristics included in Table 1 are surface area of the drainage basin, length of the main stream of the catchment, length of the drainage basin and surface area of the depositional part of the catchment. There is substantial disagreement in the literature on the surface area of the drainage basins of many major rivers. The surface area depends on the definition of a catchment. Some information sources include peripheral zones, which only contribute surface runoff in case of exceptional rainfall. Other sources are more restrictive and only regard perennial drainage as being part of the catchment area. The figures presented here are for the first, broader definition. Their rounding indicates the degree of accuracy. Sources include UNESCO reports (1972, 1974, 1979), the Atlas of World Water Balance (UNESCO, 1978), the Times Atlas of the World (1988), individual publications, and new data, digitized from topographic maps and corrected for distortions due to cartographic projection method. These same sources were used to estimate river length, while length of the drainage basin was measured from topographic maps (scale 1:1,000,000 to 1:5,000,000) with equidistant projection. Surface areas of the depositional parts of catchments, defined by the extent of Holocene fluvial sediments, were digitized from geological maps (scale 1:200,000 to 1:2,000,000). The remaining parts of a drainage basin are considered to be non-depositional. Due to the scale of the maps, significant generalizations have been made. The listed percentages are likely to be slight underestimates.

Relief

In the present data set, relief of the drainage basin is characterized in terms of the mean and maximum height of the drainage basin, the relief peakedness, the relief ratio and the slope angle of the river bed. Information on the mean elevation of catchments has been derived from Pinet and Souriau (1988) and Summerfield and Hulton (1994). Although slightly different methods were used in these two studies, their figures do not differ substantially in many cases. Where available, preference has been given to data from Summerfield and Hulton (1994). In many drainage basins, most sediment is derived from the highest portion of the catchment, thus it may be more suitable to view sediment load and sediment yield in the light of maximum elevation of the drainage area. This variable can be easily assessed from large scale topographic maps. In some cases, the elevation of the highest mapped contour has been listed, as more detailed information was not available. Relief peakedness has been calculated as the ratio of mean and maximum height, while the relief ratio is defined as the ratio of the maximum height and the length of the drainage basin. The average slope angle of the bed of the trunk stream can be calculated from the length of the stream and the maximum elevation of its head waters.

Climate

Climatic variables thought to be most relevant to this study are total annual precipitation, maximum monthly precipitation,

TABLE 1.—MORPHOMETRIC, CLIMATIC, HYDROLOGIC, TRANSPORT, AND DENUDATION DATA

RIVER	A[1] km²	A_M km²	L_r km	L_b km	S_t Mt/yr	S_D Mt/yr	E t/km²/yr	D mm/ka	H m	H_max m	P mm/yr	P_max mm/mth	T °C	T_range °C	Q m³/s	Q_max m³/s	R mm/yr	R_cf	reference[2]
Amazon	6150000	4	6299	3310	1150	223	187	69	426	6768	1490	260	18	3	200000	211838	1026	0.67	20, 6
Amudar'ya	309000	10	2620	1380	94	27	304	113	—	7459	222	41	9	28	1450	3195	148	0.70	23
Amur	1855000	12	4416	2455	52	20	28	10	571	2499	455	119	−2	46	10300	19413	175	0.38	25, 23
Apalachicola	51800	—	880	521	0.17	1	3	1	—	1458	1223	150	18	19	641	1177	390	0.32	28
Brahmaputra	610000	21	2840	1270	520	61	852	316	2734	7736	2661	664	14	14	19300	45700	998	0.38	25, 14
Brazos	114000	—	1400	1020	31	3	272	101	339	950	810	119	19	22	222	472	61	0.07	28
Burdekin	131000	7	680	520	3	—	23	9	—	1277	640	140	23	11	476	—	115	0.17	25, 23
Chao Phraya	160000	22	1200	700	11	3	69	26	—	2300	1246	283	26	7	824	—	162	0.13	8
Chari	880000	7	1400	920	4	3	5	2	—	3071	1233	263	27	6	1320	3177	47	0.04	22, 23
Colorado (Cal)	640000	—	2333	1300	150	15	234	87	1652	4730	310	47	12	24	32	—	2	—	14
Colorado (Tex)	100000	—	1450	790	13	5	130	48	—	1440	598	75	18	17	634	—	200	0.33	4, 29
Columbia	670000	—	1950	1200	15	35	22	8	1329	3748	568	84	7	24	7930	11939	373	0.62	25, 11
Colville	60900	—	662	520	6	1	98	36	469	2320	—	—	—	—	492	—	255	—	25, 11
Copper	61800	—	360	325	70	—	1133	420	1140	5952	—	—	—	—	1240	—	633	—	42, 23
Danube	815000	16	2860	1250	70	60	86	32	501	3087	960	127	7	21	6660	8921	258	0.29	28
Delaware	22900	—	518	350	0.68	1	30	11	—	1360	900	103	10	27	329	653	453	0.50	28, 23
Dnepr	504000	11	2200	1045	2.1	11	4	1	152	325	526	87	7	27	1650	3255	103	0.20	28, 5
Dnestr	72100	20	1350	695	2.5	4	35	13	—	2058	635	97	7	25	379	586	166	0.27	23
Don	422000	11	1870	770	6	14	14	5	—	367	527	74	8	27	856	2797	64	0.12	31, 34
Ebro	86800	—	930	470	21	—	242	90	402	3404	632	80	12	16	492	2475	179	0.32	28
Elbe	148000	9	1110	705	0.84	—	6	2	—	1603	694	86	7	19	690	1053	147	0.22	19
Fly	64400	24	744	475	70	—	1087	403	—	3993	—	—	—	—	4760	—	2331	—	25, 24
Fraser	220000	—	1110	736	20	11	91	34	1140	4043	300	50	10	26	3550	7014	509	0.25	23, 6
Ganges	980000	28	2510	1560	524	75	535	198	890	8848	1573	430	20	15	11600	34358	373	0.25	28, 10
Garonne	86000	6	650	330	2.2	—	26	10	—	3308	933	114	11	16	600	1010	220	0.24	17
Godavari	287000	1	1500	920	170	—	592	219	413	1300	1122	321	26	13	2920	11816	321	0.29	41
Haiho	50800	—	650	460	81	—	1595	591	—	2870	463	166	12	32	63	—	39	0.09	25
Indigirka	360000	2	1726	1120	14	2	39	14	713	3147	131	33	−16	69	1740	5568	152	0.49	17, 5
Indus	960000	10	3180	1610	250	41	260	96	1855	8611	543	132	17	25	7610	10128	250	0.56	26, 6
Irrawaddy	410000	12	2300	1420	260	92	634	235	758	5881	1878	375	20	9	13600	—	1046	0.76	25, 6
Jana	238000	—	872	815	3	1	13	5	703	3000	163	43	−12	38	920	3738	122	—	17, 5
Kemijoki	37800	3	600	320	0.15	—	4	1	—	807	437	73	−2	27	534	1605	446	0.96	15
Kizil Irmak	75800	8	1151	375	23	—	303	112	—	3916	410	65	10	23	192	292	80	0.18	28
Kolyma	647000	3	3513	1150	6	3	9	3	564	3147	249	62	−12	54	2250	10102	110	0.45	17
Krishna	256000	0	1290	860	65	—	254	94	—	1892	834	227	26	9	1607	6253	198	0.24	41
Kura	188000	2	1360	650	36	5	191	71	—	4480	668	147	8	26	515	904	86	0.13	5
Kuskokwim	116000	—	1080	700	7.5	—	65	24	602	6194	490	90	−4	38	—	—	—	—	28
Lena	2430000	4	4400	2525	12	88	5	2	602	2579	355	94	−5	47	16200	74361	210	0.60	17, 23
Liao He	170000	—	1350	515	41	—	241	89	496	2029	640	170	7	39	190	—	35	0.07	25
Limpopo	440000	—	1600	840	33	—	75	28	766	2322	520	107	21	10	160	1425	11	0.02	25
Loire	120000	5	1110	540	1.5	—	13	5	—	1885	795	84	11	15	—	—	—	—	28
Mackenzie	1448000	4	4240	2270	125	64	86	32	634	3955	390	58	−1	37	9830	—	214	0.56	18
Magdalena	260000	—	1530	1050	220	28	846	313	1203	5493	2670	311	19	11	6980	6933	847	0.32	25, 23
Mahakam	75000	20	—	420	12	—	160	59	—	2988	2000	—	—	—	2000	—	841	—	3
Mahanadi	133000	1	858	630	60	—	451	167	330	1027	1456	425	25	15	1970	—	467	0.32	7
Mekong	810000	22	4500	2950	160	60	198	73	1062	6000	1800	393	24	6	14900	—	580	0.33	25, 23
Meuse	29000	—	925	440	0.70	—	24	9	—	692	968	106	9	15	331	596	360	0.37	28
Mississippi	3344000	—	5985	2220	400	125	120	44	656	4400	612	99	9	28	18400	22730	174	0.26	21, 6, 16
Mobile	57000	—	1064	580	2.3	4	40	15	—	1360	1349	178	17	20	1590	—	880	0.65	28
Murray	910000	24	3490	1000	30	9	33	12	266	2239	582	71	17	14	698	855	24	0.05	25

TABLE 1.—Continued

RIVER	A¹) km²	A_d km²	L_r km	L_b km	S_t Mt/yr	S_d Mt/yr	E t/km²/yr	D mm/ka	H m	H_{max} m	P mm/yr	P_{max} mm/mth	T °C	T_{range} °C	Q m³/s	Q_{max} m³/s	R mm/yr	R_{cf}	reference²)
Niger	1112700	7	4160	1950	32	10	29	11	429	2918	937	254	28	8	6020	—	171	0.18	2,23
Nile	2715000	9	6670	3600	125	18	46	17	662	5110	832	177	25	7	317	1711	4	0.005	39, 25, 23
Ob	2500000	7	5570	2530	16	50	6	2	301	4506	406	68	0	41	12200	32421	154	0.38	17, 23
Oder	112000	35	909	515	0.13	7	1	0.4	—	1603	723	95	7	19	539	776	152	0.23	28, 24
Orange	1020000	0	1860	1285	91	12	89	33	1241	3482	415	75	17	14	2890	—	89	0.22	38, 23
Ord	46000	1	—	400	22	—	478	177	297	1000	530	150	25	13	165	645	113	0.21	9
Orinoco	945000	—	2740	1550	150	39	159	59	456	5493	1300	216	25	2	34900	58822	1165	0.90	28, 6
Parana	2600000	—	4500	2175	112	56	43	16	564	6720	1027	188	19	9	18000	—	218	0.21	37, 23
Pechora	322000	5	1810	760	6.1	7	19	7	147	1894	498	71	-2	36	3360	13810	329	0.67	17, 5
Po	75000	19	691	480	18	10	240	89	793	4810	1202	162	10	20	1490	1936	626	0.56	30
Red (Song Koi)	120000	9	1200	860	123	—	1025	380	420	3000	2768	470	15	12	3810	2909	1001	0.37	40
Rhein	225000	2	1360	725	0.72	17	3	1	—	4158	1121	129	7	18	2243	—	314	0.28	28
Rhone	99000	6	810	540	60	—	606	224	754	4810	960	113	9	18	1550	1898	494	0.52	12
Rio Colorado (Arg)	65000	—	1000	1460	6.9	—	106	39	—	6960	342	65	15	17	95	240	4	0.01	28
Rio Grande	670000	—	2870	1725	30	2	45	17	1279	4295	160	67	19	19	308	940	78	0.10	14
Rio Grande Santiago	125000	19	960	650	1	—	8	3	—	4577	774	201	18	7	951	—	231	0.56	14
Rio Negro (Arg)	130000	—	729	880	13	—	100	37	745	4800	423	76	12	15	285	—	50	0.05	28
Rufiji	178000	—	1400	625	17	22	96	36	912	2959	940	218	18	6	678	1170	293	0.28	25
Sacramento	73000	—	610	385	3	—	41	15	—	3187	1052	212	15	19	678	—	293	0.28	
Salween	325000	5	3060	1725	100	—	308	114	—	6070	—	—	—	20	9510	—	923	—	
San Joaquin	80100	0	560	450	1	1	12	4	—	4420	525	105	10	24	123	211	48	0.09	
Sanaga	135000	2	860	660	5.9	—	44	16	—	2000	1690	724	24	5	2069	5658	483	0.29	32
Sao Francisco	640000	6	2800	1510	6	—	9	3	609	1800	1145	232	22	15	3080	5139	152	0.13	25
Seine	78600	6	780	370	1.1	12	14	5	—	902	711	77	9	1	685	—	154	0.39	28, 24
Senegal	441000	2	1430	900	1.9	1	4	1	—	1000	665	235	29	—	761	3184	54	0.08	13, 28, 24
Sepik	81000	30	825	425	80	—	988	366	—	4500	—	—	—	35	2440	13626	950	0.59	35
Sevemaya Dvina	350000	—	—	850	4.5	14	13	5	119	200	514	72	-1	32	3360	—	303	0.10	17, 5
Shatt al Arab	1050000	19	2760	1475	103	18	98	36	669	4168	498	97	17	26	1460	—	44	0.48	23, 1
St. Lawrence	1185000	—	3060	1650	4	59	3	1	265	1917	794	99	6	31	14300	—	381	—	28, 6
Susitna	50300	—	454	370	25	—	497	184	1031	6190	1060	120	14	21	1270	2087	796	0.42	25
Susquehanna	72500	—	733	445	1.8	—	25	9	—	950	295	51	9	28	1034	1011	450	0.30	28
Syrdar'ya	219000	12	2210	1440	12	12	55	20	555	5880	317	99	27	4	581	372	84	0.19	5
Tana	91000	—	720	470	32	—	352	130	—	5200	497	84	25	25	171	—	59	—	25
Terek	43200	17	623	390	24	3	556	206	—	5642	497	69	9	37	301	1409	40	0.11	5
Ural	237000	10	2430	1020	3	3	13	5	—	1000	364	84	2	—	301	—	40	0.11	5
Uruguay	240000	—	—	1085	11	8	46	17	—	2000	1534	182	17	1	5010	7530	658	0.43	28, 23
Vistula	198000	33	1014	600	2.5	13	13	5	—	2499	740	114	7	13	1044	1910	166	0.25	28, 24
Volga	1350000	7	3350	1640	26	77	19	7	—	1638	489	68	2	32	8400	22421	196	0.42	23
Volta	394000	5	1600	980	19	3	48	18	—	500	1046	287	28	7	1270	5120	102	0.10	28, 24
Weser	46000	12	724	375	0.33	—	7	3	670	1142	848	86	7	17	313	475	215	0.26	28
Xi Jiang	464000	5	2129	1150	80	—	172	64	670	2500	1314	230	19	20	9510	15985	646	0.49	43, 36
Yangtze	1940000	—	5520	2730	480	132	247	91	1688	6800	1173	235	11	20	28500	47300	463	0.40	43, 6
Yellow (Huang He)	980000	—	4670	2070	120	226	122	45	1885	5500	484	144	11	29	1550	2858	50	0.10	27, 6
Yenisey	2580000	9	5550	2250	13	22	5	2	749	3492	439	84	-3	41	17800	77671	218	0.50	17, 6
Yukon	855000	—	3000	2140	60	65	70	26	741	6194	236	49	-5	36	6180	18132	228	0.95	25, 23
Zaire	3700000	—	4370	2020	32.8	34	9	3	740	4507	1586	277	25	3	40900	57200	349	0.22	33, 6
Zambezi	1400000	8		2020	15	34	13	1033	2606	957	231	22	8	6980		157	0.16		25, 23
	26602040	8		48	15	34	13	1033	2606	957	231	22	8	6980	157	0.16		25, 23	

1) Definitions of variables in Table 2. 2) Reference: 1, Al-Ansari et al. (1988); 2, Allen (1970); 3, Allen et al. (1979); 4, Arnborg et al. (1967); 5, Atlas Mira (1964); 6, Berner and Berner (1987); 7, Chakrapani and Subramanian (1990); 8, Chouret (1977); 9, Coleman and Wright (1975); 10, Etchanchu and Probst (1988); 11, Galloway (1976); 12, Got et al (1985); 13, Grove (1972); 14, Kanes (1970); 15, Kempe et al (1991); 16, Kesel et al (1992); 17, Lisitzin (1972); 18, Macdonald et al. (1991); 19, Markham (1992); 20, Meade et al. (1985); 21, Meade et al. (1990); 22, Meckel (1970); 23, Meybeck (1976); 24, Meybeck (1979); 25, Milliman et al. (1984); 27, Milliman et al. (1987); 28, Milliman and Syvitski (1992); 29, Naidu and Mowatt (1975); 30, Nelson (1970); 31, Nelson (1990); 32, Olivry et al. (1988); 33, Olivry (1977); 34, Palanques et al. (1990); 35, Pickup et al. (1980); 36, Pinet and Souriau (1988); 37, Prendes (1983); 38, Rooseboom and von Harmse (1979); 39, Shahin (1985); 40, UN ECAFE (1966); 41, Vaithiyanathan et al. (1988); 42, Varga et al. (1989); 43, Zhengying (1983)

TABLE 2.—DEFINITION OF VARIABLES

VARIABLE		DEFINITION
		general characteristics drainage area
A	drainage area (km²)	The surface area from which a river collects surface runoff, including the peripheral zones which only contribute runoff in case of exceptional rainfall.
A_d	percent depositional (km²)	The surface area of the part of the drainage basin covered with Holocene fluvial sediments, expressed as a percentage of the total drainage area.
L_b	basin length (km)	The maximum length of the drainage basin in the direction of the main stream.
L_r	river length (km)	The length of the flowline of the main stream of the catchment from its headwaters to the rivermouth.
		sediment
S_s	annual suspended load (Mt/yr)	The total amount of material carried in suspension by the river annually, as measured at mouth of the river.
S_d	annual dissolved load (Mt/yr)	The total amount of material carried in solution by the river annually, as measured at the mouth of the river.
E	an. specific sed. yield* (t/km²/yr)	The amount of material eroded mechanically from a unit surface area annually, averaged over the drainage area, calculated from the suspended load and the drainage area.
D	mech.denudat. rate* (mm/ka)	The average lowering of the surface per unit time resulting from mechanical denudation, calculated from the specific sediment yield.
		discharge
Q	mean flow rate (m³/s)	The long term average water discharge per unit time, at the mouth of the river.
Q_{max}	maximum flow rate (m³/s)	The average flow rate per unit time, for the month with the greatest discharge.
Q_{pk}	discharge peak.* (−)	The ratio of the average flow rate and the maximum flow rate.
R	specific runoff* (mm/yr)	The total annual height of the water column on a unit surface area which leaves the catchment as surface runoff, as calculated from the mean flow rate and the drainage area.
R_{cf}	runoff coefficient* (−)	The ratio of the total annual precipitation in the catchment and the total annual water discharge at the river mouth.
		relief
H	mean height (m)	The average height of the drainage basin as calculated by Summerfield and Hulton (1994) and Pinet and Souriau (1988).
H_{max}	maximum height (m)	The elevation of the highest point in the drainage basin.
H_{pk}	relief peakedness* (−)	The ratio of the mean height and the maximum height of the drainage basin.
H_r	relief ratio* (m/km)	The ratio of the maximum height of the drainage basin and the basin length.
α_{riv}	slope river bed (m/km)	The average slope angle of the bed of the main stream of the catchment, calculated as the ratio of the maximum height of the drainage basin and the length of the main stream.
		climate
P	tot. annual precip. (mm/yr)	The total annual precipitation, averaged over a number of meteorological stations distributed more or less equally over the erosional part of the catchment.
P_{max}	max. month. precip. (mm/mth)	The average amount of precipitation for the wettest month of the year.
P_{pk}	precip. peakedness* (−)	The ratio of the average monthly precipitation and the maximum monthly precipitation.
T	mean annual temp. (°C)	The mean annual daytime temperature, averaged over a number of meteorological stations distributed more or less equally over the erosional part of the catchment.
T_{range}	annual temp. range (°C)	The difference between the average daytime temperatures for hottest and coldest months.

*marked variables have been calculated from other variables in the data set. Some of the calculated variables have not been included in Table 2.

precipitation peakedness, mean annual temperature, and annual temperature range. Estimates for these variables have been derived from the Klimadiagramm Weltatlas (Walter and Lieth, 1967). This atlas presents the averaged results of long-term observations of thousands of long serving meteorological stations around the world. For most catchments listed here, data from between 5 and 40 stations are available. In many drainage basins, some meteorological stations are situated on high points, so that orographic effects are incorporated in the average figures presented in the atlas. The estimates presented in Table 1 are for the nondepositional part of the catchment only, as most of the sediment load measured at the river mouth is likely to have been derived from this part of the drainage basin. Precipitation peakedness has been calculated as the ratio of average monthly precipitation and maximum monthly precipitation.

Discharge

Runoff variables in Table 1 are mean flow rate, maximum flow rate, discharge peakedness, specific runoff and runoff coefficient. The mean flow rate is an annual average figure, and has been taken from a variety of sources. Estimates of the flow rate for the month with greatest discharge have, by contrast, been derived from UNESCO reports (1972, 1974, 1979) that present observations for a global network of gauging stations.

The stations have been operated in a more or less uniform way, so the set of estimates presented here may have a certain degree of internal consistency, although individual figures do not always combine with the annual flow rates quoted from other sources. Discharge peakedness is defined here as the ratio of the average monthly flow rate and the maximum monthly flow rate. Specific runoff is the total annual height of the water column on a unit surface area which leaves the catchment as surface runoff, averaged over the drainage basin. The runoff coefficient is the ratio of the total annual precipitation in the entire catchment, including the depositional part of the basin, and the total annual water discharge at the river mouth.

Sediment

Suspended sediment load estimates given in Table 1 can be used to calculate the specific sediment yield from a drainage basin. This is the amount of material eroded mechanically from a unit surface area per unit time, and leaving the catchment as suspended load, averaged over the entire drainage basin. The figures calculated here are too low, mainly for two reasons. First, bed load is not included in the calculations, although it is part of the mechanical component of erosion. Second, material stored within the drainage basin is not included. This can be up to several times greater than the load at the river mouth. Natural suspended loads have been used where available. Using an average rock density of 2700 kgm^{-2}, the specific sediment yield can be converted into the mechanical denudation rate, defined as the average lowering of the surface per unit time resulting from mechanical denudation.

CONTROLS: CLIMATE, RELIEF AND RUNOFF

Figure 1 shows a plot of specific sediment yield versus drainage area for all 97 rivers listed in Table 1. Although a slight trend in the data may be recognized marking the inverse relationship between specific sediment yield and catchment area, the overwhelming impression is one of scatter. Clearly, the potential sediment yield from a drainage basin of a given size can range over three orders of magnitude. Some authors have explained this variance in terms of precipitation characteristics, and/or topographic aspects of the drainage basin. However, there remains profound dissent about the relative importance of the various controls on denudation rate.

Several attempts have been made to analyze the relationship between denudation rate and mean annual precipitation (Langbein and Schumm, 1958; Douglas, 1967; Wilson, 1973; Ohmori, 1983). Although there is little agreement in detail, certain common elements emerge from these studies, mainly illustrating the changes in erosivity of rainfall and, through the mediation of vegetation cover, erodibility of regolith, with changes in annual precipitation rate. Many authors identify peak erosion rates in semi-arid environments, and denudation increases progressively above a mean annual precipitation rate of 1000 mm. However, some recent studies have failed to recognize mean annual precipitation as being of crucial importance. Both Ahnert (1984) and Pinet and Souriau (1988), for example, found a weak correlation between denudation rate and mean annual precipitation. This is not surprising, considering that most hillslope erosion processes are dependent on instantaneous rates of precipitation rather than mean annual rates (e.g., Caine 1980). Fournier (1960), in a study using data from 96 intermediate size and large catchments, showed that precipitation peakedness is one of the main factors governing rate of denudation, while mean drainage basin elevation and slope also play an important role. By multiple regression, Fournier (1960) found that:

$$\log E = 2.65 \log(p^2/P) + 0.46 \log h \tan a - 1.56 \quad (1)$$

where the specific sediment yield E, measured in t km^{-2} yr^{-1} is correlated with mean drainage basin elevation h, slope a, and a precipitation peakedness parameter combining the highest monthly rainfall p and the mean annual rainfall P. Although this relationship has proved to be relatively successful in regions of variable relief and rainfall (Leeder, 1991), more recently, an approach focussing on topography has been favoured. From the analysis of data from 280 rivers, Milliman and Syvitski (1992) concluded that sediment yield is a log-linear function of basin area and maximum elevation of the drainage basin. Summerfield and Hulton (1994) considered a range of topographic and climatic variables and found that mechanical denudation rates are most strongly associated with variables that express relief characteristics and runoff. Using data from the world's 33 largest drainage basins, they suggested that over 60% of the variance in total denudation is accounted for by the

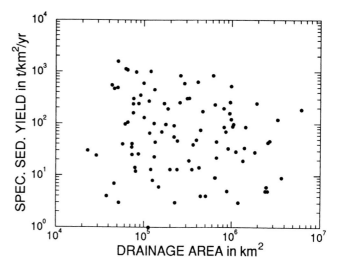

FIG. 1.—Scatter plot of specific sediment yield versus drainage area. Data from Table 1.

ratio of the maximum basin elevation and the basin length, and runoff.

Several reasons can be given for the lack of agreement on the relative importance of climatic and topographic controls on denudation. First, the type of variables considered, and their definitions, vary from study to study. Second, the number and size of drainage basins for which these variables have been described differs greatly, and the problem has often been addressed on a regional rather than a global scale. Third, the source and quality of the data used are far from uniform. And fourth, the techniques used to analyze the data are different in each case. To be able to make valid predictions about denudation rates at large, a systematic multivariate analysis of data on a large number of river systems is required.

Table 3 lists Pearson correlation coefficients between specific sediment yield and each of the estimators or control variables defined in Table 2, as calculated from the data listed in Table 1. Specific sediment yield is not particularly well correlated with any of these variables. From this it may be concluded that none of the variables defined in Table 2 can be regarded as the prime control on denudation rate. It is more likely that a significant component of the global variation in sediment yield can be explained by a combination of several controls. To test this,

TABLE 3.—PEARSONIAN COEFFICIENTS OF CORRELATION BETWEEN SPECIFIC SEDIMENT YIELD AND ESTIMATORS

ESTIMATOR	CORRELATION COEFFICIENT
drainage area	−0.1493
mean height	0.4128
maximum height	0.3391
relief peakedness	−0.1054
relief ratio	0.4427
slope river bed	0.4187
total annual precipitation	0.3463
maximum monthly precipitation	0.4340
precipitation peakedness	0.3610
mean annual temperature	0.1792
annual temperature range	−0.0575
specific runoff	0.4367
discharge peakedness	−0.0535
runoff coefficient	−0.0707

a step-wise regression has been implemented on the data listed in Table 1 and on a number of variables which can be calculated from the data in this table.

Step-wise regression routine examines a set of estimator variables, selects those that are most efficient at explaining the variance in a response variable and builds them into a model. The routine will not automatically select all estimators with a high correlation coefficient with the response variable, as some of these estimators may explain similar elements of the variance. On the other hand, estimators with a relatively low correlation coefficient with the respondent may appear in the final model, if they explain a part of the variance that is not covered by any of the other estimators. Here, the estimators are the variables that are thought to exert a certain control on the specific sediment yield, the response variable.

Using step-wise regression, five estimator variables were selected out of the 14 listed in Table 3, the combination of which is most efficient at explaining the variance of specific sediment yield on a global scale. These five are specific runoff (R), drainage area (A), maximum height (H_{max}), mean annual temperature (T) and annual temperature range (T_{range}). Specific sediment yield (E) may be expressed as:

$$\ln E = -0.416 \ln A + 4.26.10^{-4} H_{max} + 0.150T$$
$$+ 0.095 T_{range} + 0.0015R + 3.585 \qquad (2)$$

where E is in t km^{-2} yr^{-1}, A in km^2, H_{max} in m, T and T_{range} in °C, and R in mm yr^{-1}. This equation is based on data from 86 out of the total of 97 catchments included in Table 1, and no outliers have been excluded from the analysis. It explains 49.14% of the variance in specific sediment yield from these drainage basins. Using one or more additional variables does not contribute to the explanatory power of the model in a significant way. It is evident that the dimensions on either side of equation (2) don't balance. Balance could only be achieved if each estimator variable was given the same units as the response variable. Since E is in t/km²/yr, the five estimators would need to be divided or multiplied by a number of variables with different dimensions, thus removing the clarity from equation (2).

The interpretation of equation (2) is not straight forward. Not all variables included in the model do correlate particularly well with specific sediment yield, while some variables with relatively high correlation coefficients do not appear in the equation (Table 3). These variables are not necessarily unimportant to the variance in yield, and part of their influence may be reflected by controls included in equation (2). In Table 4, the Pearson correlation coefficients between all estimators used to derive the model are given. Relatively high correlation coefficients exist between several of the included and discarded estimators, while correlation coefficients between the included controls are rather lower, with the marked exception of the combination of mean annual temperature and annual temperature range. The five selected estimators cover different controls on the process-complex which determines sediment yield, with little overlap between them. Combinations of two or more of the model variables may be as adequate at explaining variance in specific sediment yield related to estimators that have not been selected, as these discarded estimators are by themselves. Thus, equation (2) represents the most effective combination of estimators, im-

plying the influence of controls that have not been explicitly incorporated in the model.

As expected from previous work (Schumm and Hadley, 1961; Milliman and Meade, 1983; Milliman and Syvitski, 1992), the inverse relationship between specific sediment yield and drainage basin surface area is an important term in equation (2). The other four terms in this equation refer to the relative importance of the weathering, hillslope erosion and fluvial transport components of the sediment routing system.

Both temperature variables in the equation are most likely to refer to the efficiency of weathering of bed rock as a function of climate. Specific sediment yield increases with mean annual temperature, possibly reflecting a general increase in chemical weathering rate and regolith thickness with temperature. The increase of specific sediment yield with temperature range may be a function of the role temperature variations play in mechanical weathering. However, the relationship between temperature and weathering is by no means simple (e.g., Yatsu, 1988), and many details are obscured in a general expression like equation (2). Maximum height is a variable that harbors elements of both local relief and large-scale topography. The positive relationship between maximum height and specific sediment yield may reflect the importance of local slope to the erosive power of processes contributing to the denudation of hillslopes. Combined with drainage area, it will bear in it an element of the relief ratio and the slope angle of the river bed, both possibly more relevant to the fluvial transport of sediment. Specific runoff determines to a certain extent the transport capacity of the fluvial system and may simultaneously refer to the amount of water available for hillslope erosion. It is relatively well correlated with mean annual precipitation and precipitation peakedness and may cover, in combination with mean annual temperature and annual temperature range, the obvious control precipitation exerts on sediment yield. Thus, the model seems to combine the influence of drainage area with weathering, hillslope erosion and fluvial transport components to explain the large-scale variance of specific sediment yield.

Equation (2) explains 49% of the variance in specific sediment yield from intermediate and large scale river catchments around the world. This implies that 51% of that variance is not readily explained in terms of the variables listed in Table 3. This may be due in part to the use of incorrect data or data that reflect a strong human influence and partly to the fact that the variables do not cover all aspects of the weathering, hillslope erosion, and fluvial transport systems that are most relevant to specific sediment yield. In addition, sediment loads may not always reflect contemporary environmental conditions, responding instead to previous episodes of landscape evolution (e.g., Church and Slaymaker, 1989). All these error sources are of significant importance and may cause considerable scatter in the dataset, but it seems unlikely that the combination of these factors is responsible for all unexplained variance.

Figure 2 is a plot of the residuals of equation (2) for the 89 drainage basins used to deduce it. The residuals represent the difference between the observed and predicted specific sediment yields. Here, the standardized residuals are expressed in standard deviation units above or below the mean. From Figure 2, it can be seen that there is a trend in the difference between observed and predicted yields, that can be described by

TABLE 4.—MATRIX OF PEARSON CORRELATION COEFFICIENTS BETWEEN THE ESTIMATORS OF SPECIFIC SEDIMENT YIELD

	A	H	H_{max}	H_{pk}	H_r	α_{riv}	P	P_{max}	P_{pk}	T	T_{range}	R	R_{cf}	Q_{pk}
A	1.0000													
H	0.0222	1.0000												
H_{max}	0.2917	0.6180	1.0000											
H_{pk}	0.4010	−0.3853	0.3342	1.0000										
H_r	−0.3187	0.2485	0.4728	0.1026	1.0000									
α_{riv}	−0.4812	0.1543	0.4368	0.0442	−0.1820	1.0000								
P	0.0579	0.1958	0.1325	0.1296	−0.0143	−0.0382	1.0000							
P_{max}	0.0909	0.2604	0.2049	0.0468	−0.0361	−0.0641	0.8649	1.0000						
P_{pk}	−0.0440	0.1345	0.1549	−0.2251	−0.0127	0.0169	−0.0939	0.3516	1.0000					
T	0.0027	0.0438	0.0275	0.0840	−0.0248	−0.0077	0.4847	0.5654	0.2301	1.0000				
T_{range}	0.0093	−0.0524	−0.0359	−0.1635	−0.0526	−0.0851	−0.5500	−0.5151	0.0174	−0.8763	1.0000			
R	−0.0002	0.1512	0.2627	0.2560	0.2759	0.2646	0.7413	0.5507	−0.2546	0.1443	−0.2549	1.0000		
R_{cf}	0.1034	−0.0393	0.1762	0.2410	−0.0317	0.0470	0.0056	−0.1253	−0.2606	−0.4643	0.3124	0.5654	1.0000	
Q_{pk}	0.8309	−0.2189	−0.2374	−0.1609	0.1455	−0.2483	−0.0092	0.4037	−0.0251	0.1296	−0.3065	−0.0946	−0.2346	1.0000

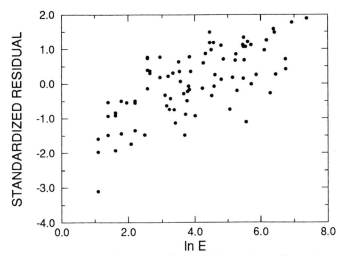

FIG. 2.—Scatter plot of the standardized residuals of the specific sediment yields, calculated using equation (2), versus the natural logarithm of the observed specific sediment yields.

$$\ln E = 0.9682\ E_p + 0.1543 \qquad (3)$$

where E_p is the specific sediment yield predicted by equation (2). The correlation coefficient between E and E_p is 0.4852. Equation (2) systematically over-predicts low specific yields and under-predicts high specific yields. Flawed measurements, human impact and time lags in the sediment routing are inducing differences between observed and predicted sediment yields, but they are probably doing so in a random fashion. Thus, they are unlikely to cause a clear trend in the model residuals. This trend is probably caused by the influence of a control not included in the model.

UPLIFT OF ROCKS

If it is accepted that the large-scale controls on weathering, hillslope denudation and fluvial erosion and transport are adequately described by data in Table 1, it must be concluded that specific sediment yield is only in part governed by the combined potential of these process systems to remove material from parts of the Earth's surface. This is reasonable as one would expect the rate of removal of material from a reference volume to be dependent not only on the erosivity of the processes acting on the surface of that volume, but also on the rate

of input of new matter (Leeder, 1991). In geological terms, the input of matter is equivalent to uplift of rocks, and it is this process which may exert the influence on sediment yield that is reflected in the unexplained part of its observed variance.

To test this hypothesis, it would be preferable to use information on the rate of uplift of rocks in the erosional parts of large drainage basins. However, there are very few reliable estimates of these rates, and most of the available measurements are for small areas only. Many observations reported as yielding rates of uplift of rock are in fact observations of exhumation (England and Molnar, 1990). At present, calculation of average rates of uplift of rocks is impossible for most intermediate and large catchments.

As regional rates of uplift of rocks are a function of the acting mechanism of uplift, tectonic setting may be used to obtain an impression of the influence of the rate of uplift of rocks on the magnitude of sediment yield. For this purpose, it is useful to make a first-order distinction between three types of mechanisms that cause uplift of rocks. These are: (1) plate convergence causing contraction and thickening, (2) active thermal processes causing doming and rifting and (3) passive crustal extension. Areas where remnants of lithospheric root zones are present underneath old mobile belts may undergo some isostatic uplift, whereas purely cratonic regions may not experience significant uplift of rocks for prolonged periods of geologic time. In many cases, two or more uplift mechanisms will operate simultaneously, either in different parts of a catchment, or in a complimentary way within one area. However, it is often possible to indicate a dominant mechanism so that a simple basin-wide classification may result. At this point it is important to stress that in any tectonic setting an important part of the observed vertical crustal movements result from isostatic compensation of the removal of material from the surface (Molnar and England, 1990). Often, the tectonic component of uplift of rocks cannot be separated from the observed strain resultant, but it may be assumed that in general, it is proportional to the total vertical motion. It is this tectonic component of vertical crustal movements that is most relevant to the present problem.

Milliman and Syvitski (1992) present figures for the specific sediment yield from 280 drainage basins of varying size, tectonic, climatic and geomorphic setting. This data set is larger than the data set in this paper, and it spans a larger range of catchment sizes, thus allowing an analysis of trends in subgroups. The yields plotted against drainage area (Fig. 3A) show a similar degree of scatter as the data represented in Figure 1. However, if the data set is subdivided according to the dominant mechanism causing uplift of rock, as outlined above, a broad zonation becomes apparent (Figs. 3B–F). Drainage basins located in regions characterised by active contractional strain have consistently high sediment yields, typically ranging from 100 to 10000 t km^{-2} yr^{-1}. Cratonic catchments on the other hand rarely have yields greater than 100 t km^{-2} yr^{-1}, while old orogens may have slightly higher yields of up to 200 t km^{-2} yr^{-1}. Although not supported by a large amount of data, an intermediate position seems to be taken by regions characterized by extension. Superimposed on this zonation is a trend caused by the inverse relationship between specific sediment yield and drainage area.

Least squares regression lines can be fit to several of the drainage basin groups (Fig. 4). For contractional settings, old orogens and cratons, sufficient data are available to allow fitting of lines over the entire range of catchment sizes. A tentative least squares fit is shown for the active extension group, as its trend coincides with those of the three larger groups. As it may be expected that the different uplift mechanisms represent a range in rates of tectonically induced uplift of rocks, from high rates in contractional settings to virtually no uplift of rocks in cratonic environments, Figure 4 illustrates the first-order influence of rate of tectonic uplift of rocks on specific sediment yield.

Within each of the groups, there is a considerable degree of scatter, and consequently the least squares regression lines should be interpreted as indicative of the order of magnitude of the specific sediment yield from a drainage basin of a certain size in a certain tectonic setting, rather than as a means for a more precise estimate. Part of the variance may be due to differences in degree of tectonic activity within each group, as can be seen from the relative position of a number of tectonic domains from which data have been used to construct Figure 4. Within each group, drainage basins with a relatively high rate of uplift of rocks consistently plot above the least squares regression line, whereas catchments with lower rates of uplift of rock project closer to or below the line. The variance is especially significant in the group of rivers draining contractional terrains, probably due to the wide range of rates of uplift of rocks represented by the data. Areas with extremely high rates of uplift, like the Southern Alps in New Zealand, with 5–8 mm yr^{-1} (Bull and Cooper, 1986; Kamp and Tippett, 1993), and the Central and Coastal Ranges in Taiwan, with 5.5–7.5 mm yr^{-1} (Lundberg and Dorsey, 1990), have specific sediment yields exceeding 10.000 t km^{-2} yr^{-1}. Rates of uplift of rocks in the Alps and Apennines, on the other hand, are much lower, probably between 0.1 and 0.7 mm yr^{-1} (Hurford et al., 1989), which is reflected in sediment yields of 100 to 600 t km^{-2} yr^{-1}. The major catchments draining the Himalayas towards the south assume an intermediate position, specific sediment yields being between 250 and 1000 t km^{-2} yr^{-1}. This is in agreement with

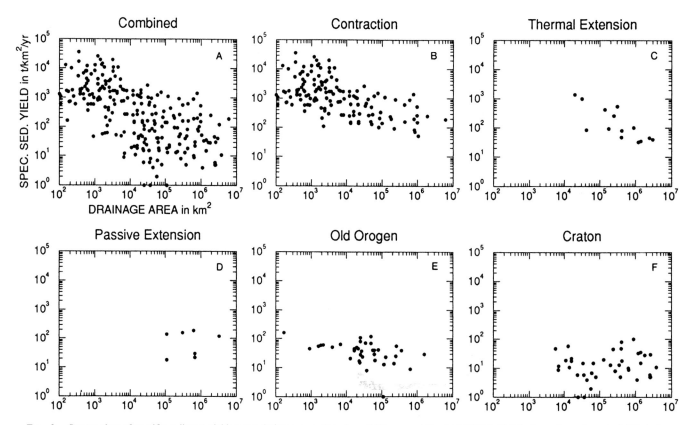

FIG. 3.—Scatter plots of specific sediment yield versus drainage area. Data from Milliman and Syvitski (1992). Plot A shows the entire set of 280 drainage basins. This set can be subdivided according to tectonic setting of the drainage basins. Plots B–F show the subsets for contractional, thermal-extensional, passive-extensional, old-orogenic, and cratonic settings respectively.

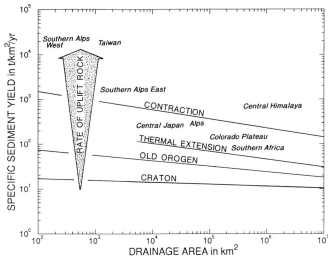

FIG. 4.—Least squares regression lines for four of the five subsets of drainage basins depicted in Figure 3. The regression line for catchments in a setting of thermal extension is tentative. The direction of increasing rate of vertical strain is indicated. Regions with relatively high rates of uplift of rock plot high with respect to the appropriate regression line, while regions with relatively lower rates of uplift of rock plot closer to, or below the regression line. Equations for the regression lines:

$$contraction: \ E = -0.0009 \, A + 2736,$$

$$thermal \ extension: \ E = -1032 \, A + 1072150,$$

$$old \ orogen: \ E = -1991 \, A + 260929,$$

$$craton: \ E = -14485 \, A + 670551,$$

where E is specific sediment yield (in t km^2 yr^{-1}) and A is the surface area of the drainage basin (in km^2).

the inferred rates of uplift of rocks in this region of 0.5 to 2.5 mm yr^{-1} (Zeitler, 1985; Searle pers. commun.) Thus, knowledge of the rate at which tectonic processes operate within a certain region allows a more accurate prediction of the specific sediment yield from that area.

As was demonstrated in the previous section, equation (2) over-estimates sediment yields from areas with relatively low denudation rates and under-estimates yields from rapidly eroding catchments. If it is accepted that, in addition to the environmental factors included in equation (2), rate of uplift of rocks is a first-order control on sediment yield, the systematic difference between observed yields and yields predicted by the equation may be interpreted in terms of the relative importance of the latter control. The over-prediction for drainage basins with small yields indicates that denudation of those catchments is limited by the input of material into the weathering-erosion-transport system. This is reasonable, as low denudation rates tend to be confined to regions with extremely slow rates of uplift of rocks. The potential for erosion, as controlled by the environmental factors included in equation (2), exceeds the rate at which tectonic processes make available material for erosion. Under-prediction of yields from rapidly eroding drainage basins may be due to the fact that here the rate of denudation is limited by the potential of surface processes to erode, rather than the rate at which material is being provided to the system by tectonic processes. Thus, there is a fundamental difference in the relative importance of weathering, hillslope erosion, fluvial

transport and uplift of rocks as controls on the rate of denudation, and therefore the rate of sediment supply, between the different tectonic regimes in which a catchment may be situated.

It is proposed here that the rate of tectonic uplift of rocks is a control on regional denudation rates of an importance similar to, or possibly greater than, that of the climatic and topographic variables governing rates of weathering, hillslope erosion, and fluvial transport. Along with providing material to the surficial denudation system, tectonic processes determine many environmental aspects of a drainage basin. Lithology, texture and structure of rocks, distribution of relief and seismic activity are all functions of the tectonic setting, while large-scale drainage patterns (Ollier, 1981; Morisawa, 1985) are also strongly influenced by the style and rate of tectonic processes. Tectonics may even interfere with atmospheric circulation patterns to induce orographic precipitation. Many of the variables presented in Table 2 bear in them an element of tectonic activity. Thus, tectonics affect sediment yield both directly and indirectly. It is not simple to separate the tectonic signal in the observed specific sediment yield from the effects of topography and precipitation: the sediment flux at the outlet of the erosion area, as well as the resulting topography of that area, are likely to reflect the way in which surface processes have interacted with the tectonic input of material. To be able to make more accurate predictions of specific sediment yield, a tectonic factor should be incorporated in the model described by equation (2). This is difficult, bearing in mind the present paucity of information.

PREDICTING SEDIMENT SUPPLY

The accuracy with which sediment supply to depositional sites can be estimated depends on the availability of information on the source area. For many present-day river systems, morphometric, climatologic and hydrologic observations can be made, thus enabling the use of equation (2) for a first-order estimate of the rate of denudation in the catchment. This estimate can then be corrected for the relative rate of uplift of rocks in the source area. In most cases, the correction factor should be between 2 and -2, where positive values apply to drainage basins with relatively high rates of vertical strain and negative values to low rates.

In studies of the ancient sedimentary record, precise information on the environmental factors governing the erosivity of processes acting in a paleo-source area is not available. This makes it impossible to employ equation (2). Instead, Figure 4 may be used to derive an order of magnitude estimate of sediment yield, if a general idea exists of the tectonic setting in which a drainage system developed. Additional knowledge of the rate of uplift of rocks and the size of the paleo-drainage system may allow a more accurate positioning with respect to the regression lines in Figure 4. It should be kept in mind that both equation (2) and Figure 4 reflect sediment yields related to modern conditions, and that over geologic time conditions have changed. Evolving vegetation, for instance, has had a changing effect on the hydrologic cycle and the erodibility of the earth's surface (Schumm, 1968). A correction may be necessary when reconstructing paleo-sediment yields.

Continental basins are commonly occupied by large river systems. For the purpose of modelling the stratigraphy of these basins, information is required on the sediment input into the basin, rather than on the output of the basin. However, in many cases only the latter is known. Figure 4 may be used to derive a first-order estimate of the sediment input into a continental basin from the sediment yield of the catchment draining the entire region. The trunkstream of a catchment is fed by a number of smaller tributary systems that carry sediment into the depositional basin. The sediment yield of the entire catchment is the resultant of the sediment flux into the depositional basin from these tributary systems and the storage of sediment in the basin. If, in Figure 4, the overall yield of a catchment is plotted, its position may be extrapolated along the trend of the regression lines in the diagram, to estimate specific sediment yields for the tributary drainage areas. Assuming that the tectonic regime in the tributary areas is similar to that of the entire catchment, the position of the tributaries with respect to the appropriate regression line in Figure 4 is comparable to that of the catchment, the only adjustment being for drainage area. Thus, when catchment-wide sediment yield is combined with surface area of the tributary systems, the sediment supply to a depositional basin may be estimated.

The methods outlined above allow estimation of sediment supply with an accuracy of well within an order of magnitude. However, both approaches are open for considerable improvement. With the advance of more detailed Digital Elevation Models, further analysis of the influence of local topography on large-scale patterns of sediment yield is likely to be facilitated. This may lead to different insights into the relative importance of factors governing the removal of material from the upstream segments of drainage basins. Furthermore, a systematic analysis of data on the sediment load of streams at the outlet of source areas may yield more precise constraints on denudation rates in upstream areas, as well as on the efficiency of present day flood plains as sediment traps. Finally, the identification of the tectonic component of uplift of rocks will allow a quantification of the control it exerts on the rate of denudation.

ACKNOWLEDGMENTS

I am indebted to Sean Hodges and Martin Shepley for their assistance in data collecting, and to Mathew Eagle for his help with the statistical analysis. Philip Allen, Sanjeev Gupta and Pete Burgess have been stimulating and critical discussion partners. Thanks to Philip also for thoroughly reading the manuscript. An earlier version of the manuscript was reviewed by Mike Blum and Stan Schumm. Their comments have been most helpful. Initial research was financed by Amerada Hess and Royal Dutch Shell Research Laboratories in Rijswijk and continued under a British Council Fellowship.

REFERENCES

AHNERT, F., 1970, Functional relationships between denudation, relief and uplift in large mid-latitude drainage basins: American Journal of Science, v. 268, p. 243–263.

AHNERT, F., 1984, Local relief and the height limits of mountain ranges: American Journal of Science, v. 284, p. 1035–1055.

AIGNER, T., BRANDENBURG, A., VAN VLIET, A., DOYLE, M., LAWRENCE, D., AND WESTRICH, J., 1991, Stratigraphic modeling of epicontinental basins: two case studies, in Dott, R. H., and Aigner, T., eds., Processes and patterns in shelf and epeiric basins: Sedimentary Geology, v. 69, p. 167–190.

AL-ANSARI, N. A., ASAAD, N. M., WALLING, D. E., AND HUSSAN, S. A., 1988, The suspended sediment discharge of the river Euphratus at Hadita, Iraq: Geografiska Annaler, v. 70A, p. 203–213.

ALLEN, G., LAURIER, D., AND TOUVENIN, J., 1979, Etude sedimentologique du Delta de la Mahakam: Paris, Compagnie Francaise des Petroles, 156 p.

ALLEN, J. R. L., 1970, Sediments of the modern Niger Delta: a summary and review, in Morgan, J. P., ed., Deltaic Sedimentation: Tulsa, Society of Economic Palaeontologists and Mineralogists Special Publication 15, p. 138–151.

ARNBORG, L., WALKER, H. J., AND PEIPPO, J., 1967, Suspended load in the Colville River, Alaska, 1962: Geografiska Annaler, v. 49A, p. 131–144.

ATLAS MIRA FISIKO GEOGRAFICHESKIY, 1964, Soviet Academy of Science.

AUBREY, W. M., 1989, Mid-Cretaceous alluvial-plain incision related to eustacy, southeastern Colorado Plateau: Geological Society of America Bulletin, v. 101, p. 443–449.

BERNER, E. K. AND BERNER, R. A., 1987, The Global Water Cycle: Englewood Cliffs, Prentice-Hall, 397 p.

BLUM, M. D., 1994, Genesis and architecture of alluvial stratigraphic sequences: a late Quaternary example from the Colorado River, Gulf Coastal Plain of Texas, in Siliciclastic Sequence Stratigraphy: Recent Developments and Applications: Tulsa, American Association of Petroleum Geologists Memoir 58, p. 259–283.

BULL, W. B. AND COOPER, A. F., 1986, Uplifted marine terraces along the Alpine Fault, New Zealand: Science, v. 234, p. 1225–1228.

BURGESS, P. M. AND ALLEN, P. A., 1996, A forward modelling analysis of the controls on sequence stratigraphic geometries, in Hesselbo, S. P. and Parkinson, D. N., eds., Sequence Stratigraphy and its Application to British Stratigraphy: London, Geological Society of London Special Publication 103, p. 9–24.

CAINE, N., 1980, The rainfall intensity-duration control of shallow landslides and debris flows: Geografisca Annaler, v. 62A, p. 23–27.

CHAKRAPANI, G. J. AND SUBRAMANIAN, V., 1990, Factors controlling sediment discharge in the Mahanadi River basin, India: Journal of Hydrology, v. 117, p. 169–185.

CHOURET, A., 1977, Regime des apports fluviatiles des materiaux solides en suspension ver le Lac Tchad: synthese des etudes de l'Orstom en Republique du Tchad: Wallingford, International Association of Hydrological Sciences Publication 122, p. 126–133.

CHURCH, M. AND SLAYMAKER, O., 1989, Disequilibrium of Holocene sediment yield in glaciated British Columbia: Nature, v. 337, p. 452–454.

CLOETINGH, S., SASSI, W., HORVATH, F., AND PUIGDEFABREGAS, C., 1993, Basin analysis and dynamics of sedimentary basin evolution — an introduction: Sedimentary Geology, v. 86, p. 1–4.

COLEMAN, J. M. AND WRIGHT, L. D., 1975, Modern river deltas: variability of processes and sand bodies, in Broussard, M. L., ed., Deltas: Houston, Houston Geological Society, p. 99–149.

DEDKOV, A. P. AND MOSZHERIN, V. I., 1992, Erosion and sediment yield in mountain regions of the world: Wallingford, International Association of Hydrological Sciences Publicaltion 209, p. 29–36.

DOUGLAS, I., 1967, Man, vegetation and the sediment yield of rivers: Nature, v. 215, p. 925–928.

EINSELE, G., 1992, Sedimentary Basins: Berlin, Springer Verlag, 628 p.

ENGLAND, P. AND MOLNAR, P., 1990, Surface uplift, uplift of rocks, and exhumation of rocks: Geology, v. 18, p. 1173–1177.

ENOS, P., 1991, Sedimentary parameters for computer modelling, in Franseen, E. K., Watney, W. L., Kendal, C. G. S., and Ross, W., eds., Sedimentary Modelling: Computer Simulations and Methods for Improved Parameter Definition: Kansas Geological Survey Bulletin, v. 233, p. 63–99.

ETCHANDU, D. AND PROBST, J. L., 1988, Evolution of the chemical composition of the Garonne River water during the period 1971–1984: Hydrological Sciences Journal, v. 33, p. 243–256.

FLINT, S. S., 1993, The application of sequence stratigraphy to ancient fluvial successions: Brisbane, 5th International Conference on Fluvial Sedimentology, Proceedings, p. K22–K32.

FOURNIER, F., 1960, Climat et erosion: la relation entre l'erosion du sol par l'eau et les precipitations atmospheriques: Paris, Presse Universitaire de France, 201 p.

GALLOWAY, W. E., 1976, Sediments and stratigraphic framework of the Copper River fan delta, Alaska: Journal of Sedimentary Petrology, v. 46, p. 726–737.

GALLOWAY, W. E., 1989, Genetic stratigraphic sequences in basin analysis I: architecture and genesis of flooding-surface bounded depositional units: American Association of Petroleum Geologists Bulletin, v. 73, p. 125–142.

GOT, H., ALOISI, J. C., AND MONACO, A., 1985, Sedimentary processes in Mediterranean deltas and shelves, in Stanley, D. J. and Wezel, F. C., eds., Geological Evolution of the Mediterranean Basin: New York, Springer Verlag, p. 355–376.

GROVE, A. T., 1972, The dissolved and solid load carried by some west African rivers: Senegal, Niger, Benue and Shari: Journal of Hydrology, v. 16, p. 277–300.

HOLEMAN, J. N., 1968, Sediment yield of major rivers of the world: Water Resources Research, v. 4, p. 737–747.

HURFORD, A. J., FISCH, M., AND JAGER, E., 1989, Unravelling the thermotectonic evolution of the Alps: a contribution from fission track analysis and mica dating, *in* Coward, M. P., Dietrich, D., and Park, R. G., eds., Alpine Tectonics: London, Geological Society of London Special Publication 45, p. 369–398.

JANSSON, M. B., 1982, Landerosion by Water in Different Climates: Uppsala, Ungi Rapport 57, Department of Physical Geography, Uppsala University, 151 p.

JANSSON, M. B., 1988, A global survey of sediment yield: Geografiska Annaler, v. 70A, p. 81–98.

JERVEY, M. T., 1988, Quantitative geological modelling of siliciclastic rock sequences and their seismic expression, *in* Wilgus, C. K., Hastings, B. S., Kendall, C. G. St. C., Posamentier, H. W., Ross, C. A., and Van Wagoner, J. C., eds., Sea-Level Changes: an Integrated Approach: Tulsa, Society of Economic Palaeontologists and Mineralogists Special Publication 42, p. 47–69.

KAMP, P. J. J. AND TIPPETT, J. M., 1993, Dynamics of Pacific Plate crust in the South Island (New Zealand) zone of oblique continent-continent convergence: Journal of Geophysical Research, v. 98, p. 105–118.

KANES, W. H., 1970, Facies and development of the Colorado River Delta in Texas, *in* Morgan, J. P., ed., Deltaic Sedimentation: Tulsa, Society of Economic Palaeontologists and Mineralogists Special Publication 15, p. 78–106.

KEMPE, S., PETTINE, M., AND CAUWET, C., 1991, Biogeochemistry of European rivers, *in* Degens, E. T., Kempe, S., and Richey, J. E., eds., Biogeochemistry of Major World Rivers, Scope 42: Chichester, Wiley, p. 169–211.

KESEL, R. H., YODIS, E. G., AND MCGRAW, D. J., 1992, An approximation of the sediment budget of the Lower Mississippi River prior to major human modification: Earth Surface Processes and Landforms, v. 17, p. 711–722.

KVALE, E. P. AND VONDRA, C. F., 1993, Effects of relative sea-level changes and local tectonics on a Lower Cretaceous fluvial to transitional marine sequence, Bighorn Basin, Wyoming, Usa, *in* Marzo, M. and Puigdefabregas, C., eds., Alluvial Sedimentation: Oxford, International Association of Sedimentologists Special Publication 17, p. 383–399.

LANGBEIN, W. B. AND SCHUMM, S. A., 1958, Yield of sediment in relation to mean annual precipitation: Transactions of the American Geophysical Union, v. 39, p. 1076–1084.

LEEDER, M. R., 1991, Denudation, vertical crustal movements and sedimentary basin infill: Geologische Rundschau, v. 80, p. 441–458.

LISITZIN, A. P., 1972, Sedimentation in the world ocean: Tulsa, Society of Economic Paleontologists and Mineralogists Special Publication 17, p. 1–218.

LUNDBERG, N. AND DORSEY, R. J., 1990, Rapid Quaternary emergence, uplift and denudation of the Coastal Range, Eastern Taiwan: Geology, v. 18, p. 638–641.

MACDONALD, R. W. AND PEDERSEN, T. F., 1991, Geochemistry of sediments of the western Canadian continental shelf: Continental Shelf Research, v. 11, p. 717–735.

MARKHAM, A., 1992, Erosion and sediment transport in Papua New Guinea. Network design and monitoring. Case study: Ok Tedi Coppermine: International Association of Hydrological Sciences Publication 210, p. 517–526.

MEADE, R. H., 1992, River-sediment inputs to major deltas, *in* Milliman, J. D. and Haq, B. I. eds., Rising Sealevel and Subsiding Coasts: Boston, Kluwer Academic Publishers, p. 63–85.

MEADE, R. H., DUNNE, T., RICHEY, J. E., SANTOS, U., AND SALATI, E., 1985, Storage and remobilization of suspended sediment in the Lower Amazon River of Brazil: Science, v. 228, p. 488–490.

MEADE, R. H., YUZYK, T. R., AND DAY, T. J., 1990, Movement and storage of sediment in rivers of the United States and Canada, *in* Wolman, M. G. and Riggs, H. C., eds., The Geology of North America, v. 1, Surface Water Hydrology: Boulder, Geological Society of America, p. 255–280.

MECKEL, L. D., 1970, Holocene sand bodies in the Colorado Delta area, northern Gulf of California, *in* Broussard, M. L., ed., Deltas, Models for Exploration: Houston, Houston Geological Society, p. 239–265.

MEYBECK, M., 1976, Total mineral dissolved transport by major world rivers: Hydrological Sciences Bulletin, v. 21, p. 265–285.

MEYBECK, M., 1979, Concentrations des eaux fluviales en elements majeurs et apports en solution aux oceans: Revue de Geology Dynamique et de Geographie Physique, v. 21, p. 215–246.

MEYBECK, M., 1988, How to establish and use world budgets of riverine materials, *in* Lerman, A. and Meybeck, M. eds., Physical and Chemical Weathering in Geochemical Cycles: Amsterdam, Kluwer Academic Publishers, p. 247–272.

MILLIMAN, J. D. AND MEADE, R. H., 1983, World wide delivery of river sediment to the oceans: The Journal of Geology, v. 91, p. 1–21.

MILLIMAN, J. D., QURAISHEE, G. S., AND BEG, M. A. A., 1984, Sediment discharge from the Indus River to the ocean, past, present and future, *in* Haq, B. U. and Milliman, J. D., eds., Marine Geology and Oceanography of the Arabian Sea and Coastal Pakistan: New York, Van Nostrand Rheinhold Co., p. 65–70.

MILLIMAN, J. D., AND SYVITSKI, J. P. M., 1992, Geomorphic/tectonic control of sediment discharge to the ocean: the importance of small mountainous rivers: The Journal of Geology, v. 100, p. 525–544.

MILLIMAN, J. D., YUN-SHAN, Q., MEI-E, R., AND SAITO, Y., 1987, Man's influence on the erosion and transport of sediment by Asian rivers: The Yellow River (Huanghe) example: The Journal of Geology, v. 95, p. 751–762.

MORISAWA, M., 1985, Rivers: New York, Longman, p. 220.

NAIDU, A. S. AND MOWATT, T. C., 1975, Depositinal environments and sediment characteristics of the Colville and adjacent deltas, northern arctic Alaska, *in* Broussard, M. L., ed., Deltas: Houston, Houston Geological Society, p. 283–309.

NELSON, B. W., 1970, Hydrography, sediment dispersal and recent historical development of the Po River, Italy, *in* Morgan, J. P., ed., Deltaic Sedimentation: Tulsa, Society of Economic Paleontologists and Mineralogists Special Publication 15, p. 152–184.

NELSON, C. H., 1990, Estimated post-Messinian sediment supply and sedimentation rates on the Ebro continental margin, Spain: Marine Geology, v. 95, p. 395–418.

OHMORI, H., 1983, Erosion rates and their relation to vegetation from the viewpoint of world-wide distribution: Bulletin of the Department of Geography University of Tokyo, v. 15, p. 77–91.

OLIVRY, J. C., 1977, Transports solides en suspension au Cameroun: Wallingford, International Association of Hydrological Sciences Publication 122, p. 134–141.

OLIVRY, J. C., BRICQUET, J. P., THIEBAUX, J. P., AND SIGHA, N., 1988, Transport de matiere sur les grandes fleuves des regions intertropicales: les premieres resultats des mesures de flux particulaires sur le basin du fleuve Congo. Wallingford, International Association of Hydrological Sciences Publication 174, p. 509–521.

OLLIER, C., 1981, Tectonics and Landforms: Harlow, Longman, 324 p.

PALANQUES, A., PLANA, F., AND MALDONADO, A., 1990, Recent influence of man on the Ebro margin sedimentation system, northwestern Mediterranean Sea: Marine Geology, v. 95, p. 247–263.

PICKUP, G., 1980, Hydrologic and sediment modelling studies in the environmental impact assessment of a major tropical dam project: Earth Surface Processes, v. 5, p. 61–75.

PINET, P. AND SOURIAU, M., 1988, Continental erosion and large scale relief: Tectonics, v. 7, p. 563–582.

POSAMENTIER, H. W. AND VAIL, P. R. 1988, Eustatic controls on clastic deposition II — sequence and systems tract models, *in* Wilgus, C. K., Hastings, B. S., Kendall, C. G. St. C., Posamentier, H. W., Ross, C. A., and Van Wagoner, J. C., eds., Sea-Level Changes: An Integrated Approach: Tulsa, Society of Economic Paleontologists and Mineralogists Special Publication 42, p. 125–154.

POSAMENTIER, H. W., JERVEY, M. T., AND VAIL, P. R., 1988, Eustatic controls on clastic deposition I — conceptual framework, *in* Wilgus, C. K., Hastings, B. S., Kendall, C. G. St. C., Posamentier, H. W., Ross, C. A., and Van Wagoner, J. C., eds., Sea-level Changes: An Integrated Approach: Tulsa, Society of Economic Paleontologists and Mineralogists Special Publication 42, p. 109–124.

PRENDES, H., 1983, A mathematical model with movable bed: It's application to the Parana River: Beijing, Proceedings 2nd International Symposium on River Sedimentation, Nanking, China, Water Resources and Electronic Power Press, p. 837–864.

PUIGDEFABREGAS, C., 1993, Controls on fluvial sequence architecture: Brisbane, 5th International Conference on Fluvial Sedimentology, Procedings, p. K42–K48.

ROOSEBOOM, A. AND VON HARMSE, H. J., 1979, Changes in the sediment load of the Orange River during the period 1929–1969: Wallingford, International Association of Hydrological Sciences Publication 128, p. 459–470.

SCHLAGER, W., 1993, Accommodation and supply—a dual control on stratigraphic sequences: Sedimentary Geology, v. 86, p. 111–136.

SCHUMM, S. A. AND HADLEY, R.F., 1961, Progress in the application of landform analysis in studies of semiarid erosion: Washington, D.C., United States Geological Survey Circular 437, 14 p.

SCHUMM, S. A., 1968, Speculations concerning paleohydrologic controls on terrestrial sedimentation: Geological Society of America Bulletin, v. 79, p. 1573–1588.

SELBY, M. J., 1993, Hillslope Materials and Processes 2nd edition: Oxford, Oxford University Press, 451 p.

SHANLEY, K. W. AND MCCABE, P. J., 1994, Perspectives on sequence stratigraphy of continental strata: American Association of Petroleum Geologists Bulletin, v. 78, p. 544–568.

SHAHIN, M., 1985, Hydrology of the Nile Basin: Developments in Water Science, v. 21, 575 p.

SINCLAIR, H. D. AND ALLEN, P. A., 1992, Vertical versus horizontal motions in the Alpine orogenic wedge: stratigraphic response in the foreland basin: Basin Research, v. 4, p. 215–232.

SUMMERFIELD, M. A. AND HULTON, N. J., 1994, Natural controls of fluvial denudation in major world drainage basins: Journal of Geophysical Research, v. 99, p. 13,871–13,884.

SWIFT, D. J. P. AND THORNE, J. A., 1991 Sedimentation on continental margins I—a general model for shelf sedimentation, in Swift, D. J. P., Oertel, G. F., Tillman, R. W., and Thorne, J. A., eds., Shelf Sand and Sandstone Bodies: Oxford, International Association of Sedimentologists Special Publication 14, p. 3–31.

TIMES ATLAS OF THE WORLD, 1988, London, Times Books.

UN ECAFE, 1966, A compendium of major international rivers in the Ecafe region: Water Resources Series 29, 74 p.

UNESCO, 1972, Discharge of selected rivers of the world, general and regime characteristics of stations selected: Studies Reports in Hydrology, no. 5, vol.1, 70 p.

UNESCO, 1974, Discharge of selected rivers of the world, mean monthly and extreme discharges (1969–1972): Studies Reports in Hydrology, no. 5, v. 3, part 2, 124 p.

UNESCO, 1978, Atlas of World Water Balance: New York, Unesco.

UNESCO, 1979, Discharge of selected rivers of the world, mean monthly and extreme discharges (1972–1975): Studies Reports in Hydrology no. 5, v. 3, part 3, 104 p.

VAITHIYANATHAN, P., RAMANATHAN, A., AND SUBRAMANIAN, V., 1988, Erosion, transport and deposition of sediments by tropical rivers in India: Wallingford, International Association of Hydrological Sciences Publication 174, p. 561–574.

VAN WAGONER, J. C., MITCHUM, R. M., CAMPION, K. M., AND RAHMANIAN, V. D., 1990, Siliciclastic sequence stratigraphy in well logs, core, and outcrops: concepts for high-resolution correlation of time and facies: Tulsa, American Association of Petroleum Geologists Methods in Exploration Series, 7, 55 p.

VARGA, S., BRUK, S., AND BABIL-MLADENOVIC, M., 1989, Sedimentation in the Danube and tributaries upstream from the Iron Gates (Djerdap) Dam: beijing, Proceedings of the 4th International Symposium on River Sedimentation, p. 1111–1118.

WALLING, D. E. AND WEBB, B. W., 1983, The dissolved loads of rivers: a global overview: Wallingford, International Association of Hydrological Sciences Publication 141, p. 3–20.

WALTER, H. AND LIETH, H., 1967, Klimadiagramm Weltatlas: Jena, Gustav Fischer Verlag.

WILSON, L., 1973, Variations in mean annual sediment yield as a function of mean annual precipitation: American Journal of Science, v. 273, p. 335–349.

YATSU, E., 1988, The Nature of Weathering: Tokyo, Sozosha, 624 p.

ZEITLER, P. K., 1985, Cooling history of the NW Himalaya, Pakistan: Tectonics, v. 4, p. 127–151.

ZHENGYING, Q., 1983, The problems of river control in China: Proceedings of the 2nd International Symposium on river sedimentation, Nanking, China: Beijing, Water Resources and Electronic Power Press, p. 8–19.

CYCLIC VARIABLES CONTROLLING FLUVIAL SEQUENCE DEVELOPMENT: PROBLEMS AND PERSPECTIVES

FRANK G. ETHRIDGE
Colorado State University, Fort Collins, CO 80523
LESLI J. WOOD*
Amoco Production Co., Houston, TX
AND
S. A. SCHUMM
Colorado State University, Fort Collins, CO 80523

ABSTRACT: The sequence stratigraphic model, with its initial emphasis on eustatically driven controls on sedimentary sequences, has generated considerable interest in the ultimate controls on alluvial successions. Arguments regarding controls on alluvial sequences have been ongoing for many decades. Proposed controls include global (eustatic) or local base-level fluctuation, climate, tectonics, and sediment supply. Since sediment supply is in general a function of one or more of the other three controls, the possible number can be reduced to three.

Most models and explanations for fluvial successions are too simplistic. They attempt to explain these successions on the basis of a single controlling factor. Additionally the models commonly fail to take into account modern geomorphic concepts of complexity and ignore the fact that controls other than base-level fall can produce incision and that base-level lowering does not always result in incision and rejuvenation of a fluvial system. They also fail to realize that base-level fluctuations may have their greatest effect only in the lower reaches of a fluvial system and that the amounts of sediment produced by incision alone cannot account for the volume of sediment observed in most stratigraphic sequences. For these reasons it is difficult to justify the application of systems tracts designations to upstream portions of fluvial valley-fill successions.

Field and analog experiment studies demonstrate the difficulty of distinguishing various controls in fluvial systems because (1) similar erosional and depositional features and sequences can be produced by different processes and vice versa and (2) different levels of sensitivity may result in a minor, a major, or no response of a system to an extrinsic perturbation.

CURRENT MODELS AND THEIR SHORTCOMINGS

Early versions of the sequence stratigraphic model were based on the premise that all significant sedimentary cyclicity is eustatically controlled (Posamentier et al., 1988) or, at least, controlled by relative changes in sea level (Van Wagoner et al., 1990). This model, while being criticized by numerous researchers (Miall, 1986, 1991, 1992; Galloway, 1989; Walker, 1990; Wescott, 1993; and Schumm, 1993) has created a revolution in the way that sedimentary rocks are subdivided, correlated and mapped. This revolution has led to a better understanding of the origin and distribution of hydrocarbon reservoirs. It has also generated renewed interest in the ultimate controls on sedimentary sequences, especially fluvial and valley-fill succession. Interest in the origin and evolution of fluvial valleys and their fill, in light of modern geomorphic concepts, has led to an appreciation that factors other than eustasy can also give rise to valley formation and filling, and that eustatic changes do not necessarily lead to valley formation or valley filling (Schumm, 1993; Posamentier and James, 1993; Shanley and McCabe, 1994; Schumm and Ethridge, 1994; and Zaitlin et al., 1994). Even Lane (1955), who is often quoted in arguments regarding the effect of base-level control on fluvial incision and aggradation, recognized that changes in other factors such as water and sediment discharge and particle size were also important controls. In fact, the first two historical cases that he discusses are analogous to changes that might result from climate or tectonic controls rather than downstream base-level controls.

The relative and absolute effects of sea level (base level), tectonic and climatic changes on alluvial architecture and sequences continue to be a subject of considerable debate especially for non-marine deposits (Shanley and McCabe, 1994). Sediment supply is also listed as an influencing factor by several authors, however, since sediment supply is really a function of one or a combination of the three factors listed, it will not be regarded here as a major allocyclic control. Tectonics, other than basin subsidence and climate are sometimes considered as local factors that must be incorporated into the sequence stratigraphic model before it can be applied to a specific basin (Posamentier et al., 1988; Posamentier and Weimer, 1993). Schlager (1993) and Ryer (1994), however, argue that rates of change in accommodation (largely controlled by sea-level changes) and sediment supply (largely controlled by tectonic and climatic changes) operate as dual controls on stratigraphic sequences (Fig. 1). Cloetingh (1991) argues, from a numerical modeling viewpoint, that glacial eustatic fluctuations and tectonics sometimes operate on similar time scales and that both are of importance in controlling patterns of erosion and sedimentation. Leeder (1993) goes even further when he suggests that all three

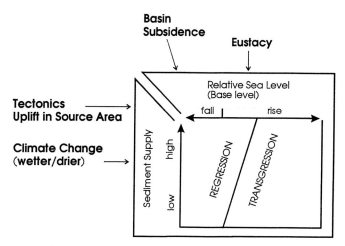

FIG. 1.—Diagram relating transgressions and regressions to changes in relative sea level (base level) and sediment supply. Major allocyclic controls are shown above and to the left of the main diagram (modified after Curray, 1964; Ryer, 1994).

*Present address: Bureau of Economic Geology, University of Texas at Austin, Austin, TX 78713
Relative Role of Eustasy, Climate, and Tectonism in Continental Rocks, SEPM Special Publication No. 59

controls overlap in scale and time to control the meso-architecture of sedimentary deposits (Fig. 2).

Numerous observations and data from Late Quaternary fluvial sequences suggest that the importance of sea-level changes diminish with increasing distance inland from the shoreline (Saucier, 1981, 1991; Koss, 1992; Schumm, 1993; Posamentier and James, 1993; Koss et al., 1994; Shanley and McCabe, 1994; and Zaitlin et al,. 1994). These observations and the continuing debate over the relative importance of all three controls on the nature and architecture of fluvial deposits have prompted this review of cyclic variables that control fluvial sequence development. The objectives of this paper are to: review evidence presented for and against major external (allocyclic) controls on fluvial successions including valley fills, point out some of the problems associated with the interpretation of these controls in the rock record, and suggest areas of future research.

BASE LEVEL, CLIMATE, AND TECTONIC CONTROL

General

Base level, climatic, and/or tectonic changes have all been used to explain the origin and/or evolution of Holocene and ancient fluvial systems and valley-fill deposits. Table 1 gives a long, but by no means, exhaustive list of examples along with inferred controls. Of the 31 references listed, 11 attribute changes in fluvial character in whole or in part to base level, 12 to climate and 17 to tectonics. Twenty-three of the references attribute changes to a single control. It is probable that all of the fluvial systems described in these papers were affected to some degree by more than one of the three controls and that each time the system was perturbed by an allocyclic control or controls, the response was complex. The complex nature of the response of fluvial systems to changes in one or more allocyclic control is discussed in Schumm (1977, 1993), Schumm and Brakenridge (1987), Wescott (1993), and Schumm and Ethridge

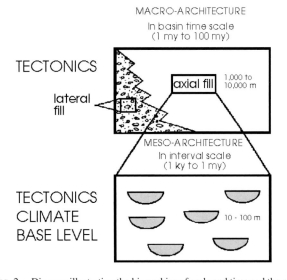

Fig. 2.—Diagram illustrating the hierarchies of scale and time and the major allocyclic controls that determine the architecture of fluvial successions at the macro- and meso-architectural levels. Ranges of time scale of formation and thickness of deposits are listed above and within the two large boxes (modified after Leeder, 1993).

(1994). As an example numerous fluvial systems show evidence of a complicated erosional and depositional history resulting from a single base-level change event. It is easy to imagine how some or all of these erosional and depositional events might be attributed to a separate allocyclic control.

Base Level (Sea Level)

There is no consensus on how a stream profile will vary with a change in base level (Butcher, 1990). Schumm (1993) concludes that the effect of base-level change depends on the rate, amount and direction of change, river character and dynamics and erodibility of the sediment source area.

Posamentier and Vail (1988) concluded that the dominant response of fluvial systems to a rapid base-level fall is rejuvenation and incision, whereas a slow base-level fall would result in a general, widespread denudation of the landscape and a lack of significant stream rejuvenation and incision. Pittman et al. (1991), in proposing an alternate model for the development of sedimentary sequences, argue that onlap can even occur during sea-level fall if the rate of fall is less than the rate of subsidence at the shoreline minus the rate of sedimentation. Butcher (1990), in an examination of the knickpoint concept, concluded that relative sea-level rise will generally cause alluviation, and sea-level fall incision. However, given the competing factors of tectonism and climate almost any scenario is possible. Posamentier et al. (1992) suggested three different possible responses to a relative sea-level fall depending on the difference in angle between the fluvial profile and the surface exposed by the fall (Fig. 3A). Schumm (1993) carried this discussion further by suggesting that the fluvial response to a base-level change will depend not only on the difference in the angles, if any, but also the magnitude of the difference (Fig. 3B). Under certain situations, channels can adjust simply by a change in pattern (Fig. 3B–b, c and d). Inferred changes from meandering to anastomosing channels the Rhine-Meuse delta were probably caused by a rapidly rising base level between 7500 and 4000 yr BP (Tornqvist, 1993). Apparently these changes in planform pattern affected depositional geometries some 65–km upstream from the mouth of the present river system. Saucier (1991) suggested that smaller coastal-plain streams, like the Pearl and Sabine in Louisiana, respond to sea-level changes primarily by lengthening or shortening with accompanying changes in sinuosity and channel shape and without significant entrenchment and valley filling. He concluded that late Wisconsin sea-level changes probably had little or no effect on these rivers upstream of the modern deltaic and chenier plain.

Even if the nature of the downstream response of a fluvial system to base-level change can be determined, other considerations such as the relation between downstream and upstream responses must be considered. In most fluvial systems, upstream and downstream reaches are out-of-phase with respect to aggradation and degradation, as demonstrated by analog experiments (Germanoski, 1991; Ware et al., 1992; Schumm, 1993) and studies of modern and ancient systems (Saucier 1981, 1991). In the lower Mississippi River valley, at or shortly before maximum glaciation, rapid aggradation occurred in the northern portion in response to massive amounts of outwash. This period of rapid aggradation in the northern portion of the fluvial valley corresponded with lowstand and a time of max-

TABLE 1.—PARTIAL LIST OF RESEARCH PAPERS, IN WHICH THE AUTHORS INFERRED ONE OR MORE ALLOCYCLIC PROCESS AS RESPONSIBLE FOR FLUVIAL OR VALLEY FILL DEPOSITS OR FOR CHANGES IN THESE DEPOSITS THROUGH TIME

AGE	LOCATION	BASE LEVEL	CLIMATE	TECTONICS	REFERENCE
Upper Cretaceous-Lower Eocene	Alberta, Canada			X	McLean & Jerzykiewicz (1978)
Middle Jurassic	England	X		X	Nami & Leeder (1978)
Devonian	South Wales	X	X		Allen & Williams (1982)
Miocene	Pakistan		X	X	Behrensmeyer & Tauxe (1982)
Upper Triassic	Utah & Arizona			X	Blakey & Gubitosa (1984)
Eocene-Oligocene	South Dakota		X		Retallack (1986)
Middle Eocene	Wyoming			X	Groll & Steidtmann (1987)
Lower Eocene/Plio-Pleistocene	Wyoming/Idaho			X	Kraus & Middleton (1987)
Miocene	Spain			X	Nicholas (1987)
Cretaceous	Wyoming			X	Shuster & Steidtmann (1987)
Devonian	New York			X	Willis & Bridge (1988)
Early Cretaceous	New Mexico	X			Aubry (1989)
Lower Jurassic	North Sea	X	X	X	Nystuen et al. (1989)
Holocene	U.S. Great Plains		X		Hall (1990)
Devonian	Greenland		X		Olsen (1990)
Holocene	Iowa		X		Van Nest & Bettis (1990)
Upper Cretaceous	Utah	X			Shanley & McCabe (1991)
Middle Jurassic	England	X			Eschard et al. (1992)
Holocene	Alaska		X		Ashley & Hamilton (1993)
U. Cretaceous/Paleocene	France		X	X	Cojan (1993)
Neogene	Nebraska	X	X		Diffendal & Lowrie (1992)
Carboniferous	Wales	X		X	Hartley (1993a)
Holocene	Arizona		X		Herford (1993)
Lower Cretaceous	Wyoming	X			Kvale & Vondra (1993)
Triassic	Spain			X	López-Gómez & Arche (1993)
Jurassic	England	X			Mjøs & Prestholm (1993)
Upper Cretaceous	Wyoming			X	Martinsen et al. (1993)
Plio-Quaternary	Spain			X	Mather (1993)
Mississippian	Alaska			X	Melvin (1993)
Devonian	Greenland		X		Olsen & Larsen (1993)
Middle Cretaceous	Wyoming-Utah	X		X	Ryer (1994)

imum exposure and incision of the Gulf of Mexico continental shelf and development of shelf-margin deltas (Saucier, 1991, Table 1).

In general, the zone of incision caused by base-level lowering propagates further upstream than the zone of deposition caused by base-level rise, and the average depth of incision associated with base-level fall is much greater than the average thickness of deposits caused by the subsequent rise in base level. These observations have been made during experimental studies of base-level change (Ethridge et al., 1991; Koss, 1992) and in modern fluvial valleys (Schumm et al., 1984). This is because the incision occupies only part of the valley floor. Flow, which previously spread widely over the flood-plain during floods, is now concentrated in the deeper channel and headward erosion can be rapid. However, during base-level rise not only is the channel filled, but the slope of the entire valley floor is modified by deposition, which requires a much longer period of time.

The upstream distance that is effected by base-level change has been the subject of considerable debate (Table 2). Lane (1955) observed significant channel floor aggradation in the Little Colorado River 70–km upstream from Imperial Dam. In contrast, Leopold and Bull (1979) concluded that base-level changes affect the vertical position of rivers only locally and to a minor extent. For rivers along the tectonically active coast of northern California (Merritts et al., 1994) relative sea-level changes affected only the lower few tens of kilometers. The middle and upper portions of these rivers are dominated by long-term uplift. Fisk (1944) concluded that Pleistocene sea-level lowering caused bedrock incision far upstream in the Mississippi River valley. In contrast Saucier (1981, 1991) states that the direct influence of late Wisconsin sea-level fall ex-

tended up valley only to the latitude of Baton Rouge, Louisiana; a distance of 370 km above the Head of Passes (Schumm, 1993). Blum (1992, 1994a, b) concludes that for another large extrabasinal stream, the Colorado River system in Texas, there is no correlation between base-level changes and periods of deposition, incision or flood-plain abandonment and soil formation upstream of the Holocene alluvial-deltaic plain; a distance of less than 100 km.

Computer modeling studies also appear to give conflicting results in terms of the upstream effects of base-level change. Paola (1991) suggests that for reasonable input parameters, it is expected that fluctuations on a scale of 100,000 yr will be felt upstream for a distance of 100 km. On the other hand, Humphry (1991) suggests that most fluvial systems are ineffective in transmitting base-level changes upstream in time frames ranging from 10,000 to 100,00 yr because of the nonlinearity of the stream profile and climatic variability.

Suggestions by Aubry (1989) for Cretaceous deposits of New Mexico and Colorado and Diffendal and Lowrie (1992) for Neogene deposits of the Great Plains that there is a direct correlation between sea-level fluctuations and inland fluvial stratigraphy are not supported by the data and observations described above. Schumm (1993) makes a geometrical argument, in the case of the Cretaceous deposits, that adjustments to such a minor change in channel gradients over long distances could easily be accommodated by changes in channel pattern without the need for significant cutting and filling. These arguments would be even more compelling for the Neogene deposits of the Great Plains because of the greater distances involved. Even if correlations based on limited age dates are reasonable, it

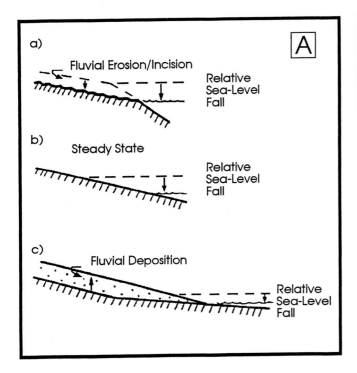
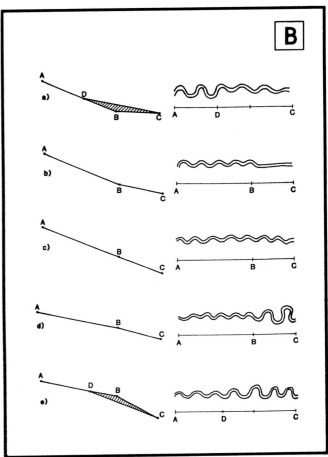

FIG. 3.—(A) Three possible scenarios of fluvial response to a relative sea-level fall: (a) exposure of a shelf surface steeper than the graded alluvial profile results in fluvial incision; (b) exposure of a shelf surface with the same gradient as the graded alluvial profile results in a steady state condition with no fluvial erosion or deposition; (c) exposure of a shelf surface with a lower gradient than the graded alluvial profile results in deposition (modified from Posamentier et al., 1992). (B) Effects of sea-level (base-level) change across continental shelves with different inclination. In (a) and (b), the angle of the continental shelf is less than that of the stream channel. In (c), the angles are the same, and in (d) and (e), the angle of the continental shelf is greater than that of the stream channel. In (a), the decrease in slope is large, the stream channel cannot adjust to this change by a pattern change and deposition results. In (b), the decrease in slope is small and the stream channel adjusts with a decrease in sinuosity. In (c), the stream channel extends across the surface without aggradation or degradation. In (c), the increase in slope is small and the stream channel adjusts with an increase sinuosity. In (d), the increase in slope is large and deposition occurs (from Schumm, 1993).

doesn't necessarily imply a similarity of causes as discussed below under convergence, divergence and sensitivity.

Tectonics

It has long been recognized that tectonic uplift and subsidence can affect the form, character and location of fluvial systems and that tectonic uplift can cause incision and valley formation (Schumm et al., 1987). Sloss (1988, 1991) concluded that tectonics is a major factor in the development of cratonic sequences. Leeder (1993) suggested that tectonics is the dominant factor at the basin-wide scale and controls the macro-architecture of terrigenous clastic deposits (Fig. 2). There is little dispute over these conclusions, however, at the meso-architecture scale (Fig. 2) where the controls of tectonics, climate and base level overlap and interfere, there continues to be considerable debate over the relative and absolute importance of eustasy and tectonics. Generalizations regarding the relative depth of incision along a fluvial system (Posamentier and

James, 1993), downfilling versus backfilling, and sediment delivery versus time (Schumm, 1993) need testing and quantification before they can be used to interpret the rock record, especially at an outcrop or well log scale.

Clastic wedges and, in particular, their fluvial/alluvial-fan components were once used to date tectonic activity in foreland basins, where the flux of clastic sediments into the basin was believed to coincide with thrusting and uplift in the orogenic belt. Recent studies by Leeder and Gawthorpe (1987), Heller et al. (1988) and Blair and Billodeau (1988), however, have demonstrated that just the opposite is true and they proposed a new two-stage model. In this model, progradation of a clastic wedge across the basin most commonly occurs during times of reduced rates of thrusting when the foredeep does not act as a sediment trap. During active tectonism, coarse-grained fluvial and alluvial-fan deposits are concentrated in a narrow, foredeep zone immediately adjacent to the uplifted terrain. Mack and Seager (1990) and Lang (1993) have applied this concept to fluvial sequences in New Mexico and Australia respectively.

TABLE 2.—UPSTREAM EFFECT OF BASE-LEVEL (SEA-LEVEL) CHANGE ON FLUVIAL SYSTEMS

REFERENCE	AGE	LOCATION	DISTANCE OF UPSTREAM EFFECT
Fisk (1944)	Holocene	Miss. valley	suggested that the entire alluvial valley up to Cairo was probably affected
Lane (1955)	Holocene	Little Colorado River	base-level rise as result of dam caused a rise of the stream bed 70-km upstream in 7 yr
Saucier (1981, 1991)	Holocene	Miss. valley	prism of Holocene back-swamp deposits extends only up to Baton Rouge, Louisiana, a distance of 200 km from the head of passes
Paola et al. (1991)	not applicable	computer model	fluctuations on a scale of 100,000 yr will be felt a distance of 100 km
Humphry (1991)	not applicable	computer model	ineffective in transmitting base-level changes upstream based on non-linearity of profile and climate changes
Blum (1992, 1994a,b)	Holocene	Colorado River, TX	sea-level fluctuations effective up to apex of alluvial-deltaic plain, less than 100-km upstream
Törnqvist (1993)	Holocene	Rhine-Meuse delta	change in planform from meandering to anastomosing as a result of rapid sea-level rise over a distance of 65 km
Merritts et al. (1994)	Pleistocene and Holocene	three rivers, northern CA	eustatically-driven depositional wedge extends tens of km upstream on all three rivers
Koss et al. (1994)	not applicable	analog model	only the exposed shelf and coastal plain were affected by sea-level fluctuations; no significant effect on the drainage basin

TABLE 3.—GENERAL CRITERIA FOR DISTINGUISHING EUSTATIC AND TECTONIC CONTROLS ON THE DEVELOPMENT OF SEQUENCES AND/OR CLASTIC WEDGES. CRITERIA, AS LISTED, FAVORS TECTONIC CONTROLS. (A) MODIFIED FROM MØRK ET AL. (1989); EMBRY (1990) AND CLOETINGH (1991). (B) MODIFIED FROM RYER (1994).

(A)
- sediment source areas vary greatly from one sequence to the next
- sediment regimes change drastically and abruptly across sequence boundaries
- faults terminate at sequence boundaries
- significant changes in subsidence and uplift patterns within basins across sequence boundaries
- sequence boundaries absent in parts of basins
- significant differences in the magnitude and extent of some of the subaerial unconformities and time-equivalents recognized by others.

(B)
- localized distribution of clastic wedge
- relatively thick wedges, especially near the thrust front
- lack of significant unconformities
- widespread conglomerates

Criteria for differentiating eustatic and tectonic controls on sequences and clastic wedges in general are summarized by Mørk et al. (1989) and Embry (1990) for the Canadian Sverdrup basin and the Barents shelf and Ryer (1994) for the Western Cretaceous Interior (Table 3). While they take different approaches to the problem and have worked in different types of basins some similarities occur in their criteria. Mørk et al. (1989) and Embry (1990) suggest that sequences once thought to be controlled entirely by eustasy are now thought to have been dominated by tectonic controls. Ryer (1994) describes clastic wedges that are inferred to have been dominated by tectonics, eustasy and by a combination of the two controls.

Heward (1978), Erikkson (1978), Steel and Asheim (1978), Anadón (1986), Mack and Seager (1990), Crews and Ethridge (1993), and Hartley (1993b) all argue for tectonic control on coarse-grained basin-margin alluvial-fan deposits ranging in

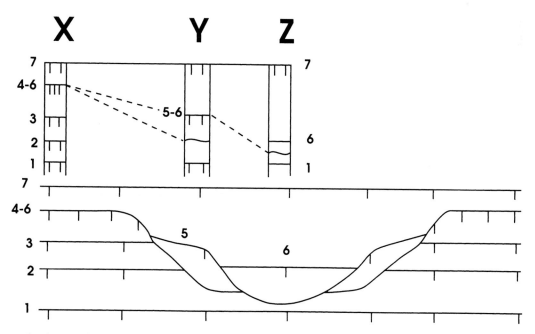

FIG. 4.—Diagram showing complex soil stratigraphies generated by simple phases of valley incision and filling (Note: the complex cutting and filling represented by units 4–6 might result from a single event such as a base-level change). The numbers 1 to 7 designate various geomorphic surfaces. The letters X through Z designate outcrop exposures. Short vertical lines indicate paleosols. Greater horizontal density of these lines indicates greater length of time for pedogenesis (modified from Wright, 1992). The difficulty of reconstructing the complicated two-dimensional stratigraphy of valley incision and filling from limited outcrop exposures is seen in this comparison.

TABLE 4.—VARIOUS INTERPRETATIONS OF CLIMATIC CONTROL ON ANCIENT FLUVIAL SUCCESSIONS (FROM RETALLACK, 1986; MAIZELS, 1987; HALL, 1990; DIFFENDAL AND LOWRIE, 1992; ASHLEY AND HAMILTON, 1993; HEREFORD, 1993; LÓPEZ-GOMEZ AND ARCHE, 1993; LEOPOLD, 1994).

- Change from moist to dry climate = incision
- Change to a moister climate = incision
- Catastrophic flooding = local incision
- cold climate = eroding and depositing
- Climate with wet and dry seasons = reactivation surfaces

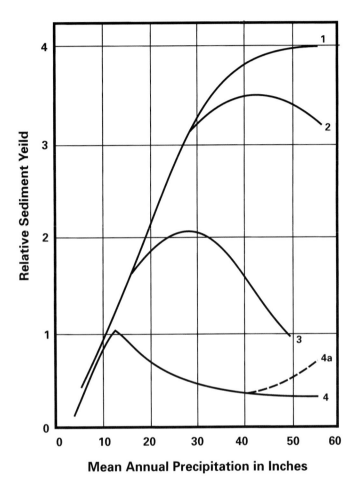

FIG. 5.—Hypothetical series of curves illustrating the relationship between precipitation and relative sediment yield during geologic time. Curve 1, before the appearance of land vegetation; curve 2 after the appearance of primitive vegetation on the earth's surface; curve 3 after the appearance of flowering plants and conifers; curve 4 after the appearance of grasses. The curves show only the relative differences to be expected for a given rock type (from Schumm, 1968).

age from Pliocene to Precambrian. Arguments in favor of tectonism includes the coarse-grained nature of the deposits, thick coarsening-upward/fining-upward sequences, proximity to actual or inferred uplifted terrane, evidence of unroofing, relatively high sedimentation, syntectonic intraformational unconformities, and/or that complex heterogeneity suggested by short-term climatic fluctuations are not present. A tectonic argument is certainly easier to make for coarse-grained, basin-margin fans than for other fluvial systems, however, a climatic argument has been made for some fan deposits.

McLean and Jerzykiewicz (1978), Blakey and Gubitosa (1984), Kraus and Middleton (1987), Groll and Steidtmann (1987), Nicholas (1987), Shuster and Steidtmann (1987), Lawrence and Williams (1987), Willis and Bridge (1988), López and Arche (1993), Martinsen et al. (1993), Mather (1993), Melvin (1993) argue that the dominant control on observed cyclicity or vertical changes in ancient fluvial successions ranging in age from Pleistocene to Devonian was tectonism. Evidence cited for these conclusions include reversals or changes in the dominant flow directions as determined by paleocurrent data, changes in the rate of subsidence as determined from subsidence curves, relation between major bounding surfaces and inferred basin margin faulting, and similarity of observed geometries and stacking patterns to results from numerical models that evaluate avulsion/subsidence ratios (Allen, 1979; Bridge and Leeder, 1979). Some of authors suggest that tectonism was the dominant control because they could not find direct evidence to support climatic or eustatic changes and some admit that controls other than tectonism could have produced similar results in fluvial deposits.

Climate

Climate has been suggested as a major control on modern and ancient alluvial fan successions (Talbot and Williams, 1979; Lettis, 1985; Ponti, 1985; Dorn et al., 1987; Frostick and Reid, 1989; Blair et al., 1990; Wright, 1992). Wright (1992) discusses the importance of soils in the recognition of climate as the major control on fan development. He discusses six fan models, with the climatically controlled fans being those in

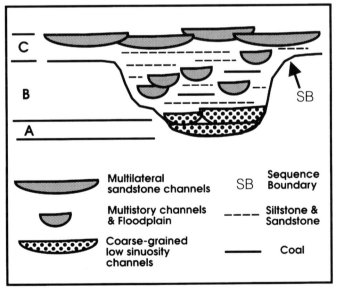

FIG. 6.—Hypothetical cross section of a fluvial succession showing the relationships between various portions of the succession and systems tracts (based in part on diagrams by Wright and Marriott, 1993; Boyd and Diessel, 1994; Zaitlin et al., 1994). In this cross section letters A, B, and C replace the Lst, Tst and Hst designations commonly illustrated. If a particular valley-fill succession can be demonstrated to have formed within the zone of sea-level influence (the downstream subsiding alluvial-deltaic plain) and key surfaces can be correlated basinward, then the system tract designations may be appropriate. In most other cases, use of the systems tracts designations is inappropriate.

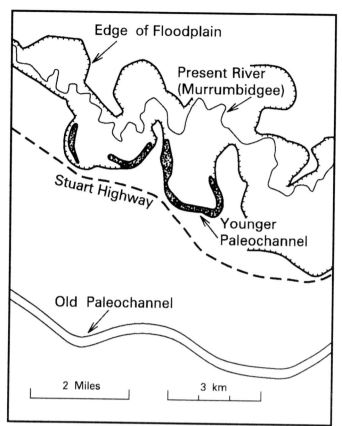

FIG. 7.—Riverine Plains near Darlington Point, New South Wales, Australia. The sinuous Murrumbidgee River flows to the west (left) at the top of the figure. The irregular floodplain contains large meander scars and oxbow lakes (younger paleochannel). The oldest paleochannel crosses the lower part of the figure (modified from Schumm and Brakenridge, 1987).

which soils were strongly developed and covered the entire fan during periods of stability. The basic premise is that during colder periods increased river discharge led to entrenchment and flooding of valley floors to form perennial lakes. In this case, large areas of the piedmont and valley border slopes became stabilized with vegetation (Wright, 1992).

However, the interpretation of climatic control on ancient alluvial fans is difficult for several reasons. Blair and McPherson (1994) provide data from modern alluvial fans to invalidate the concept of wet and dry alluvial fans. Their Table 1 suggest that debris-flow deposits are common in both desert and non-desert alluvial fans. Wright (1992) suggested a potential problem in stating that soil distribution and development on the so-called climatically-controlled fans resembles that of a fully entrenched fan. The apparently similar patterns may result from both tectonic uplift and climate change. A second problem involves the correlation and dating of paleosols. A number of studies of Late Cenozoic alluvial fans suggest climate control as a major factor in fan development. This conclusion is based on the assumption that paleosols with the same degree of development, response to in situ shear-wave velocities and amount of dissection are synchronous through the extent of preserved fan deposits. Without absolute age dates, which are, for the most part absent, this assumption is unwarranted. A third

problem is the difficulty involved in differentiation of auto-cyclic and allocyclic controls. Weaver (1984) and Schumm et al. (1987) proposed an autocyclic, threshold model for alluvial fan evolution. Because of the interaction of allocyclic controls such as climate change and autocyclic changes such as episodic fan head trenching, it might be expected, as observed in east-central Idaho, that various fans in the same general area would show dramatically different responses to a regional climate change (Weaver, 1984). To complicate matters even more, Fros-trick and Reid (1989) suggested climatic control for Pleistocene fans of the Dead Sea on the basis of sheet-like character and interbedding of coarse-grained fan deposits and fine-grained lake deposits and the absence of coarsening- and fining-upward sequences or significant trenching.

Climatic variation has also been suggested as the major control on some fluvial successions (Behrensmeyer and Tauxe, 1982; Retallack, 1986; Maizels, 1987; Hall, 1990; Van Nest and Bettis, 1990; Olsen, 1990; Wright, 1992; Ashley and Hamilton, 1993; Hereford, 1993; López-Gómez and Arche, 1993; Olsen and Larsen, 1993; Trewin, 1993). The wide variety of interpretations (Table 4) leaves one confused as to just what are the effects of major climatic changes on alluvial systems, because in many cases tributary incision can cause main stream deposition (Schumm 1965) as a result of climate change. Paleosol type, development and distribution should provide critical evidence for the interpretation of climatic factors. Wright (1992), however, demonstrated that the complex soil stratigraphies developed during valley incision and fill will be difficult to reconstruct with limited exposures (Fig. 4). Schumm (1965, 1993) concluded that the geomorphic effect of major climatic changes will depend on the nature of the pre-change and the post-change climates. For example, at present, a change from semiarid to arid under natural conditions will result in decreased sediment delivery with time whereas a change from humid to semiarid will result in increase in sediment delivery. This is the result of the dramatic vegetation change from sparse desert vegetation to grasslands of the semiarid to humid forests with increasing precipitation (Langbein and Schumm, 1958). It should be remembered also that vegetation evolved through geologic time with considerable significance for sediment yields. For example, before the appearance of terrestrial vegetation, erosion and sediment yield rates should have increased with precipitation to some maximum value (Fig. 5). As primitive land plants and eventually flowering plants appeared, erosion and sediment yield rates should have decreased above some relatively high value of precipitation. Only with the appearance of grasses is there a peak of sediment yield at about 330 mm (13 in; Fig. 5; Schumm, 1968). Perhaps the best that can be said at this point is that correlation between climate change and fluvial activity is possible, but prediction of the geomorphic effect is difficult at best (Van Nest and Bettis, 1990).

FLUVIAL VALLEY FILLS AND SYSTEMS TRACTS

Van Wagoner et al. (1990) suggested a correlation between fluvial architectural patterns and systems tracts. More recently attempts have been made to relate various parts of a valley-fill succession to lowstand (LST), transgressive (TST) and highstand (HST) system tracts (Shanley and McCabe, 1991, 1993, 1994; Vail et al., 1991; Mjøs and Prestholm, 1993; Wright and

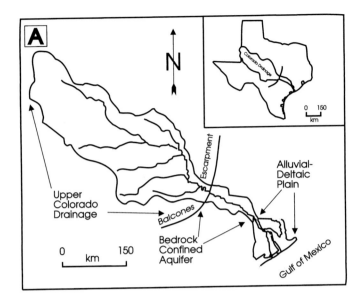

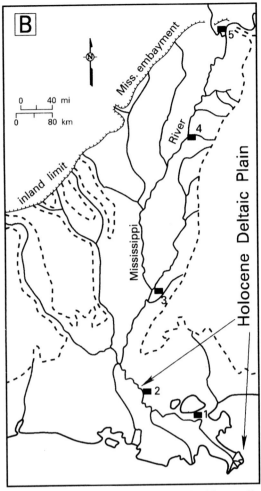

FIG. 8. (A) Map of Colorado River in Texas, showing upper part of drainage basin above the Balconies Escarpment, the bedrock-confined lower valley and the lower Colorado alluvial-deltaic plain (modified from Blum, 1994a). (B) Map of the lower Mississippi River, showing the Holocene deltaic plain and the Pleistocene bedrock-confined lower valley (that portion between of 2 to 5; dashed lines define the limits of the Pleistocene valley). Number refer to cities along the modern Mississippi River (1 = New Orleans, LA; 2 = Baton Rouge, LA; 3 = Vicksburg, MS; 4 = Memphis, TN; and 5 = Cairo, IL; modified from Saucier, 1991). Note that in both examples the portion of the fluvial system directly affected by sea-level fluctuations is restricted to the subsiding deltaic or alluvial-deltaic plain.

TABLE 5.—RELATIVE WIDTH/DEPTH RATIOS OF EXPERIMENTAL, INCISED VALLEYS ILLUSTRATING THE CONCEPT OF CONVERGENCE (DATA FROM WOOD ET AL. 1993, 1994A, B).

➤ Relationship between rates of base-level change and w/d ratio: Faster rates of change = higher ratio.
➤ Relationship between shelf angle and w/d ratio: Lower shelf angle = higher ratio.
➤ Relationship between magnitude of base-level change and w/d ratio: Lower magnitude = higher ratio.

Marriott, 1993; Zaitlin et al., 1994; Boyd and Diessel, 1994). Publications on this subject note that many Holocene and ancient valley fills have a somewhat similar succession of deposits that grade upward from coarse-grained, low-sinuosity fluvial deposits to isolated channel or point bar deposits encased in extensive floodplain fines (Fig. 6). Some authors suggest that the succession may be capped by extensive fluvial sheet sands. Explanations for this succession usually center around the occurrence of higher gradients following relative sea-level lowstand and incision and changes in the rate at which accommo-

dation is created throughout the filling and overtopping of valleys. Potential problems associated with this interpretation are illustrated with examples from the Riverine Plain of Australia, the Mississippi and Colorado River valleys, and the Canterbury Plains of New Zealand.

The lowest alluvium in many modern valley fills is coarse grained and probably was deposited by a high-gradient low-sinuosity channel. These deposits generally change upward to deposits formed in large meandering streams and finally to those of smaller, modern, meandering channels (Schumm and Brakenridge, 1987). This environmental succession is observed in glaciated and in unglaciated regions. The Riverine Plain of New South Wales, Australia (Fig. 7) displays a similar overall history in an area where tectonics and sea-level change were negligible and climate change from relatively dry to wet to intermediate was probably responsible for the changing channel patterns (Schumm and Brackenridge, 1987). Thermoluminescence chronology (Page et al., 1991) has clarified age relations but has not altered the basic history.

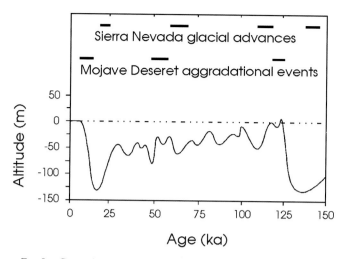

FIG. 9.—Comparison of times of insolation maxima during the past 150 ky, major glacial advances in the Sierra Nevada and times of aggradation events in the Mojave Desert (modified from Bull, 1991).

As noted above, the effects of Holocene sea-level rise are restricted to the area downstream from the subsiding alluvial-deltaic plain on both the Colorado River in Texas (Fig. 8A) and the Mississippi River (Fig. 8B; Saucier 1981, 1991; Blum, 1994a, b). In the alluvial valleys upstream of this area, the complex sequence of erosion, deposition and soil formation are related to climatically-controlled changes in sediment supply, which are out of phase with glacio-eustatically-driven base-level changes. Leckie (1994) interprets incision and valley formation to be actively occurring along some coarse-grained rivers of the southern Canterbury Plains, South Island of New Zealand during the modern highstand. Both upstream and downstream areas are undergoing incision but for entirely different reasons. In upstream areas, incision is a result of tectonic uplift of the Southern Alps and near the coast it occurs as a result of coastal retreat. Other areas of the coastline, such as those north of Banks Peninsula, a major bedrock headland, are undergoing progradation, and fluvial incision is only occurring in the upstream portions of rivers.

Based on the above discussion, it would appear that attempts to define ancient valley-fill deposits by association with their chronostratigraphically-equivalent marine systems tracts may be valid in some instances for those that formed basinward of the apex of the actively subsiding fluvio-deltaic plain but probably incorrect for the more landward portions of most, if not all, fluvial valleys. We should not, as some models suggest, assume that the nature of erosional surfaces and valley-fill deposits reflect only the adjustment of a graded-river profile to base-level changes (Boyd and Diessel, 1994; Zaitlin et al., 1994) or that the rates of sediment supply remain constant (Posamentier et al., 1988) when interpreting ancient fluvial valley-fill deposits. Legarreta et al. (1993), Martinsen (1994) and Olsen et al. (1995) all recognize regionally extensive discontinuity surfaces in ancient fluvial deposits but argue for caution in the application of systems tract terminology and the interpretation of specific allocyclic controls. Instead, they focus interpretations of fluvial architecture and stacking patterns on the rate of relative base-level change (i.e., the rate of creation or destruction of accommodation space).

CONVERGENCE, DIVERGENCE AND SENSITIVITY

In geomorphology, the term **convergence** is used to describe a situation where different processes and/or causes produce similar results (Schumm, 1991, p. 58). This concept was noted by Cojan (1993) in a study of large-scale Upper Cretaceous/Paleocene sequences of fluvial and lacustrine rocks in the Provence basin of southern France. She concluded that these sequences resulted from variation in sediment supply, stage fluctuations, tectonic activity or climatic conditions and that any one or a combination of these factors can lead to the observed sequential arrangement. A number of other authors including Nami and Leeder (1978), Allen and Williams (1982), Behrensmeyer and Tauxe (1982), Nystuen et al. (1989), Hartley (1993a), Diffendal and Lowrie (1992), López-Gómez and Arche (1993), Olsen and Larsen, (1993) and Ryer (1994) conclude that two or more controls may have been important in origin and evolution of ancient fluvial systems. The concept was clearly observed in a series of flume experiments (Wood et al. 1993; Wood et al., 1994a, b). In these analog experiments it was determined that incised valley width/depth ratios increased under conditions of increasing rates of base-level change, decreasing shelf angles and magnitudes of base-level change (Table 5).

Divergence is the opposite of convergence and refers to similar processes and/or causes producing different effects (Schumm, 1991, p. 62). The most obvious example of divergence, as related to sequence stratigraphy, is that base-level lowering can result in aggradation, erosion, change in channel form or no change depending on the magnitude and difference between the slope of the channel and the slope of the newly exposed surface as discussed above (Fig. 3B).

Sensitivity is the susceptibility of landforms to alter in response to external change (Schumm, 1991). Many examples of sensitivity are found in the geomorphic literature, but only two will be reviewed here. Van Nest and Bettis (1990) noted that the presence of significant valley-fill deposits in a small drainage basin in central Iowa contrasted with the absence of deposits of the same age in western Iowa valleys. This happened in spite of the fact that the direction and magnitude of Holocene climate change was similar in western and central Iowa. A second example by Bull (1991) illustrates the complex behavior of hillslopes in the Mojave Desert and the San Gabriel Mountains. A comparison of the Holocene sea-level fluctuation curve, Sierra Nevada glacial advances and Mojave Desert aggradation events in a period of 150,000 yr (Fig. 9) show good agreement between sea-level fluctuations and geomorphic responses to climatic change in many study areas. This observation then begs the question as to why there are not more aggradational events in the Mojave Desert and San Gabriel Mountain streams. The answer lies in the sensitivity of the hillslope subsystem. Major aggradational events occurred only when large amounts of bedload sediment were rapidly eroded from fully-stocked hillslope sediment reservoirs (Bull, 1991). Without this readily available sediment source, aggradation was impossible. Therefore, sea-level rise did not necessarily trigger an aggradational response.

AREAS OF POSSIBLE FUTURE RESEARCH

In a general sense, the problem of developing predictive models of fluvial systems is not to choose between tectonics, climate or eustasy as the dominant control on siliciclastic sedimentary sequences but to assess the relative and possibly the absolute, importance of these variables and their interaction in space and time (Cleotingh, 1991). Posamentier and Weimer (1993) and Shanley and McCabe (1994) suggest that three outstanding problems in sequence stratigraphy are whether or not distinct surfaces and deposits exist in non-marine successions, whether or not they can be correlated with coastal and marine successions and to what extent stratigraphic expression of sea-level changes are masked or replaced by those of tectonic and climatic changes.

Criteria for establishing the presence of incised valleys in the rock record are now well established. Criteria for distinguishing between the relative importance of tectonics, climate and eustasy, especially in fluvial successions and valley-fill deposits, are not well defined and much addition research is needed. This research should at the very least involve multi-disciplinary studies of entire Holocene fluvio-deltaic systems. Studies such as those by Blum (1992; 1994a, b) for the Colorado River in Texas and Saucier (1981, 1991) are good examples. These authors argue that surfaces and systems tracts cannot be correlated between valley-fill deposits in the subsiding, alluvial-deltaic plain and interior fluvial valley-fill successions. In contrast, Shanley and McCabe (1991, 1994), Wright and Marriott (1993), and Zaitlin et al. (1994) appear to suggest that these correlations can be made, but it is not always clear if the deposits that they have examined were formed outside of a subsiding coastal plain affected by sea-level fluctuations. Analog experiments and numerical models, which permit assessment of the effects of changes with a single extrinsic control, will continue to provide concepts that can be tested in detailed outcrop and subsurface studies.

SUMMARY AND CONCLUSIONS

Early versions of the sequence stratigraphic model were based on the assumption that relative changes in sea level (base level) were the most important control on cyclicity in the rock record. Research into the origin, distribution and internal heterogeneity of fluvial and valley-fill successions has demonstrated that tectonic and climatic changes may be of equal or even greater importance.

A review of relevant literature suggests that many studies of ancient fluvial and valley-fill systems conclude that one control is the dominant factor in explaining observed changes. While it is certainly true that any one control can significantly affect these systems, it is more likely that two or more of these controls had a major role in shaping ancient sedimentary sequences. To date, the relative importance of these controls is not well understood except for the vague consensus that the effect of sea-level change diminishes in a landward direction. In fact there is field and experimental evidence to suggest that erosional and depositional cycles in inland fluvial and valley-fill systems may be out of phase with cycles in the same systems in the coastal zone or that they may be in phase but because of entirely different controls.

Limited data suggests that the application of systems tract designations to various portions of ancient valley-fill or fluvial systems is fraught with problems, especially for systems that developed outside of a subsiding fluvio-deltaic coastal zone.

Convergence, divergence and sensitivity are factors that must be taken into account in future attempts to interpret the relative importance of major controls on fluvial and valley-fill systems.

ACKNOWLEDGMENTS

Most of the literature review was made while the senior author was a visiting Professor at the Department of Geology and Petroleum Geology, King's College, University of Aberdeen, Scotland. Debt is acknowledged to the faculty and staff and especially Professor Brian Williams for their support and for many stimulating discussions. This manuscript has benefited greatly from constructive reviews of an earlier version by Dale Leckie and John Van Wagoner.

REFERENCES

ALLEN, J. R. L., 1979, Studies in fluviatile sedimentation: an elementary geometrical model for the connectedness of avulsion-related channel sand bodies: Sedimentary Geology, v. 24, p. 253–267.

ALLEN, J. R. L. AND WILLIAMS, B. P. J., 1982, The architecture of an alluvial suite: the rocks between the Townsend Tuff and Pickard Bay Tuff Beds (Early Devonian), southwest Wales: Philosophical Transactions of the Royal Society of London, v. B297, p. 51–89.

ANADÓN, PERE, CABRERA, L., COLOMBO, F., MJARZO, M., AND RIBA, O., 1986, Syntectonic intraformational unconformities in alluvial fan deposits, eastern Ebro basin margins (NE Spain), in Allen, P. A. and Homewood, P., eds., Foreland Basins: Oxford, International Association of Sedimentologists Special Publication 8, p. 259–271.

ASHLEY, G. M. AND HAMILTON, T. D., 1993, Fluvial response to Late Quaternary climatic fluctuations, central Kobuk valley, northwestern Alaska: Journal Sedimentary Petrology, v. 63, p. 814–827.

AUBREY, W. M., 1989, Mid Cretaceous alluvial-plain incision related to eustasy, southeastern Colorado Plateau: Geological Society America Bulletin, v. 101, p. 521–533.

BEHRENSMEYER, A. K. AND TAUXE, L., 1982, Isochronous fluvial systems in Miocene deposits of northern Pakistan: Sedimentology, v. 29, p. 331–352.

BLAIR, T. C. AND BILODEAU, W. L., 1988, Development of tectonic cyclothems in rift, pull-apart and foreland basins: sedimentary response to episodic tectonism: Geology, v. 16, p. 517–520.

BLAIR, T. C. AND MCPHERSON, J. G., 1994, Alluvial fans and their natural distinction from rivers based on morphology, hydraulic processes, sedimentary processes, and facies assemblages: Journal of Sedimentary Research, v. A64, p. 450–489.

BLAIR, T. C., CLARK, J. S., AND WELLS, S. G., 1990, Quaternary continental stratigraphy, landscape evolution and application to archeology: Jarilla piedmont and Tularosa graben floor, White Sands Missile Range, New Mexico: Geological Society America Bulletin, v. 102, p 749–759.

BLAKELY, R. C. AND GUBITOSA, R., 1984, Controls of sandstone body geometry and architecture in the Chinle Formation (Upper Triassic), Colorado Plateau: Sedimentary Geology, v. 38, p. 51–86.

BLUM, M. D., 1992, Modern depositional environments and recent alluvial history of the lower Colorado River, Gulf coastal plain, Texas: Unpublished Ph.D. Dissertation, University of Texas at Austin, Austin, 304 p.

BLUM, M. D., 1994a, Genesis and architecture of alluvial stratigraphic sequences: A Late Quaternary example from the Colorado River, Gulf Coastal Plain of Texas, in Weimer, P. and Posamentier, H. W., eds., Siliciclastic Sequence Stratigraphy: Recent Developments and Applications: Tulsa, American Association Petroleum Geologists Memoir 58, p. 259–279.

BLUM, M. D., 1994b, Late Quaternary sedimentation, lower Colorado River, Gulf Coastal Plain of Texas: Geological Society of America Bulletin, v. 106, p. 1002–1016.

BOYD, R. AND DIESSEL, C., 1994, The application of sequence stratigraphy to non-marine clastics and coal, in Posamentier H. W. and Mutti, E., eds., Second High-Resolution Sequence Stratigraphy Conference: Tremp, International Union of Geological Sciences, p. 13–20.

BRIDGE, J. S. AND LEEDER, M. R., 1979, A simulation model of alluvial stratigraphy: Sedimentology, v. 26, p. 617–644.

BULL, W. B., 1991, Geomorphic Responses to Climatic Change: New York, Oxford University Press, N.Y., N.Y., 326 p.

BUTCHER, S. W., 1990, The nickpoint concept and its implications regarding onlap to the stratigraphic record, *in* Cross, T. A., ed., Quantitative Dynamic Stratigraphy: Englewood Cliffs, Prentice Hall, p. 375–385.

CLOETINGH, S., 1991, Tectonics and sea-level changes: A controversy?, *in* Muller, D. W., McKenzie, J. A., and Weissert, H., eds., Controversies in Modern Geology: Evolution of Geological Theories in Sedimentology, Earth History, and Tectonics: London, Academic Press, p. 249–277.

COJAN, I., 1993, Alternating fluvial and lacustrine sedimentation: tectonic and climatic controls (Provence Basin, S. France, Upper Cretaceous/Paleocene), *in* Marzo, M. and Puigdefabregas, C., eds., Alluvial Sedimentation: Oxford, International Association of Sedimentologists Special Publication 17, Blackwell Scientific Publishing, p. 425–438.

CREWS, S. G. AND ETHRIDGE, F. G., 1993, Laramide tectonics and humid alluvial fan sedimentation, NE Uinta Uplift, Utah and Wyoming: Journal Sedimentary Petrology, v. 63, p. 420–436.

CURRAY, J. R., 1964, Transgressions and regressions, *in* Miller, R. L., ed., Papers in Marine Geology Shepard Commemorative Volume: New York, McMillan, p. 175–203.

DIFFENDAL, R. F., JR. AND LOWRIE, A., 1992, Application of sea level stratigraphy to Neogene Great Plains stratigraphy: Transactions of the Gulf Coast Association of Geological Societies, v. 42, p. 121–129.

DORN, R. I., DENIRO, M. J. AND AJIE, H. O., 1987, Isotopic evidence for climatic influence on alluvial-fan development in Death Valley, California: Geology, v. 15, p. 108–110.

EMBRY, A. F., 1990, A tectonic origin for depositional sequences in extensional basins — Implications for basin modeling, *in* Cross, T. A. ed., Quantitative Dynamic Stratigraphy: New York, Prentice-Hall, p. 491–501.

ERIKSSON, K. A., 1978, Alluvial and destructive beach facies from the Archean Moodies Group, Barberton Mountain Land, South Africa and Swaziland, *in* Miall, A. D., ed., Fluvial Sedimentology: Calgary, Canadian Society of Petroleum Geologists Memoir 5, p. 287–311.

ESCHARD, R., RAVENNE, C., HOUEL, P., AND KNOX, R., 1992, Three dimensional reservoir architecture of a valley-fill sequence and a deltaic aggradational sequence: influences of minor relative sea-level variations (Scalby Formation, England), *in* Miall, A. D. and Tyler, N., eds., The Three-Dimensional Facies Architecture of Terrigenous Clastic Sediments and Its Implications for Hydrocarbon Discovery and Recovery: Tulsa, Sepm (Society for Sedimentary Geology) Concepts in Sedimentology and Paleontology Series 3, p. 133–147.

ETHRIDGE, F. G., SCHUMM, S. A., WOOD, L. J., KOSS, J. E., AND WARE, J. V., 1991, Sequence stratigraphy: What can we learn from experimental studies of base-level change? (Extended abs.), *in* Leckie, D. A., Posamentier, H. W. and Lovell, R. W. W., eds., 1991 Nuna Conference on High-Resolution Sequence Stratigraphy: Calgary, Geological Association of Canada, Program/Proceedings/ Guidebook, p. 13–14.

FISK, H. N., 1944, Geological investigation of the alluvial valley of the lower Mississippi River: Vicksburg, United States Army Corps of Engineers Mississippi River Commission, 79 p.

FROSTICK, L. E. AND REID, I., 1989, Climatic versus tectonic controls of fan sequences: lessons from the Dead Sea, Israel: Journal of the Geological Society of London, v. 146, p. 527–538.

GALLOWAY, W. E., 1989, Genetic stratigraphic sequences in basin analysis: Architecture and genesis of flooding-surface bounded depositional units: American Association of Petroleum Geologists Bulletin, v. 73, p. 125–142.

GERMANOSKI, D., 1991, Flume study of the relative impact of falling versus rising base-level in steep braided streams (abs.): San Diego, Geological Society America Abstracts with Programs, 1991 Annual Meeting, p. A240.

GROLL, P. E. AND STEIDTMANN, J. R., 1987, Fluvial response to Eocene tectonism, the Bridger Formation, southern Wind River Range, Wyoming, *in* Ethridge, F. G., Flores, R. M., and Harvey, M. R., eds., Recent Developments in Fluvial Sedimentology: Contributions from the Third International Fluvial Sedimentology Conference: Tulsa, Society of Economic Paleontologists and Mineralogists Special Publication 39, p. 263–268.

HALL, S. A., 1990, Channel trenching and climatic change in the southern U.S. Great Plains: Geology, v. 18, p. 342–345.

HARTLEY, A. J., 1993a, A depositional model for the Mid-Westphalian A to late Westphalian B coal Measures of South Wales: Journal of the Geological Society of London, v. 150, p. 1121–1136.

HARTLEY, A. J., 1993b, Sedimentological response of an alluvial system to source area tectonism: the Seilao Member of the Late Cretaceous to Eocene Purilactis Formation of northern Chile, *in* Marzo, M. and Puigdefabregas, C., eds., Alluvial Sedimentation: Oxford, International Association of Sedimentologists Special Publication 17, Blackwell Scientific Publishing, p. 489–500.

HELLER, P. L., ANGEVINE, C. L., WILSON, N. S., AND PAOLA, C., 1988, Two-stage stratigraphic model for foreland basin sequences: Geology, v. 16, p. 501–504.

HERFORD, R., 1993, Entrenchment and Widening of the Upper San Pedro River, Arizona: Boulder, Geological Society America Special Paper 282, 46 p.

HEWARD, A. P., 1978, Alluvial fan sequence and megasequence models: with examples from Westphalian D – Stephanian B coalfields, northern Spain, *in* Miall, A.D., ed., Fluvial Sedimentology: Calgary, Canadian Society of Petroleum Geologists Memoir 5, p. 669–702.

HUMPHREY, N. F., 1991, Response of alluvial river systems to base-level changes on intermediate time scales (order 10,000 to 1000,000 years) (abs.): Geological Society of America Abstracts with Programs, p. A160.

KOSS, J. E., 1992, Effects of base-level change on fluvial and coastal plain systems: an experimental approach: Thesis, Colorado State University, Fort Collins, 157 p.

KOSS, J. E., ETHRIDGE, F. G., AND SCHUMM, S. A., 1994, An experimental study of the effects of base-level change on fluvial, coastal plain and shelf systems: Journal of Sedimentary Research, v. B64, p. 90–98.

KRAUS, M. J. AND MIDDLETON, L. T., 1987, Contrasting architecture of two alluvial suites in different structural settings, *in* Ethridge, F. G., Flores, R. M., and Harvey, M. R., eds., Recent Developments in Fluvial Sedimentology: Contributions from the Third International Fluvial Sedimentology Conference: Tulsa, Society of Economic Paleontology and Mineralogy Special Publication 39, p. 253–262.

KVALE, E. P. AND VONDRA, C. F., 1993, Effects of relative sea-level changes and local tectonics on a Lower Cretaceous fluvial to transitional marine sequence, Bighorn Basin, Wyoming, U.S.A., *in* Marzo, M. and Puigdefabregas, C., eds., Alluvial Sedimentation: Oxford, International Association Sedimentologists Special Publication 17, Blackwell Scientific Publishing, p. 383–399.

LANE, E. W., 1955, The importance of fluvial morphology in hydraulic engineering: American Society of Civil Engineering, Proceedings, v. 81, p. 1–17.

LANG, S. C., 1993, Evolution of Devonian alluvial systems in an oblique-slip mobile zone — an example from the Broken River Province, northeastern Australia: Sedimentary Geology, v. 85, p. 501–535.

LANGBEIN, W. B. AND SCHUMM, S. A., 1958, Yield of sediment in relation to mean annual precipitation: American Geophysical Union, Transactions, v. 39, p. 1076–1084.

LAWRENCE, D. A. AND WILLIAMS, B. P. J., 1987, Evolution of drainage systems in response to Acadian deformation: the Devonian Battery Point Formation, eastern Canada, *in* Ethridge, F. G., Flores, R. M., and Harvey, M. R., eds., Recent Developments in Fluvial Sedimentology: Contributions from the Third International Fluvial Sedimentology Conference: Tulsa, Society of Economic Paleontologists and Mineralogists Special Publication 39, p. 287–300.

LECKIE, D., 1994, Canterbury Plains, New Zealand implications for sequence stratigraphic models: American Association of Petroleum Geologists Bulletin, v. 78, p. 1240–1256.

LEEDER, M. R., 1993, Tectonic controls upon drainage basin development, river channel migration and alluvial architecture: implications for hydrocarbon reservoir development and characterization, *in* North, C. P. and Prosser, D. J., eds., Characterization of Fluvial and Aeolian Hydrocarbon Reservoirs: London, The Geological Society of London, Special Publication 73, p. 7–22.

LEEDER, M. R. AND GAWTHORPE, R. L., 1987, Sedimentary models for extensional tilt-block/half graben basins: Geological Society of London Special Publication 28, p. 139–152.

LEGARRETA, L., ULIANA, M. A., LAROTONDA, C. A., AND MECONI, G. R., 1993, Approaches to nonmarine sequence stratigraphy – theoretical models and examples from Argentine basins, *in* Eschard, R. and Doliqez, B., eds., Subsurface Reservoir Characterization from Outcrop Observations: Paris, Editions Technip, p. 125–143.

LEOPOLD, L. B., 1994, A View of the River: Cambridge, Harvard University Press, 298 p.

LEOPOLD, L. B. AND BULL, W.B., 1979, Base level, aggradation, and grade: Proceedings American Philosophical Society, v. 123, p. 168–202.

LETTIS, W. R., 1985, Late Cenozoic stratigraphy and structure of the west margin of the central San Joaquin Valley, California, *in* Weide, D. L., ed., Soils and Quaternary Geology of the Southwestern United States: Boulder, Geological Society of America Special Publication 203, p. 97–114.

LÓPEZ-GÓMEZ, J. AND ARCHE, A., 1993, Architecture of the Canizar fluvial sheet sandstones, Early Triassic, Iberian Ranges, eastern Spain, *in* Marzo, M. and Puigdefabregas, C., eds., Alluvial Sedimentation: Oxford, International Association Sedimentologists Special Publication 17, Blackwell Scientific Publishing, p. 363–381.

MACK, G. H. AND SEAGER, W. R., 1990, Tectonic control on facies distribution of the Camp Rice and Palomas Formations (Pliocene-Pleistocene) in the southern Rio Grande rift: Geological Society America Bulletin, v. 102, p. 45–53.

MAIZELS, J. K., 1987, Large-scale flood deposits associated with the formation of coarse-grained, braided terrace sequences, in Ethridge, F. G., Flores, R. M., and Harvey, M. R., eds., Recent Developments in Fluvial Sedimentology: Contributions from the Third International Fluvial Sedimentology Conference: Tulsa, Society of Economic Paleontologists and Mineralogists Special Publication 39, p. 135–148.

MARTINSEN, O. J., 1994, Subtle but significant changes in fluvial style in the Erickson Sandstone (Campanian), Rock Springs Uplift, Wyoming: implications for stratigraphic analysis of non-marine rocks, in Posamentier, H. W. and Mutti, E., eds., Second High-Resolution Sequence Stratigraphy Conference: Tremp, International Union of Geological Sciences, p. 111–119.

MARTINSEN, O. J., MARTINSEN, R. S., AND STEIDTMAN, J .R., 1993, Mesaverde Group (Upper Cretaceous), southeastern Wyoming: Allostratigraphy versus sequence stratigraphy in a tectonically active area: American Association of Petroleum Geologists Bulletin, v. 77, p. 1351–1373.

MATHER, A. E., 1993, Basin inversion: some consequences for drainage evolution and alluvial architecture: Sedimentology, v. 40, p. 1069–1089.

MCLEAN, J .R. AND JERZYKIEWICZ, T., 1978, Cyclicity, tectonics and coal: some aspects of fluvial sedimentology in the Brazeau-Paskapoo formations, Coal Valley area, Alberta, Canada, in Miall, A. D., ed., Fluvial Sedimentology: Calgary, Canadian Society of Petroleum Geologists Memoir 5, p. 441–468.

MELVIN, J., 1993, Evolving fluvial style in the Kekiktuk Formation (Mississippian), Endicott field area, Alaska: base level response to contemporaneous tectonism: American Association of Petroleum Geologists Bulletin, v. 77, p. 1723–1744.

MERRITTS, D. J., VINCENT, K. R., AND WOHL, E. E., 1994, Long river profiles, tectonism, and eustasy: a guide to interpreting fluvial terraces: Journal of Geophysical Research, v. 99, p. 14031–14050.

MIALL, A. D., 1986, Eustatic sea level changes interpreted from seismic stratigraphy: A critique of the methodology with particular reference to the North Sea Jurassic record: American Association Petroleum Geology Bulletin, v. 70 p. 131–137.

MIALL, A. D., 1991, Stratigraphic sequences and their chronostratigraphic correlation: Journal of Sedimentary Petrology, v. 61, p. 497–505.

MIALL, A. D., 1992, Exxon global cycle: An event for every occasion?: Geology, v. 20, p. 787–790.

MJØS, R. AND PRESTHOLM, E., 1993, The geometry and organization of fluvio-deltaic channel sandstones in the Jurassic Saltwick Formation, Yorkshire, England: Sedimentology, v. 40, p. 919–935.

MØRK, A., EMBRY, A. F. AND WEITSCHAT, W. 1989, Triassic transgressive-regressive cycles in the Sverdrup basin, Svalbard and the Barents shelf, in Collinson, J. D., ed., Correlation in Hydrocarbon Exploration: London, Norwegian Petroleum Society, Graham and Trotman, p. 113–130.

NAMI, M. AND LEEDER, M. R., 1978, Changing channel morphology and magnitude in the Scalby Formation (M. Jurassic) of Yorkshire, England, in Miall, A. D., ed., Fluvial Sedimentology: Calgary, Canadian Society of Petroleum Geologists Memoir 5, p. 431–440,

NICHOLS, G. J., 1987, Structural controls on fluvial distributary systems — the Luna system, northern Spain, in Ethridge, F. G., Flores, R. M., and Harvey, M. R., eds., Recent Developments in Fluvial Sedimentology: Contributions from the Third International Fluvial Sedimentology Conference: Tulsa, Society of Economic Paleontologists and Mineralogists Special Publication 39, p. 269–277.

NYSTUEN, J. P., KNARUD, R., KNUT, J., AND STANLEY, K. O., 1989, Correlation of Triassic to Lower Jurassic sequences, Snorre field and adjacent areas, northern North Sea, in Collinson, J. D., ed., Correlation in Hydrocarbon Exploration: London, Norwegian Petroleum Society, Graham and Trotman, p. 273–289.

OLSON, H., 1990, Astronomical forcing of meandering river behavior: Milankovitch cycles in the Devonian of East Greenland: Palaeogeography, Palaeoclimatology, Palaeoecology, v. 79, p. 99–116.

OLSON. H. AND LARSEN, P.–H., 1993, Structural and climatic controls on fluvial depositional systems: Devonian, North-East Greenland, in Marzo, M. and Puigdefabregas, C., eds., Alluvial Sedimentation: Oxford, International Association of Sedimentologists Special Publication 17, Blackwell Scientific Publications, p. 401–423.

OLSEN, T. STEEL, R., HØGSETH, K. SKAR, T. AND RØE, S. L., 1995, Sequential architecture in a fluvial succession: Sequence stratigraphy in the Upper Cretaceous Mesaverde Group, Price, Utah: Journal of Sedimentary Research, v. B65, p. 265–280.

PAGE, K. J., NANSON, G. C., AND PRICE, D. M. , 1991, Thermoluminescence chronology of late quaternary deposition on the Riverine Plain of south-central Australia: Australian Geographer, v. 22, p. 14–23.

PAOLA, C., HELLER, P. L., AND ANGEVINE, C. L., 1991, The response distance of river systems to variations in sea level (abs.): Geological Society of America Abstracts with Programs, p. A170.

PITTMAN, W. D. AND GOLOVOCHENKO, X., 1991, Modeling sedimentary sequences, in Muller, D. W., McKenzie, J. A. and Weissert, H., eds., Controversies in Modern Geology: Evolution of Geological Theories in Sedimentology, Earth History and Tectonics: London, Academic Press, p. 279–390.

PONTI, D. J., 1985, The Quaternary alluvial sequence of the Antelope Valley, in Weide, D. L., ed., Soils and Quaternary Geology of the Southwestern United States: Boulder, Geological Society of America Special Paper 203, p. 79–96.

POSAMENTIER, H. W. JERVEY, M. T., AND VAIL, P. R., 1988, Eustatic controls on clastic deposition I – conceptual framework, in Wilgus, C. K., Hastings, B. S., Kendall, C. G. St. C., Posamentier, H. W., Ross, C. J., and Van Wagoner, J. C. , eds., Sea-Level Changes: An Integrated Approach: Tulsa, Society Economic Paleontologists and Mineralogists Special Publication 42, p. 109–124.

POSAMENTIER, H. W. AND VAIL, P. R., 1988, Eustatic controls on clastic deposition II – sequence and systems tract models, in Wilgus, C. K., Hastings, B. S., Kendall, C. G. St. C., Posamentier, H. W., Ross, C. J., and Van Wagoner, J. C., eds., Sea-Level Changes: An Integrated Approach: Tulsa, Society Economic Paleontologists and Mineralogists Special Publication 42, p. 125–154.

POSAMENTIER, H. W., ALLEN, G. P., JAMES, D. P., AND TESSON, M., 1992, Forced regressions in a sequence stratigraphic framework: concepts, examples and exploration significance: American Association of Petroleum Geologists Bulletin, v. 76, p. 1687–1709.

POSAMENTIER, H. W. AND JAMES, D. P., 1993, An overview of sequence stratigraphic concepts: Uses and abuses, in Posamentier, H. W., Summerhayes, C. P., Haq, B. U. and Allen, G. P., eds., Sequence Stratigraphy and Facies Associations: Oxford, International Association of. Sedimentologists, Special Publication 18, p. 3–18.

POSAMENTIER, H. W. AND WEIMER, P., 1993, Siliciclastic sequence stratigraphy and petroleum geology — Where to from here?: American Association Petroleum Geology Bulletin, v. 77, p. 731–742.

RETALLACK, G. J., 1986, Fossil soils as grounds for interpreting long-term controls on ancient rivers: Journal Sedimentary Petrology, v. 56, p. 1–18.

RYER, T. A., 1994, Interplay of tectonics, eustasy, and sedimentation in the formation of Mid-Cretaceous clastic wedges, Central and Northern Rocky Mountain regions, in Dolson, J. C., Hendricks, M. L. and Wescott, W. A., eds., Unconformity-Related Hydrocarbon Exploration and Exploitation in Sedimentary Sequences: Denver, Rocky Mountain Association of Geologists, p. 35–44.

SAUCIER, R. T., 1981, Current thinking on riverine processes and geologic history as related to humane settlement in the southeast: Geoscience and Man, v. 22, p. 7–18.

SAUCIER, R. T., 1991, Quaternary geology of the lower Mississippi valley: geomorphology, stratigraphy and chronology, in Morrison, R. B., ed., The Geology of North America, v. K–2, Quaternary Nonglacial Geology: Conterminous U.S.: Boulder, The Geological Society of America, p. 550–564.

SCHLAGER, W., 1993, Accommodation and supply — A dual control on stratigraphic sequences: Sedimentary Geology, v. 86, p. 111–136.

SCHUMM, S. A., 1965, Quaternary paleohydrology, in Wright, H. E., Jr. and Frey, D. G., Eds., The Quaternary of the United States: Princeton, Princeton University Press, p. 783–794.

SCHUMM, S. A., 1968, Speculations concerning paleohydrologic controls of terrestrial sedimentation: Geological Society of America Bulletin, v. 79, p. 1572–1588.

SCHUMM, S. A., 1977, The Fluvial System: New York, Wiley and Sons, 338 p.

SCHUMM, S. A., 1981, Evolution and response of the fluvial system: sedimentologic implications, in Ethridge, F. G. and Flores, R. M., eds., Recent and Ancient Nonmarine Depositional Environments: Models for Exploration: Tulsa, Society of Economic Paleontologists and Mineralogists Special Publication 31, p. 19–29.

SCHUMM, S. A., 1991, To Interpret the Earth Ten Ways to be Wrong: New York, Cambridge University Press, 133 p.

SCHUMM, S. A., 1993, River response to base-level changes: Implications for sequence stratigraphy: Journal of Geology, v. 101, p. 279–294.

SCHUMM, S. A., HARVEY, M. D., AND WATSON, C. C., 1984, Incised Channels: Morphology Dynamics and Control: Littleton, Water Resources Publication, 200 p.

SCHUMM, S. A. AND BRAKENRIDGE, G. R., 1987, River responses, in Ruddiman, W. F. and Wright, H. E., Jr., eds., North America and Adjacent Oceans During the Last Deglaciation: Boulder, Geological Society of America, The Geology of North America, v. K–3 p. 221–240.

SCHUMM, S. A., MOSLEY, M. P., AND WEAVER, W. E., 1987, Experimental Fluvial Geomorphology: New York, John Wiley and Sons, 413 p.

SCHUMM, S. A. AND ETHRIDGE, F. G., 1994, Origin, evolution and morphology of fluvial valleys, in Dalrymple, R. W., Boyd, R., and Zaitlin, B. A., eds., Incised-Valley Systems: origin and Sedimentary Sequences: Tulsa, Sepm (Society for Sedimentary Geology) Special Publication 51, p.11–27.

SHANLEY, K. W. AND MCCABE, P. J., 1991, Predicting facies architecture through sequence stratigraphy – an example from the Kaiparowits Plateau, Utah: Geology, v. 19, p. 742–745.

SHANLEY, K. W. AND MCCABE, P. J, 1993, Alluvial architecture in a sequence stratigraphic framework: a case history from the Upper Cretaceous of southern Utah, Usa, in Flint, S. S. and Bryant, I. D., eds., The Geological Modeling of Hydrocarbon Reservoirs and Outcrop Analogues: Oxford, International Association of Sedimentologists Special Publication 15, p. 21–55.

SHANLEY, K. W. AND MCCABE, P. J., 1994, Perspectives on the sequence stratigraphy of continental strata: American Association of Petroleum Geologists Bulletin, v. 78, p. 544–568.

SHUSTER, M. W. AND STEIDTMANN, J. R., 1987, Fluvial-sandstone architecture and thrust-induced subsidence, northern Green River basin, Wyoming, in Ethridge, F. G., Flores, R. M., and Harvey, M. R., eds., Recent Developments in Fluvial Sedimentology: Contributions from the Third International Fluvial Sedimentology Conference: Tulsa, Society of Economic Paleontologists and Mineralogists Special Publication 39, p. 279–285.

SLOSS, L. L., 1988, Forty years of sequence stratigraphy: Geological Society of America Bulletin, v.100, p. 1661–1665.

SLOSS, L. L., 1991, The tectonic factor in sea level change: a counter veiling view: Journal of Geophysical Research, v. 96, p. 6609–6617.

STEEL, R. AND ASHEIM, S. M., 1978, Alluvial sand deposition in a rapidly subsiding basin (Devonian, Norway), in Miall, A. D., ed., Fluvial Sedimentology: Calgary, Canadian Society of Petroleum Geologists Memoir 5, p. 385–412.

TALBOT, M. R. AND WILLIAMS, M. A. J., 1979, Cyclic alluvial fan sedimentation and the flanks of fixed dunes, Janjari, central Niger: Catena, v. 6, p. 43–62.

TÖRNQVIST, T. E., 1993, Holocene alteration of meandering and anastomosing fluvial systems in the Rhine-Meuse delta (central Netherlands) controlled by sea-level rise and subsoil erodibility: Journal of Sedimentary Petrology, v. 63, p. 683–693.

TREWIN, N. H., 1993, Controls on fluvial deposition in mixed fluvial and aeolian facies within the Tumblagooda Sandstone (Late Silurian) of Western Australia: Sedimentary Geology, v. 85, p. 387–400.

VAIL, P. R., AUDERMARD, F., BOWMAN, S. A., EISNER, P. N., PEREZ-CRUS, C., 1991, The stratigraphic signatures of tectonics, eustasy and sedimentology — an overview, in Einsele, G., Ricken, W., and Seilacher, A., Eds., Cycles and Events in Stratigraphy: Berlin, Springer-Verlag, p. 617–659.

VAN NEST, J. A. AND BETTIS, E. A., III, 1990, Postglacial response of a stream in central Iowa to changes in climate and drainage basin factors: Quaternary Research, v. 33, p. 73–85.

VAN WAGONER, J. C., MITCHUM, R. M., CAMPION, K. M. AND RAHMANIAN, V. D., 1990, Siliciclastic Sequence Stratigraphy in Well Logs, Cores, and Outcrops: Concepts for High-Resolution Correlation of time and Facies: Tulsa, American Association Petroleum Geologists Methods in Exploration Series 7, 55 p.

WALKER, R. G., 1990, Facies modeling and sequence stratigraphy: Journal Sedimentary Petrology, v. 60, p. 777–786.

WARE, J. V., ETHRIDGE, F. G., AND SCHUMM, S. A., 1992, The effect of base-level fluctuations on meandering channels: An experimental approach (abs.): Sepm (Society for Sedimentary Geology) 1992 Theme Meeting, p. 64.

WEAVER, W. E., 1984, Experimental study of alluvial fans: Unpublished Ph.D. Dissertation, Colorado State University, Fort Collins, 423 p.

WESCOTT, W. A., 1993, Geomorphic thresholds and complex response of fluvial systems − Some implications for sequence stratigraphy: American Association Petroleum Geology Bulletin , v. 77, p. 1208–1218.

WILLIS, B. J. AND BRIDGE, J. S., 1988, Evolution of Catskill river systems, New York State, in McMillan, N. J., Embry, A. F. and Glass, D. J., eds., Devonian of the World, Volume II: Sedimentation: Calgary, Canadian Society of Petroleum Geologists, p. 85–106.

WOOD, L. J., ETHRIDGE, F. G., AND SCHUMM, S. A., 1993, The effect of rate of base-level fluctuation on coastal plain, shelf, and slope depositional systems: An experimental approach, in Posamentier, H. W., Summerhayes, C. P., Haq, B. U., and Allen, G. P., eds., Sequence Stratigraphy and Facies Associations: Oxford, International Association of Sedimentologists Special Publication 18, p. 43–53.

WOOD, L. J., ETHRIDGE, F. G. AND SCHUMM, S. A., 1994a, An experimental study of the influence of subaqueous shelf angle on coastal-plain and shelf deposits, in Weimer, P. and Posamentier, H. W., eds., Siliciclastic Sequence Stratigraphy Recent Developments and Applications: Tulsa, American Association of Petroleum Geologists Memoir 58, p. 381–391.

WOOD, L. J., KOSS, J. E. AND ETHRIDGE, F. G., 1994b, Simulating unconformity development and unconformable stratigraphic relationships through physical experiments, in Dolson, John, J. C., Hendricks, M. L., and Wescott, W. A., eds., Unconformity-Related Hydrocarbon Exploration and Exploitation in Sedimentary Sequences: Denver, Rocky Mountain Association of Geologists, p. 23–34.

WRIGHT, V. P., 1992, Paleopedology — stratigraphic relationships and empirical models, in Martin, I. P. and Chesworth, W. eds., Weathering, Soils, and Paleosols: Amsterdam, Elsevier, Development in Earth Surface Processes 2, p. 457–499.

WRIGHT, V. P. AND MARRIOTT, S. B., 1993, The sequence stratigraphy of fluvial depositional systems: The role of floodplain sediment storage: Sedimentary Geology, v. 86, p. 203–210.

ZAITLIN, B. A., DALRYMPLE, R. W., BOYD, R., AND LECKIE, D., eds., 1994, The Stratigraphic Organization of Incised Valley Systems: Implications to Hydrocarbon Exploration and Production with Examples from the Western Canada Sedimentary Basin: Calgary, Canadian Society of Petroleum Geologists, Calgary, 158 p.

QUATERNARY ALLUVIAL PLAIN CONSTRUCTION IN RESPONSE TO GLACIO-EUSTATIC AND CLIMATIC CONTROLS, TEXAS GULF COASTAL PLAIN

MICHAEL D. BLUM
Department of Geosciences, 214 Bessey Hall, University of Nebraska – Lincoln, Lincoln, Nebraska 68588
AND
DAVID M. PRICE
Thermoluminescence Dating Laboratory, School of Geosciences, University of Wollongong, Wollongong, New South Wales 2522 Australia

ABSTRACT: The Texas Gulf Coastal Plain consists of a series of low-gradient, fan-shaped alluvial plains emanating from each major river valley. The majority of alluvial plain surfaces have been mapped as Pleistocene Beaumont Formation or younger unnamed strata, and interpreted to represent eustatically-controlled deposition during the oxygen isotope stage 5 and modern interglacial highstands. Reevaluation of preexisting data combined with reexamination of Beaumont and younger strata of the Colorado River suggests the stratigraphic and geochronologic framework needs revision, and processes of alluvial plain deposition are more complex than previous interpretations have inferred. As a result, Beaumont and younger strata provide an opportunity to examine alluvial plain construction within a sequence-stratigraphic framework and discuss some key characteristics and the heirarchal nature of eustatically-controlled versus climatically-controlled components of alluvial plain depositional sequences.

Mapping from satellite imagery, field documentation of geomorphic and stratigraphic relationships, consideration of the stratigraphic significance of surface and buried soils, and a number of radiocarbon and thermoluminescence ages suggests that Beaumont and younger alluvial plains consist of multiple cross-cutting and/or superimposed valley fills of widely varying age, and may represent the last 300–400 ky or more. Valley fills become partitioned by initial lowering of sea-level below interglacial highstand positions, when channels rapidly incise and valley axes become fixed in place as they extend across the subaerially-exposed shelf. While shorelines remain basinward of highstand positions, the remainder of the alluvial plain is characterized by non-deposition and soil development. During this time, multiple episodes of lateral migration, aggradation, degradation, and/or flood-plain abandonment with soil formation occur within incised and extended valleys in response to climatic controls on discharge and sediment supply. This creates a composite basal valley fill unconformity, as well as multiple smaller-scale allostratigraphic units within the valley fill. With late stages of transgression and highstand valleys fill at paces set by upstream controls on sediment delivery. As valley filling nears completion, veneers of flood basin sediments spread laterally, which buries soils developed on downdip margins of the alluvial plain. Complete valley filling during highstand is one of several processes that promotes avulsion, with relocation of valley axes before the next sea-level fall, such that successive 100–ky valley fills have a distributary pattern, and successive increments of geologic time occur lateral to each other.

INTRODUCTION

River systems of the Gulf Coastal Plain have long been used as natural laboratories for the study of fluvial processes and facies, and development of facies models for interpretation of the ancient stratigraphic record. Classic, widely cited examples include Fisk's (1944, 1947) documentation of meandering-stream facies of the lower Mississippi River, Bernard and Major (1963) and Bernard et al.'s (1970) studies of the Brazos River of Texas, which resulted in development of a general model for fine-grained meandering streams, and McGowan and Garner's (1970) work on the Colorado River of Texas, which served a similar purpose for coarse-grained, chute-dominated meandering streams. Development of facies models through the study of modern systems of the Gulf Coast and elsewhere was an important step forward, because it was possible to link general system controls with depositional processes, a variety of autocyclic mechanisms, and resultant facies patterns.

With the advent of sequence stratigraphic concepts, many turned their attention to the responses of depositional systems to allocyclic controls. First-generation conceptual models suggested: (a) that strata could be subdivided and correlated using regionally-significant unconformities that result from basinward shifts in systems tracts during a relative sea-level fall; and (b) that stratal architecture reflected primarily on trends in the generation of accommodation space, also a function of relative sea-level change (e.g., Jervey et al., 1988; Posamentier and Vail, 1988; Posamentier et al., 1988). Because of the emphasis on base-level change and translation of the shoreline, these models may be well suited to coastal and marine strata, and to inter-bedded marine and continental deposits, but questions have been raised concerning their general applicability to alluvial and

other fully non-marine successions (Blum, 1990, 1993; Miall, 1991a; Schumm, 1993; Wescott, 1993; Wright and Marriot, 1993; Shanley and McCabe, 1994). Alluvial successions should record the coupling of upstream controls on discharge regimes and sediment supply, which are mostly independent of base-level change, with base-level controls on accommodation and preservation. It follows that full integration of alluvial successions into a sequence stratigraphic paradigm can only be achieved when features and characteristics commonly recognized within alluvial deposits, for example hierarchies of bounding unconformities (e.g., Miall, 1988, 1991b), systematic lateral and vertical changes in facies assemblages and stratal geometries (e.g., Shanley and McCabe, 1991, 1993, 1994; Van Wagoner et al., 1990; Gibling and Bird, 1994), and various types of paleosols (e.g., Kraus and Bown, 1986; 1988; Retal-lack, 1986; Wright, 1992; Wright and Marriot, 1993; Willis and Behrensmeyer, 1994), can be explained by robust models that account for interactions between multiple sets of controls.

It seems unlikely that the significance of key features and characteristics within alluvial successions can be fully appreciated from studies of ancient deposits alone, since dating methods are imprecise, and correlations between fluvial deposits and allocyclic controls are tenuous at best. Interpretive models for many ancient alluvial successions might instead be developed from Quaternary analogs, since key features can be compared with independently-derived records of changes in forcing mechanisms. Prospective study areas should possess a well-preserved stratigraphic record, occur in an area where allocyclic controls are well-known, and occur within a subsiding basin so the record has a high preservation potential. Here again the Gulf Coastal Plain can serve as a natural laboratory. The Coastal

Relative Role of Eustasy, Climate, and Tectonism in Continental Rocks, SEPM Special Publication No. 59
Copyright © 1998, SEPM (Society for Sedimentary Geology), ISBN 1-56576-042-5

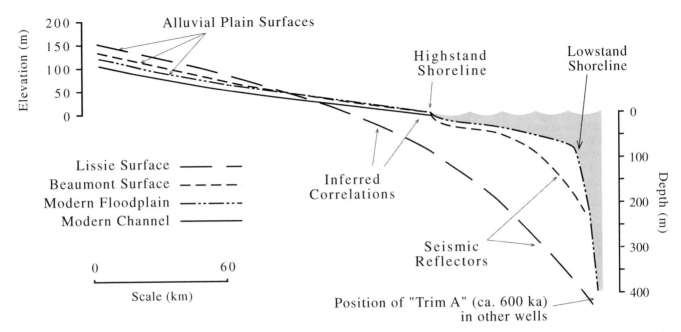

FIG. 4.—Longitudinal profiles of Texas Coastal Plain morphostratigraphic units and inferred correlations with widespread seismic reflectors in the marine record offshore. After Winker (1979), as updated in DuBar (1991). The first downhole appearance of the foram *Trimosina denticulata*, "Trim A", placed at approximately 600 Ka by Armentrout and Clements (1990), occurs just above the reflector correlated to the top of Lissie in offshore wells elsewhere in the Gulf.

landward to shore-parallel reworking of deposits during periods of transgression, (c) qualitative and quantitative differences in sediment input from different fluvial systems, and (d) the importance of structural controls on partitioning of the stratigraphic record. They differ, however, in terms of correlation of key surfaces used to subdivide shelf successions into depositional sequences, as well as chronostratigraphic interpretations and inferred relationships to causal mechanisms.

Reexamination of the fully non-marine updip Quaternary record has lagged behind research in the offshore, so much so that significant differences of interpretation within an overall modern paradigm are not yet possible. A few small-scale, site-specific studies suggest Beaumont and younger strata are more complex than originally envisioned and may span a longer and/ or different time period (e.g., Van Siclen, 1985, 1991; Aronow et al., 1991; Paine, 1991). But the lack of systematic reinvestigation of fully non-marine strata makes it difficult to move beyond simplistic and largely untested sea-level control models, with alluvial deposition constrained to periods of transgression and highstand. Such a model was satisfactory when the concept of four long Pleistocene interglacials was accepted, but needs reevaluation today. Willis, Lissie, Beaumont, and post-Beaumont strata are now, for example, thought to represent the entire Plio-Pleistocene through Holocene (DuBar et al., 1991), yet studies of oxygen isotopes in marine sediments clearly show that middle to late Pleistocene glacial-interglacial cycles occurred with a periodicity of about 100 ka (see Fig. 5), and the time represented by full interglacial highstands was only 10–15% of each cycle (e.g., Imbrie et al., 1984; Williams et al., 1988; Porter, 1989). Moreover, rates of sediment delivery to the coastal plain are not directly related to sea-level change, and fluvial systems may be more sensitive to upstream controls than previously envisioned.

Colorado River System

The Colorado River is an extrabasinal fluvial system (drainage area of 110,000 km²) with major tributaries draining the Southern High Plains and Edwards Plateau of west and central Texas. As the river emerges from a canyon at Balcones Escarpment, the drainage basin narrows, and the lower Colorado River transects the Coastal Plain for 280 km until discharging into the Gulf (Fig. 6). On the Inner Coastal Plain, the lower Colorado River flows within a well-defined bedrock valley that transects Upper Cretaceous carbonates, then progressively younger and less steeply dipping Tertiary siliciclastics (Barnes, 1979, 1981, 1982, 1983, 1987). The bedrock valley contains terraces of Pliocene(?) through Holocene age that document relatively short-term episodes of sediment storage superimposed on a late Cenozoic trend of bedrock valley deepening (Blum and Valastro,

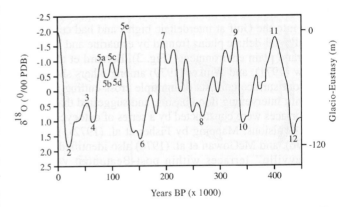

FIG. 5.—Record of changes in oxygen isotope composition of foraminifera in marine sediments, which provides a proxy record of changes in global ice volume and the corresponding glacio-eustatic component of sea-level change (after Imbrie et al., 1984).

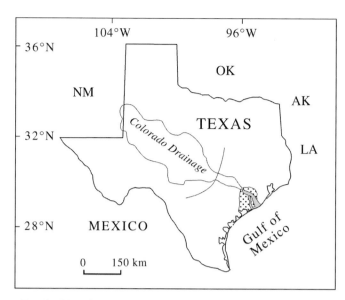

FIG. 6.—Map of Texas and the Colorado River drainage basin. Also shown are Pleistocene alluvial-plain deposits (combined Lissie and Beaumont—darker pattern) and the post-Beaumont incised valley (lighter pattern).

1994). Farther downstream, the Colorado channel emerges onto a constructional Quaternary alluvial plain consisting of Lissie, Beaumont, and post-Beaumont strata. Based on warping of presumed isotope stage 5 interglacial shorelines, total subsidence on the Colorado alluvial plain along the present-day coast has been estimated at 3–5 m over the last 120 ky (Winker, 1979), an average of 0.025–0.04 mm/yr. The shelf is really an extension of the fluvially-dominated coastal plain, since it remains subaerially-exposed for considerable portions of each glacio-eustatic cycle, and alluvial-deltaic facies comprise a significant portion of shelf sediments (Suter and Berryhill, 1985; Suter, 1987; Morton and Price, 1987; Anderson et al., 1992, 1996).

Due to the above physiographic setting, most sediment for the Colorado alluvial plain comes from above Balcones Escarpment (92% of drainage area). The bedrock valley on the Inner Coastal Plain stores sediment over short time periods but mostly acts as a conduit that transfers material to the alluvial plain and farther into the basin. Hence rates of sediment delivery to the alluvial plain depend on processes operating in the upper Colorado drainage, and alluvial-plain deposition should reflect interactions between controls on sediment supply and sea-level change.

POST-BEAUMONT STRATA

Studies of post-Beaumont deposits of Colorado River provide insight on relationships between stratigraphic sequences in the continental interior and the alluvial plain. Since many of these deposits are within the radiocarbon window, episodes of fluvial activity can be correlated with independent records of climate change in the continental interior and glacio-eustasy in the Gulf without use of a pre-existing model. Much of this work is published elsewhere (Blum, 1990, 1993; Blum and Valastro, 1994; Blum et al., 1994, 1995) and only summarized below.

Post-Beaumont Stratigraphic Record

Post-Beaumont strata of the Colorado River have been subdivided into a series of unconformity-bounded allostratigraphic

units (Fig. 7), with chronological control afforded by more than 40 radiocarbon ages (Blum and Valastro, 1994). In the degradational bedrock valley, at least two distinct terraces with underlying alluvial deposits correspond to the "Deweyville" terraces identified elsewhere on the Gulf Coastal Plain (e.g., Bernard, 1950; Bernard and LeBlanc, 1965; Saucier and Fleetwood, 1970; Brown et al., 1976; McGowan et al., 1975; Autin et al., 1991; DuBar et al., 1991; Blum et al., 1995); the youngest of these has been refered to as the Eagle Lake Alloformation (ELA), and consists of up to 10 m of sediments that underlie a terrace and well-developed soil profile at 17–20 m above the present-day channel. ELA deposition occurred around 20–14 Ka, and was followed by bedrock valley incision about 14–12 Ka, then deposition of a complex Holocene valley fill referred to as the Columbus Bend Alloformation (CBA). Columbus Bend Allomembers 1 and 2 (CBA–1 and CBA–2) rest on Tertiary bedrock at or near the present low water channel, and underlie a terrace at 12–14 m above the modern channel, with total thicknesses exceeding 12 m. CBA–1 was deposited from 12–5 Ka, whereas CBA–2 was deposited from 5–1 Ka. Columbus Bend Allomember 3 (CBA–3) consists of channel and flood-plain depositional environments from the last 600–1000 yr. Unconformities within the CBA represent episodes of flood-plain abandonment and soil formation, with continued lateral migration of the channel, but little to no bedrock valley cutting.

Sedimentary facies differ significantly between late Pleistocene "Deweyville" units, including ELA, and the Holocene CBA (Blum and Valastro, 1994; Blum et al., 1995), which suggests changes in sedimentation styles through time. Late Pleistocene units contain few vertical accretion flood-plain deposits, and point bar/chute bar gravel and sand extend to the top of many sections (Fig. 8A). This indicates that floods during the late Pleistocene were for the most part contained within bankfull channel perimeters, and flood-plain construction proceeded by lateral rather than vertical accretion. By contrast, vertical accretion facies occur throughout the CBA, and increase in thickness and volumetric significance through time (Fig. 8B). Hence deep overbank floods and flood-plain construction by vertical accretion became more important through time, and was a most important process during the late Holocene.

Allostratigraphic units within the bedrock-confined Colorado valley correlate with allostratigraphic units defined for major valley axes of the upper Colorado drainage (e.g., Blum et al., 1994). Major discontinuities in alluvial sequences also correlate with climatic changes on the Edwards Plateau (Toomey et al., 1993), which suggests that changes in discharge regimes and sediment supply drove fluvial adjustments that produced basic elements of the allostratigraphic framework (Blum, 1993; Blum and Valastro, 1994). Individual stratigraphic units represent periods when the supply of sediment along valley axes exceeded transport rates, and excess sediments were placed into storage, whereas unconformities represent time periods when sediment supply was limited relative to transport rates, and/or flood magnitudes decreased. For example, unconformities between ELA and CBA developed in response to decreases in sediment supply and the resultant deep incision of bedrock valleys. By contrast, unconformities within the CBA developed in response to decreased flood magnitudes, which resulted in flood-plain abandonment and soil formation with coeval lateral migration by the channel but no additional bedrock incision.

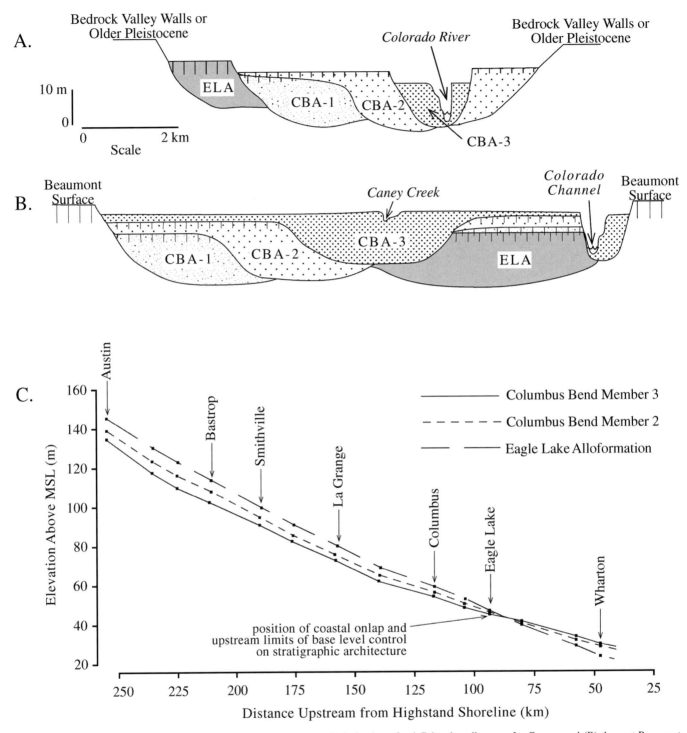

FIG. 7—Schematic cross-sections of post-Beaumont valley fill for (A) the bedrock-confined Colorado valley near La Grange, and (B) the post-Beaumont incised valley fill near Wharton, illustrating downstream persistence of allostratigraphic units, bounding unconformities, and soils (vertical lines) with downstream changes in stratigraphic architecture. (C) Longitudinal profiles for the Ela and different members of the Cba. Upstream from Eagle Lake, longitudinal profiles for Cba–1 and Cba–2 are essentially the same, and plotted as such. Downstream from that point, longitudinal profiles for the Ela and Cba–1 are essentially the same and are plotted as such (from Blum, 1993; Blum and Valastro, 1994).

Changes in sedimentation styles have been attributed to changes in drainage basin hydrology that accompanied the transition from glacial to interglacial conditions. During the last glacial cycle much of the bedrock landscape within the upper Colorado drainage was covered by deep soils, which required long periods of time to form and therefore reflect climatic and vegetational conditions that promoted long-term landscape stability. Such stability may have been the norm for most of the

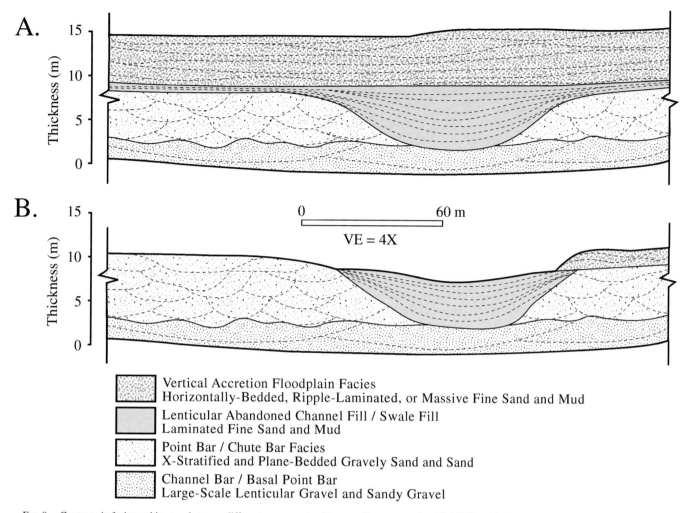

Vertical Accretion Floodplain Facies
Horizontally-Bedded, Ripple-Laminated, or Massive Fine Sand and Mud

Lenticular Abandoned Channel Fill / Swale Fill
Laminated Fine Sand and Mud

Point Bar / Chute Bar Facies
X-Stratified and Plane-Bedded Gravely Sand and Sand

Channel Bar / Basal Point Bar
Large-Scale Lenticular Gravel and Sandy Gravel

FIG. 8.—Contrasts in facies architecture between different components of the post-Beaumont valley fill. (A) Simplified composite section through late Holocene Cba–2, which contains a large proportion of vertical accretion flood-plain facies. (B) Simplified composite section through the late Pleistocene Ela, dominated by channel-related sands with few vertical accretion flood-plain facies (from Blum, 1992).

100–ky glacial period, when global and regional temperatures were cooler than interglacial conditions of today, eustatic sea-level oscillated between 15 and 65 m below present (see Porter, 1989), and precipitation in the southcentral United States most likely resulted from midlatitude cyclones. Post-glacial sea-level rise and increases in sea surface temperatures, coupled with increased land surface temperatures during summer months, promoted increases in the inland penetration of warm, moist tropical air and corresponding increases in the frequency of high-intensity convectional and tropical storms. This in turn favored degradation of soils such that by late Holocene time the upland landscape consisted of exposed limestone bedrock (Toomey et al., 1993). As a result, rates of runoff to stream channels would have been at a minimum through the long and complex glacial period, when upland landscapes were covered by deep soils, and reached a maximum during the late Holocene interglacial when the present bedrock landscape was exposed. By comparison with the glacial period, interglacial fluvial systems have flashy discharge regimes, with floods that frequently exceed bankfull channels, and they deposit thick packages of vertical accretion facies in flood-plain settings (Blum et al., 1994, 1995; Blum and Valastro, 1994).

Post-Beaumont allostratigraphic units, component facies, unconformities, and paleosols persist through the degradational bedrock valley to the aggradational Quaternary alluvial plain. However, stratigraphic architecture changes in the downstream direction as a result of the last glacio-eustatic cycle (see Fig. 7). In this part of the valley, meanderbelt sand-dominated "Deweyville" sediments were deposited during the long and complex falling sea level stage that accompanied the last glacial period, within a valley that was incised below Beaumont alluvial-plain surfaces and extended basinward to shelf-phase and shelf-edge deltas (Abdullah and Anderson, 1991; Anderson et al., 1992, 1996). In contrast to the bedrock valley farther upstream, this part of the valley records sediment bypass from about 14–12 Ka, but no further valley cutting because relative base-level was rising at the time. Deposition of CBA–1 also occurred within deeper parts of the incised valley during the post-glacial transgression, while the shoreline was mostly still basinward and lower in elevation than present. By contrast, flood-plain mud-rich CBA–2 and CBA–3 were deposited during the late Holocene sea-level highstand during a period of widespread valley aggradation, and flood-plain facies associ-

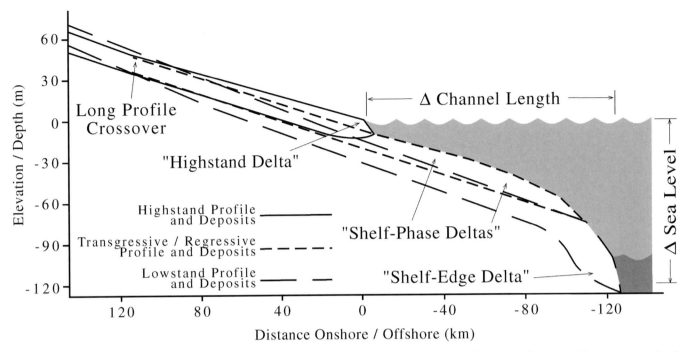

FIG. 9.—Schematic long profiles of allostratigraphic units within post-Beaumont valley fill on the coastal plain, projected to temporally equivalent sea-level positions. Exercise shows probable thickness and vertical relations between different components of an incised valley sequence.

ated with these units onlap and bury paleosols that define the upper boundaries to older allostratigraphic units within the incised valley. Depth of burial for the ELA and CBA–1 clearly increases in the downstream direction, which shows that meanderbelts were graded to shorelines well out on a subaerially-exposed shelf during sea-level lowstand and the last transgression, whereas CBA–2 and CBA–3 meanderbelts were graded to the late Holocene sea-level highstand. In far downstream reaches, CBA–3 includes the Caney Creek meanderbelt as well as the modern Colorado channel. Avulsion from the aggradational Caney Creek course, with occupation of the modern channel, occurred in response to near complete filling of the incised valley during the present highstand (Blum, 1993).

Key Characteristics of Post-Beaumont Incised Valley Fill

Data summarized above permits an outline of the scale and key characteristics of a fully non-marine incised-valley sequence. The post-Beaumont incised valley is a distinct feature that was initially partitioned by glacio-eustatic lowering of sea-level below the elevation of Beaumont alluvial plains, and the valley fill is bounded by a composite basal unconformity that can be traced up and out of the valley to deeply weathered soils that developed on Beaumont surfaces during the last glacial cycle (see Fig. 7b). Projections of long profiles from the ELA to the lowstand shelf-edge delta suggest total thickness of the valley fill ranges from 15–20 m at updip limits of the alluvial plain to 35 m or so at the highstand shoreline (Fig. 9), whereas the valley fill extends from 15–40 km along strike. Multiple smaller-scale allostratigraphic units occur within the valley fill, with each defined by a basal unconformity that can be traced to a paleosol that developed during short-term (10^3 to 10^4 yr) episodes of channel incision and/or flood-plain abandonment.

These smaller-scale allostratigraphic units within the valley fill may be 10–20 m in thickness and extend 2–10 km along strike. The post-Beaumont incised-valley fill as a whole records fluvial responses to a 100–ky glacial-interglacial cycle, whereas the composite nature of the valley fill unconformity results from development of the smaller-scale allostratigraphic units that record fluvial responses to shorter-term climatic and glacio-eustatic fluctuations within the 100–ky cycle.

Systematic changes in facies architecture occur within the post-Beaumont incised valley, as meanderbelt sand-dominated allostratigraphic units dominate deeper parts of the valley fill and flood-plain mud-rich units dominate the upper part. Similar patterns have been discussed in Fisk's (1944) classic paper on the lower Mississippi alluvial valley, in conceptual sequence stratigraphic models (e.g., Posamentier and Vail, 1988; Posamentier et al., 1988; Wright and Marriot, 1993), and in case studies of ancient strata (e.g., Van Wagoner et al., 1990; Shanley and McCabe, 1993; Gibling and Bird, 1994). A common explanation would be that increases in slope during base-level fall and lowstand result in increased stream power, an ability to transport coarse material, and a resultant facies motif dominated by laterally amalgamated sand bodies with few overbank fines. By contrast, decreases in slope during base-level rise and highstand produce decreases in slope and stream power, increased storage of flood-plain facies, and a resultant facies motif dominated by isolated meanderbelt sands encased in overbank fines.

While it seems clear that stacking patterns for allostratigraphic units within the post-Beaumont valley fill of the Colorado River represent long profile adjustments to glacio-eustatically controlled changes in base-level (e.g., Fig. 7), changes in slope had little to do with component facies and facies architecture within each unit. Instead, these are attributes that can be traced far upstream into the bedrock valley and reflect pri-

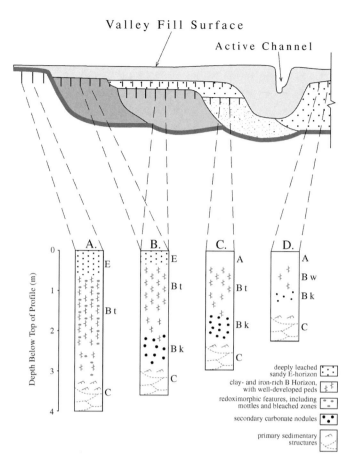

FIG. 10.—Schematic illustration of differences in soil profiles developed in alluvial sediments of Colorado River, partly as a function of duration of subaerial exposure and soil formation. (A) Profile developed in sandy facies of Pleistocene Beaumont Formation, soil forming interval >100 ky. (B) Profile characteristic of sandy facies of Eagle Lake Alloformation or similar falling stage and lowstand units within the post-Beaumont valley fill, soil forming interval 10–15 ky. (C) Profile characteristic of sandy facies of Cba–1 within the post-Beaumont valley fill, soil forming interval 3–5 ky. (D) Profile characteristic of sandy facies of Cba–2 within the post-Beaumont valley fill, soil forming interval <1 ky.

marily on upstream controls on discharge regimes and sediment supply. As such, specific causal relations in the Colorado system may have little portability beyond this particular region and the Late Quaternary. However, at a more fundamental level, such changes may record feedback between the climate and oceanographic systems, such that flood-plain mud-poor depositional systems would characterize time periods when regional land and ocean temperatures are significantly cooler, climatically-controlled global sea-level is low, and channels and valleys are extended across subaerially-exposed continental shelves. By contrast, flood-plain mud-rich systems would characterize periods when global and regional temperatures are warmer, and continental shelves are flooded. The volumetric significance of flood-plain facies is further enhanced by forced storage of sediments within the coastal plain as channels shorten and valleys fill during late stages of transgression and highstand. Resultant changes in facies architecture would then correlate to, or covary with, periods of sea-level change, but the issue of causality is more complicated and reflects on interactions between the different sets of controls.

Paleosols play an important role in differentiation of the post-Beaumont valley as a whole, as well as subdivision of allostratigraphic units within the valley fill. Differences in paleosols within alluvial successions reflect a number of factors, for example facies, climate and vegetation, drainage conditions, rates of deposition and/or the duration of non-deposition and soil development, and the degree of post-burial diagenetic modification (e.g., Retallack, 1986; Kraus and Bown, 1986, 1988; Wright, 1992; Willis and Behrensmeyer, 1994). Most important from a stratigraphic point-of-view may be differences in the duration of periods of non-deposition and the degree of development of pedogenic characteristics, which may provide an index of the time scales over which different allostratigraphic units form. For the Colorado system, Beaumont surface soils can be readily distinguished from paleosols within the incised valley because of the degree of soil development acquired during 100 ky or more of subaerial exposure while the shoreline was basinward of, and lower in elevation than, full interglacial highstand positions. Similarly, paleosols that define the upper boundary to allostratigraphic units within the valley fill vary in the degree of expression of key characteristics, for example depth of leaching and degree of development of clay-rich B horizons, partly as a function of the duration of episodes of flood-plain abandonment and soil development during the last glacial cycle (Fig. 10). Also important is the issue of preservation potential and eventual recognition in the stratigraphic record. Well-developed paleosols that bound allostratigraphic units deep within the valley fill have a high preservation potential and should be easily recognized. By contrast, less well-developed paleosols within the upper 2–3 m of the valley fill may be difficult to recognize and differentiate as individual profiles after deep weathering during successive glacial cycles overwhelms many original characteristics, or they may be removed by development of a later valley fill.

BEAUMONT ALLUVIAL PLAINS

Studies of post-Beaumont strata, combined with reexamination of previous work on the Beaumont Formation suggests the stratigraphy, chronology, and processes of Texas Gulf Coastal Plain alluvial successions are more complex than existing interpretations infer. The areal extent of Beaumont alluvial plains is, for example, much greater than the same rivers have constructed during the present highstand. Moreover, several workers (e.g., Winker, 1979; Aronow et al., 1991; Paine, 1991) have noted paleosols within Beaumont deposits, indicating significant periods of non-deposition and soil development before burial by more Beaumont deposits. Finally, Winker's (1979; see also DuBar et al., 1991) correlation of the onlapping contact between Beaumont and Lissie onshore and a regional seismic reflector offshore suggests the Beaumont Formation of Colorado River is 40 m or more in thickness at the highstand shoreline, and 3–4 times that at the shelf edge (see Fig. 4). Elsewhere in the offshore Gulf, this reflector occurs below the first downhole appearance of *Trimosina denticulata*, placed at approximately 600 Ka by Armentrout and Clements (1990). These data and observations suggest the Beaumont Formation represents a large three-dimensional wedge comprised of multiple cross-cutting and superimposed valley fills, and that Beaumont deposition must span multiple Pleistocene 100-ky glacial-interglacial cycles.

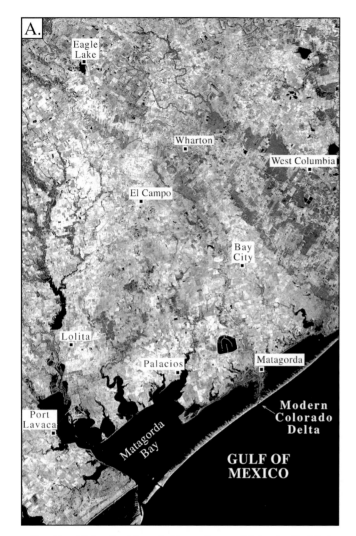

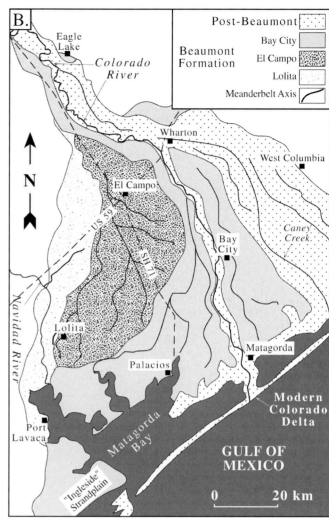

FIG. 11.—(A) Landsat Thematic Mapper image of Colorado alluvial plain. (B) Preliminary geologic map of Beaumont and younger strata of the Colorado alluvial plain, subdividing the Beaumont surface into multiple large-scale cross-cutting valley fills. Modified and updated from Blum and Price (1994).

With these ideas in mind, studies were initiated to develop a revised stratigraphic framework for Beaumont strata of the Colorado alluvial plain, based on: (1) mapping and differentiation of valley fills from LANDSAT imagery; (2) documentation of stratigraphic relations within and between valley fills from shallow cores and available outcrops; (3) documentation of the stratigraphic significance of paleosols; (4) documentation of facies architecture within valley fills; and (5) development of a chronostratigraphic framework with thermoluminescence (TL) ages.

Beaumont Valley Fills

Cross-cutting relationships identifiable in satellite imagery suggest the Beaumont alluvial plain of the Colorado River can be differentiated into three principal valley fills (Fig. 11). Each has an areal extent similar to the post-Beaumont valley, and consists of a well-defined meanderbelt axis or axes with clearly identifiable meanderbelt sands that become anastamosing and/ or distributary in the downstream direction, and are flanked by floodbasin muds. The oldest, informally designated the Lolita

valley fill, occurs along the western flanks of the Colorado alluvial plain, and is partially occupied by present-day Navidad River. A valley fill of intermediate age is centered over, and informally named after, the town of El Campo. The youngest and most clearly defined is referred to as the Bay City valley fill and has been reoccupied by the modern Colorado channel. Its western boundary is partly defined by Tres Palacios River, a small stream that has eroded headward from a drowned channel and meanderbelt (Tres Palacios Bay) into lower topography along the contact between floodbasin depositional environments and the El Campo valley fill. The post-Beaumont valley fill defines the eastern margins of the Bay City unit.

Stratigraphic relations between valley fills also have been inferred from the relative degree of soil development on alluvial plain surfaces and the superposition of deeply-weathered paleosols. Shallow cores indicate that soils developed on sandy meanderbelts of the El Campo valley fill are more deeply leached and well-developed than soils developed in sandy facies of the Bay City valley fill. Similar observations apply to vertisols developed in muddy flood-plain and flood-basin facies from the two valleys, which suggests they have been stable

FIG. 12.—Stratigraphic relations between Lolita valley fill and overlying flood-plain/flood-basin facies of the El Campo valley fill. (A) Exposure along Navidad River. TL dates of 323 ± 51 and 307 ± 37 Ka obtained from this locality. (B) Exposure father downdip along Lavaca Bay, illustrating the top of Lolita valley fill is roughly parallel to the present coastal plain surface.

and part of the Bay City valley fill is exposed along banks of the modern channel for 50–60 km. Exposures in reaches dominated by meanderbelt traces in satellite imagery typically consist of 5–10 m of point bar sands capped by a well-developed Beaumont surface soil, or in some cases a weakly developed paleosol overlain by Beaumont floodbasin facies and a well-developed vertisol (Fig. 13A). This contrasts markedly with exposures common to reaches dominated at the surface by flood-basin facies. The upper parts of most sections consist of 3–5 m of interbedded vertical accretion fine sand and mud capped by a typical Beaumont vertisol, but flood-plain and flood-basin facies often unconformably overlie paleosols that defines the upper boundary to a buried allostratigraphic unit (Figs. 13B, 14). These paleosols are most analogous to those that define the top of ELA within the post-Beaumont valley fill, suggesting subaerial exposure and soil development for 15 ky or more before burial. Depth of burial for these allostratigraphic units increases in the downstream direction which shows that meanderbelts had a gradient somewhat steeper than the highstand coastal plain, were deep within an incised valley, and were graded to falling stage or lowstand shorelines that were well out on a subaerially-exposed shelf.

FIG. 13.—Stratigraphic relations within the Bay City valley fill. (A) Vertisol developed in floodbasin facies from the youngest Beaumont highstand meanderbelt axis unconformably overlying weakly-developed paleosol in meanderbelt sand from another highstand meanderbelt. Beaumont surface soil is in turn overlain unconformably by modern flood-plain facies. TL date of 119 ± 9 Ka obtained from this location. (B) Paleosol (arrow) that bounds an allostratigraphic unit deposited deeper within the Bay City valley fill during a time period when sea-level was much lower than highstand conditions.

features in the landscape and subject to soil formation for substantially different periods of time. In a few localities, stratigraphic relations between well-developed paleosols that define the upper boundary to valley fills are well-exposed. The best example occurs in bluffs along Navidad River, where a 10–m–thick section of meanderbelt sands and muds from the Lolita valley fill, capped by a deeply weathered paleosol developed in sandy facies, is overlain unconformably by flood-basin facies and a thick well-developed vertisol associated with the El Campo valley fill (Fig. 12A). The same stratigraphic relations occur in cores farther updip, and in exposures farther downdip along the margins of Lavaca Bay (Fig. 12B). Depths of burial for the upper boundary to the Lolita valley fill remain relatively constant for some 70 km in the downdip direction, indicating meanderbelts had a gradient comparable to the highstand coastal plain, and were graded to sea-level positions similar to maximum present-day or isotope stage 5 interglacial highstands.

Stratigraphic relationships within individual valley fills resemble those of the post-Beaumont valley fill. Clear examples occur within the Bay City unit, where a Beaumont meanderbelt was reoccupied by the Colorado River during the last few hundred years following abandonment of the Caney Creek channel,

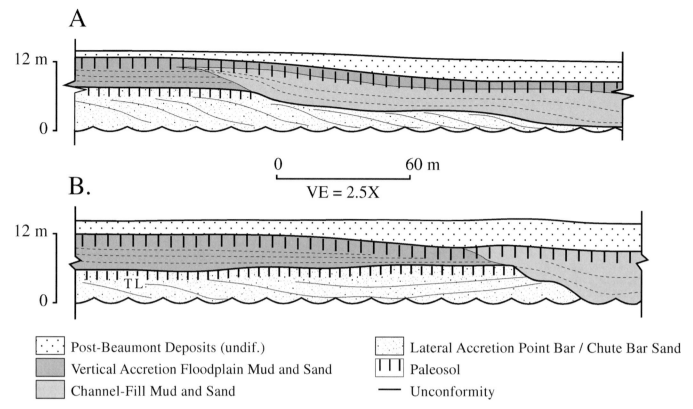

Post-Beaumont Deposits (undif.)

Vertical Accretion Floodplain Mud and Sand

Channel-Fill Mud and Sand

Lateral Accretion Point Bar / Chute Bar Sand

Paleosol

— Unconformity

FIG. 14.—Interpreted photomosaics of outcrops within the Bay City valley fill, illustrating stratigraphic relations between channel facies from isotope stage 6 lowstand, flood-plain and channel-fill facies from isotope stage 5 highstand, and modern veneer. (A) Location shown in Figure 13B. (B) A similar exposure farther downstream. TL date of 155 ± 15 Ka obtained from this location as indicated.

Although deeper parts of the Bay City valley fill are not exposed, outcrop data suggests facies architecture resembles that of the post-Beaumont valley fill. Buried allostratigraphic units are dominated by meanderbelt sands and channel-related facies, with fines concentrated in lenticular abandoned channel fills, and there is a clear lack of vertical accretion flood-plain facies. By contrast, allostratigraphic units within the upper part of the succession contain meanderbelt sands surrounded by, and encased within, thick flood-plain and flood-basin muds.

Beaumont Geochronology

Lack of chronological control on alluvial plain deposition has been a persistent problem since Beaumont strata are older than the radiocarbon window, and techniques suitable for use in deposits of middle to late Pleistocene age remain elusive. Recently, workers in Australia have achieved success with thermoluminescence (TL) dates on fluvial sands (Nanson et al., 1988, 1992, 1995; Patton et al., 1993). TL measures the time elapsed since sand-sized grains were subject to solar radiation during transport (see Berger, 1988), and, most importantly, TL can provide numerical age estimates for sandy deposits within the middle to late Pleistocene period. To establish an initial chronostratigraphic framework for Beaumont strata, samples were collected from measured sections in 3 distinct stratigraphic contexts: (1) the two youngest meanderbelts within the Bay City valley fill (see Fig. 13A); (2) meanderbelt sands within buried allostratigraphic units of the Bay City valley fill (see

Fig. 14B); and (3) meanderbelt sands from exposures of the Lolita valley fill along Navidad River (see Fig. 12). TL measurements were made at the TL Dating Laboratory of the School of Geosciences, University of Wollongong (Australia) using a combined additive/regenerative method (see Readhead, 1984), based on extraction and analysis of the 90– to 125–μ quartz grain fraction (Table 1).

The two youngest meanderbelts from the Bay City valley fill grade to former sea-level positions equivalent to or slightly below modern, and can therefore be inferred to represent the maximum isotope stage 5e interglacial highstand (ca. 129–120 Ka; Chen et al., 1991) or shortly thereafter. Point bar sands from the older of the two produced TL ages of 119 ± 10 Ka (W–1696) and 119 ± 9 Ka (W–1697), crevasse splay sands from the older meanderbelt produced a TL age of 115 ± 7 Ka (W–1624), and point bar sands from the younger of the two produced a TL age of 102 ± 6 Ka (W–1625). Such dates agree well with stratigraphic relations, and place the youngest part of the Beaumont alluvial plain in the stage 5 interglacial highstand. By contrast, point bar sands from a buried allostratigraphic unit in the Bay City valley fill yielded TL ages of 155 ± 15 Ka (W–1695) and 96.4 ± 6.2 Ka (W–1623). The first date is consistent with stratigraphic relations and suggests deposition during the stage 6 glacial period, but the second date is younger than dates on overlying strata from highstand meanderbelts and the two units are separated by a paleosol indicating a significant period of non-deposition prior to burial. We initially estimated the discrepancy was due to the sample's position near the pres-

TABLE 1.—THERMOLUMINESCENCE ANALYTICAL RESULTS. THE COMBINED SPECIFIC ACTIVITY OF THE U AND TH DECAY CHAINS WAS MEASURED BY THICK ALPHA SOURCE COUNTING OVER A 42MM SCINTILLATION SCREEN, AND ASSUMES SECULAR EQUILIBRIUM (SEE READHEAD, 1984). UNCERTAINTY LEVELS REPRESENT ONE STANDARD DEVIATION. ESTIMATED AGES REFER TO AGES ESTIMATED FROM STRATIGRAPHIC RELATIONS AND LONGITUDINAL PROFILES DEFINED IN THE FIELD AND FROM INTERPRETATION OF SATELLITE IMAGERY (SEE TEXT). CHRONOSTRATIGRAPHIC ASSIGNMENT REFERS TO CORRESPONDING OXYGEN ISOTOPE STAGES IN WILLIAMS ET AL. (1988).

SAMPLE #	W-1623	W-1624	W-1625	W-1695	W-1696	W-1697	W-1698	W-1699
Plateau Region (°C)	300–500	300–500	300–500	300–500	300–500	300–500	300–500	300–500
Analysis Temp. (°C)	350/375	350/375	350/375	350/375	350/375	350/375	350/375	350/375
Paleodose (grays)	198 ± 12	132 ± 6	169 ± 8	197 ± 17	112 ± 6	131 ± 9	302 ± 45	251 ± 26
K content (% by AES)	1.45 ± 0.005	0.55 ± 0.005	1.15 ± 0.005	0.75 ± 0.005	0.335 ± 0.005	0.55 ± 0.005	0.405 ± 0.005	0.335 ± 0.005
Moisture Content (% weight)	5.0 ± 3 (25 ± 3)*	3.2 ± 3	2.5 ± 3	2.1 ± 3	0.1 ± 3	3.4 ± 3	3.8 ± 3	1.9 ± 3
Specific Activity (Bg/kg U + Th)	21.6 ± 0.7	21.1 ± 0.7	14.4 ± 0.4	15.8 ± 0.4	20.1 ± 0.6	18.8 ± 0.6	18.1 ± 0.6	14.9 ± 0.5
Annual Radiation Dose (µGy/yr)	2055 ± 49 (1681 ± 40)*	1151 ± 50	1656 ± 49	1271 ± 50	940 ± 51	1102 ± 49	934 ± 49	817 ± 50
TL Age (Ka)	96.4 ± 6.2 (118 ± 8)*	115 ± 7	102 ± 6	155 ± 15	119 ± 9	119 ± 10	323 ± 51	307 ± 37
Estimated Age	stage 6 glacial	stage 5 interglacial	stage 5 interglacial	stage 6 glacial	stage 5 interglacial	stage 5 interglacial	stage 9 interglacial	stage 9 interglacial
Geochronological Position	problematic	stage 5e interglacial	stage 5c interglacial	stage 6 glacial	stage 5e interglacial	stage 5e interglacial	stage 9 interglacial	stage 9 interglacial

*Values listed in parentheses for sample W-1623 were recalculated using saturation moisture content of 25 ± 3%, since sample had been below the water table for much of its history, rather than the value of 5.0 ± 3% measured in the laboratory.

ent-day low-water channel and the likelihood that much of its history was spent in the phreatic zone. Recalculation assuming a saturation moisture content of 25 ± 3% produced a TL age of 118 ± 8 Ka (see Table 1), which is still problematic given stratigraphic relations. It is possible the concentration of radioactive salts, which strongly affects TL dose rates, was substantially different from what can be measured for the sample today, so the calculated TL age remains considerably lower than what would be expected from stratigraphic relations. Finally, samples collected from point bar sands in exposures of the Lolita valley fill along Navidad River produced TL ages of 323 ± 51 Ka (W–1698) and 307 ± 37 Ka (W–1699). Large error terms result from approaching TL saturation and the limits of the technique in this environment, but TL ages correspond to the isotope stage 9 interglacial highstand approximately 337 to 302 Ka (Morrison, 1991).

In sum, stratigraphic relations and TL ages provide an initial, revised chronostratigraphic framework for the Beaumont alluvial plain of Colorado River. The oldest component exposed at the near surface consists of the Lolita valley fill, which TL ages suggest should represent the isotope stage 10 to 9 glacial-interglacial cycle, whereas the youngest part of the Beaumont alluvial plain consists of the Bay City valley fill, which TL ages constrain to the isotope stage 6 to 5 glacial-interglacial cycle. The El Campo valley fill is undated but intermediate in age between the Lolita and Bay City valley fills, and should represent the stage 8 to 7 glacial-interglacial cycle.

CONCEPTUAL MODEL FOR GULF COAST ALLUVIAL PLAINS

Studies of Beaumont and post-Beaumont strata suggest a general model that describes Texas Gulf Coast alluvial plain deposition in response to interacting glacio-eustatic and climatic controls. These studies also point to features and characteristics that may be significant to sequence stratigraphic interpretations of the fully non-marine components of alluvial successions, as well as their relative scale, hierarchical nature, and genetic significance. Hence Gulf Coast alluvial plains may provide a good analog for interpretation of alluvial successions in ancient passive margin and other settings where repetitive, high-frequency glacio-eustatic and climatic controls dominated sedimentary rhythms.

Beaumont and younger alluvial plains consist of multiple cross-cutting and/or superimposed valley fills of widely varying age (Fig. 15), with valley-fill complexes exposed at the surface representing the last 3–400 ky or more. Valley fills probably become partitioned by initial glacio-eustatic lowering of base-level below interglacial highstand positions, when channels incise through highstand aggradational/progradational wedges and valley axes become somewhat fixed in place as they extend across the newly subaerial shelf (Fig. 16A). While base-level remains lower in elevation than highstand aggradational/progradational wedges, and shorelines remain basinward of highstand positions, the rest of the alluvial plain is characterized by non-deposition and soil development. However, rather than the complete sediment bypass of the coastal plain envisioned in early sequence models, multiple episodes of lateral migration, aggradation, degradation, and/or abandonment of flood plains with soil formation, occur within incised and extended valleys during the falling stage and lowstand. This creates a composite basal valley-fill unconformity, a strongly time-transgressive equivalent to the classically-defined Exxon-type sequence boundary envisioned for the marine offshore, as well as multiple smaller-scale allostratigraphic units within the valley fill (Figs. 16B, C). With transgression and highstand, and related channel and valley shortening, incised valleys fill at paces set by upstream controls on sediment delivery (Fig. 16C). As filling nears completion veneers of flood-plain and floodbasin sediments spread laterally and bury deeply weathered soils developed on subsiding downdip margins of the older alluvial plain (Fig. 16D). Total thickness of an individual valley fill ranges from 15–20 m at updip limits of the alluvial plain to 35 m or so at the highstand shoreline, whereas they may extend 15–40 km along strike.

Stratigraphic evolution of individual valley fills probably reflects interactions between climatically-controlled changes in discharge regimes and sediment supply, and glacio-eustasy. Such a view departs from standard interpretations in two respects. First, the channel incises following base-level fall below the highstand depositional shoreline, where the major break in slope exists, but the lag time for fluvial response may be insignificant compared to the maximum rates (mm/yr) and duration of base-level fall itself. Second, changes in slope forced by relative base-level lowering are insufficient in magnitude to trigger complete sediment bypass of the coastal plain for any significant length of time. Instead, through most of the 100–ky glacial periods when the shoreline is basinward of highstand positions, and lower in elevation than highstand aggradational/progradational wedges, channels are graded to shoreline posi-

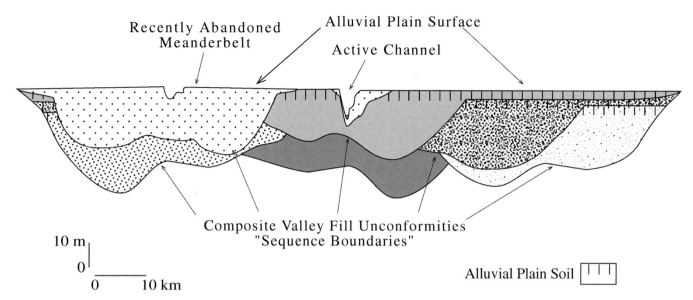

FIG. 15.—Conceptual cross-section through Quaternary alluvial plain, illustrating geometry, composite basal unconformities, and bounding alluvial-plain soil profiles for 100-ky glacio-eustatically controlled incised valley fills and depositional sequences.

tions on the shelf or at the shelf-edge. The coastal plain part of the fluvial system then becomes an extension of the bedrock-confined valley farther upstream, with similar sediment transport capabilities, and undergoes multiple episodes of net deposition and sediment storage, punctuated by valley incision and/or sediment bypass, during an individual 100-ky glacio-eustatic cycle. Periods of active channel aggradation, lateral migration, and/or flood-plain construction within the incised valley represent large volumes of sediment delivered from upstream sources, or increased flood magnitudes that result in widespread vertical accretion. By contrast, unconformities within the valley fill sequence represent periods of channel incision and/or sediment bypass due to decreased sediment supply, abandonment of flood plains due to decreased flood magnitudes, or meanderbelt avulsion and relocation within the incised valley. In each case, soils develop on abandoned flood plains and represent the upper boundary to smaller-scale allostratigraphic units.

Episodes of deposition, production of unconformities, and periods of soil development within the valley fill probably reflect changes in discharge regimes and sediment supply, as controlled by climate changes in the continental interior. However, stratigraphic architecture within an individual valley fill clearly reflects the trend of glacio-eustatically driven base-level change. Small-scale allostratigraphic units deposited during the long and complex falling stage and lowstand, when the shoreline was lower in elevation and farther basinward, occur deep within the incised valley as a flight of downward-stepping terraces with underlying sandbodies. Indeed, these units may represent the manifestation of the concept of forced regression and the "forced regressive systems tract" within incised valleys, and correlate to the series of basinward- and downward-stepping falling stage shorelines discussed by previous workers (e.g., Posamentier et al., 1992; Hunt and Tucker, 1992; Helland-Hansen and Gjelberg, 1994). After weathering and soil development, falling stage and lowstand sandbodies may be partially cut out, but mostly onlapped and buried, by stratigraphic units deposited during transgressive and highstand conditions, and

may comprise the bulk of reservoir-quality sands within the incised-valley fill. Such sandbodies may be 10–20 m in thickness and extend 2–10 km along strike.

Changes in facies architecture within an incised-valley fill may correlate with trends in relative base-level change but are not directly caused by them. Facies patterns, sand body geometry, and sand body interconnectedness within the most recent valley-fill sequence of the Colorado alluvial plain reflect interactions between climatic controls on discharge regimes and sediment supply, as well as forced channel and valley gradient changes that accompany glacio-eustatically-driven base-level change. Coarse-grained meanderbelts with few vertical accretion deposits dominated the long and complex glacial periods when base-level and the shoreline were in mid-shelf or farther basinward positions. During this time, feedback between the climate and oceanographic systems mitigated against deep, overbank floods and generation of thick overbank fines. Thus allostratigraphic units deposited during glacial periods occur deep within the incised valley and consist of laterally amalgamated sands and gravels with minimal heterogeneity. During the interglacial period, increases in sea surface size and temperature, coupled with warmer land surface temperatures, produced high-intensity tropically-derived storms that resulted in erosion of soils from uplands, and consequent increases in the depth and flashiness of floods. As a result, allostratigraphic units deposited coincident with sea-level highstand contain an abundance of flood-plain deposits, and sand bodies tend to be less amalgamated. These characteristics are greatly enhanced in downdip portions of the incised valley by a forced shortening of channels, flattening of gradients, and forced updip storage of sediments in response to base-level rise and landward translation of the shoreline.

Finally, channel shortening, valley filling, and reductions in depositional slope during the late stages of transgression and highstand is one set of events that promotes channel avulsion and development of an anastamosing and/or distributary pattern of meanderbelt sands with intervening flood-plain and flood-

A. Late Highstand - complete filling of incised valley
- burial of older alluvial plain surface and soil
- avulsion to new meanderbelt position

B. Falling Stage
and Lowstand
- abandonment of alluvial plains, partitioning of incised valley
- soil development on abandoned alluvial plain
- multiple episodes of deposition and soil formation within
  incised valley, and development of the "falling stage systems tract"

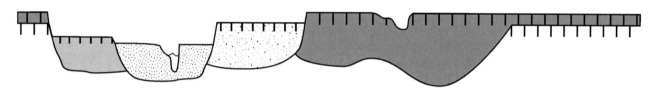

C. Transgression and
Early Highstand
- initial stages of aggradation within incised valley through
  multiple episodes of deposition and soil formation
- continued soil development on alluvial plain

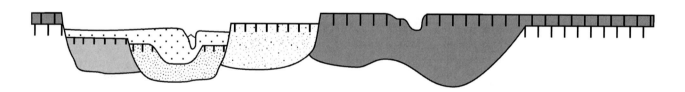

D. Late Highstand - complete filling of incised valley through multiple avulsions
  and widespread flood plain/flood basin aggradation
- burial of soils on older abandoned alluvial plain surface
- avulsion to new meanderbelt position

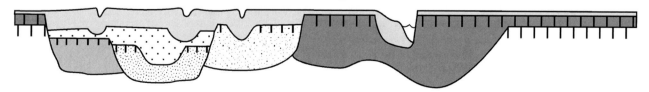

Fig. 16.—Model for evolution of alluvial plains and component incised valley fills.

basin muds. For high sediment yield extrabasinal fluvial systems, which completely fill incised valleys and construct large alluvial-deltaic headlands (see Winker, 1979, Galloway, 1981), avulsion may result in relocation of the valley axis before the next base-level fall, such that successive incised valleys have a distributary pattern, and successive 100–ky increments of geologic time occur lateral to each other. At the opposite end of the spectrum, low sediment yield basin margin and intrabasinal systems may not completely fill their incised valleys, and avulsion may remain confined by valley walls through the entire highstand, and until the next base-level fall occurs. In this type of setting, low net accommodation produced by subsidence over the course of a 100–ky glacial-interglacial cycle, the repetitive magnitude and frequency of glacio-eustatic cycles, and repetitive forced changes in channel and valley gradients combine to insure that development of the next incised valley will cannibalize much of the record from the previous cycle.

In closing, studies described above, combined with recent investigations of the Colorado system on the now-submerged shelf (Anderson et al., 1996), provide an opportunity to develop a fully longitudinal view of fluvial response to external controls. As described elsewhere (Blum, 1993; Blum and Valastro, 1994), the Late Cenozoic geologic history of the bedrock valley on the Inner Gulf Coastal Plain is dominated by long-term valley incision through Cretaceous and Tertiary components of the basin fill, in response to a tectonically-controlled base-level of erosion. Cycles of lateral migration, valley widening, and sediment storage, followed by renewed valley incision with terrace formation, were superimposed on this long-term trend, and probably record climatically-controlled changes in discharge and sediment supply. By contrast, as this paper suggests, stratigraphic records from the fully non-marine alluvial plain reflect intereactions between the 100–ky glacio-eustatic cycles of base-level change, which serve to partition incised valleys, and shorter-term climatically-controlled changes in discharge and sediment supply which play a major role in determining how they fill. For the now-submerged shelf, subject to repeated transgressions and regressions of the shoreline during each 100–ky glacial-interglacial cycle, Anderson et al. (1996) show that sediment supply from continental interiors plays an important role when comparing different river systems, but eustasy coupled with higher rates of subsidence controls facies patterns and the overall stratigraphic evolution of individual valley fills.

ACKNOWLEDGMENTS

Research on post-Beaumont strata was supported by the Gulf Coast Association of Geological Societies, the Geological Society of America, and the United States National Science Foundation. Studies of Beaumont strata have been supported by donors to the Petroleum Research Fund of the American Chemical Society, Exxon Production Research, Union Pacific Resources, the National Science Foundation, and the School of Geosciences of the University of Wollongong, Australia. We thank Charles D. Winker, John B. Anderson, Whitney J. Autin, and Torbjorn Tornquist for comments and/or constructive reviews of the manuscript.

REFERENCES

ABDULAH, K. C. AND J. B. ANDERSON, 1991, Eustatic controls on the evolution of the Pleistocene Brazos-Colorado deltas, Texas, *in* Coastal Depositional Systems in the Gulf of Mexico: Quaternary Framework and Environmental Issues: Gulf Coast Section of the Society of Economic Paleontologists and Mineralogists Foundation, Proceedings of the 12th Annual Research Conference, p. 1–7.

ANDERSON, J. B., THOMAS, M. A., SIRINGIN, F. P., AND SMYTH, W. C., 1992, Quaternary evolution of the Texas Coast and Shelf, *in* Fletcher, C. H. III AND WEHMILLER, J. F., eds., Quaternary Coasts of the United States: Marine and Lacustrine Systems: Tulsa, Society of Economic Paleontologists and Mineralogists Special Publication 48, p. 253–265.

ANDERSON, J. B., ABDULAH, K., SARZALEJO, S., SIRINGIN, F., AND THOMAS, M. A., 1996, Late Quaternary sedimentation and high-resolution sequence stratigraphy of the East Texas Shelf. *in* De Batist, M. and Jacobs, P., eds., Geology of Siliciclastic Shelf Seas: London, Geologic Society of London Special Publication 117, p. 95–124.

ARMENTROUT, J. M. AND CLEMENT, J. F., 1990, Biostratigraphic calibration of depositional cycles: a case study in High Island-Galveston East Breaks Area, offshore Texas, *in* Armentrout, J. M. and Perkins, B. F., eds., Sequence Stratigraphy as an Exploration Tool: Concepts and Practices in the Gulf Coast: Gulf Coast Section of the Society of Economic Paleontologists and Mineralogists Foundation, Proceedings of the 11th Annual Research Conference. p. 21–51.

ARONOW, S., 1971, Nueces River delta plain of Pleistocene Beaumont Formation: American Association of Petroleum Geologists Bulletin, v. 55, p. 1231–1248.

ARONOW, S., NECK, R. W., AND MCCLURE, W. L., 1991, The Caroline Street local fauna: a late Pleistocene freshwater molluscan/vertebrate fauna from Houston, Harris Co., Texas: Gulf Coast Association of Geological Societies Transactions, v. 41, p. 17–28.

AUTIN, W. J., BURNS, S. F., MILLER, B. J., SAUCIER, R. T. AND SNEAD, J. J., 1991, Quaternary geology of the Lower Mississippi Valley, *in* Morrison, R. B., ed., Quaternary Non-Glacial Geology of the Conterminous United States, v. K–2, The Geology of North America: Boulder, Geological Society of North America, p. 547–582.

BARNES, V. E., 1979, Geologic Atlas of Texas: The Seguin Sheet: Austin, Bureau of Economic Geology, University of Texas at Austin.

BARNES, V. E., 1981, Geologic Atlas of Texas: The Austin Sheet: Austin, Bureau of Economic Geology, University of Texas at Austin.

BARNES, V. E., 1982, Geologic Atlas of Texas: The Houston Sheet: Austin, Bureau of Economic Geology, University of Texas at Austin.

BARNES, V. E., 1983, Geologic Atlas of Texas: The San Antonio Sheet: Austin, Bureau of Economic Geology, University of Texas at Austin.

BARNES, V. E., 1987, Geologic Atlas of Texas: The Beeville-Bay City Sheet: Austin, Bureau of Economic Geology, University of Texas at Austin.

BERNARD, H. A., 1950, Quaternary Geology of Southeast Texas: Unpublished Ph.D. Dissertation, Louisiana State University, Baton Rouge, 164 p.

BERGER, G. W., 1988, Dating Quaternary events by luminescence, *in* Easterbrook, D. J., ed., Dating Quaternary Sediments: Boulder, Geological Society of America Special Paper 227. p. 13–50.

BERNARD, H. A. AND MAJOR, C. F. Jr., 1963, Recent meanderbelt deposits of the Brazos River: an alluvial "sand" model (abs.): American Association of Petroleum Geologists Bulletin, v. 47, p. 350.

BERNARD, H. A. AND LEBLANC, R. J., 1965, Resume of the Quaternary Geology of the Northwestern Gulf of Mexico Province, *in* Wright, H. E. and Frey, D. G., eds., The Quaternary of the United States: Princeton, Princeton University Press. p. 137–185.

BERNARD, H. A., MAJOR, C. F., PARROTT, B. S., AND LEBLANC, R. J. SR., 1970, Recent Sediments of Southeast Texas: Field Guide to the Brazos Alluvial-Deltaic Plain and the Galveston Barrier Island Complex: Austin, The University of Texas at Austin Bureau of Economic Geology Guidebook 11, 132 p.

BLUM, M. D., 1990, Climatic and eustatic controls on Gulf Coastal Plain fluvial sedimentation: an example from the Late Quaternary of the Colorado River, Texas, *in* Armentrout, J. M. and Perkins, B. F., eds., Sequence Stratigraphy as an Exploration Tool: Concepts and Practices in the Gulf Coast: Gulf Coast Section of the Society of Economic Paleontologists and Mineralogists Foundation, Proceedings of the 11th Annual Research Conference, p. 71–83.

BLUM, M. D., 1992, Modern Depositional Environments and Recent Alluvial History of the Lower Colorado River, Gulf Coastal Plain of Texas: Unpublished Ph.D. Dissertation, University of Texas at Austin, Austin, 286 p.

BLUM, M. D., 1993, Genesis and architecture of incised valley fill sequences: a Late Quaternary example from the Colorado River, Gulf Coastal Plain of Texas, *in* Weimer, P. and Posamentier, H. W., eds., Siliciclastic Sequence Stratigraphy: Recent Developments and Applications: Tulsa, American Association of Petroleum Geologists Memoir 58. p. 259–283.

BLUM, M. D. AND PRICE, D. M., 1994, Glacio-eustatic and climatic controls on Pleistocene alluvial plain deposition, Texas Coastal Plain: Transactions of the Gulf Coast Association of Geological Societies, v. 44, p. 85–92.

BLUM, M. D. AND VALASTRO, S. JR., 1994, Late Quaternary sedimentation, Lower Colorado River, Gulf Coastal Plain of Texas: Geological Society of Americ Bulletin. v. 106, p. 1002–1016.

BLUM, M. D., TOOMEY, R. S. III, AND VALASTRO, S. Jr. , 1994, Fluvial response to Late Quaternary climatic and environmental change, Edwards Plateau of Texas: Paleogeography, Paleoclimatology, and Paleoecology. v. 108, p. 1–21.

BLUM, M. D., MORTON, R. A., AND DURBIN, J. M., 1995, "Deweyville" terraces and deposits of the Texas Gulf Coastal Plain: Transactions of the Gulf Coast Association of Geological Societies, v. 45, p. 53–60.

BROWN, L. F., BREWTON, J. L., MCGOWEN, J. H., EVANS, T. J., FISHER, W. L., AND GROAT, C. G., 1976, Environmental Geologic Atlas of the Texas Coastal Zone: Corpus Christi Area: Austin, University of Texas at Austin Bureau of Economic Geology, 123 p.

CHEN, J. H., CURRAN, H. A., WHITE, B., AND WASSERBURG, G. J., 1991, Precise chronology of the last interglacial period: 234U–230Th data from fossil coral reefs in the Bahamas: Geological Society of America Bulletin, v. 103, p. 85–97.

DOERING, J. A., 1935, Post-Fleming surface formations of southeast Texas and south Louisiana: American Association of Petroleum Geologists Bulletin, v. 19, p. 651–688.

DOERING, J. A., 1956, Review of Quaternary surface formations of the Gulf Coast Region: American Association of Petroleum Geologists Bulletin, v. 40, p. 1816–1862.

DUBAR, J. R., EWING, T. E., LUNDELIUS, E. L., OTVOS, E. G., AND WINKER, C. D., 1991, Quaternary geology of the Gulf of Mexico Coastal Plain, in Morrison, R. B. ed., Quaternary Non-Glacial Geology of the Conterminous United States: Boulder, Geological Society of America, Geology of North America Volume K–2, p. 583–610.

DUESSEN, A., 1914, Geology and Underground Waters of the Southeastern Part of the Texas Coastal Plain: Washington, United States Geological Survey Water Supply Paper 335, 365 p.

DUESSEN, A., 1924, Geology of the Coastal Plain of Texas west of Brazos River: Washington, United States Geological Survey Professional Paper 126, 145 p.

FISHER, W. L., MCGOWAN, J. H., BROWN, L. F. JR., AND GROAT, C. G., 1972, Environmental Geologic Atlas of the Texas Coastal Zone: Galveston – Houston Area: Austin, University of Texas at Austin Bureau of Economic Geology, 91 p.

FISK, H. N., 1944, Geological Investigations of the Alluvial Valley of the Lower Mississippi River: Vicksburg, Mississippi River Commission, United States Army Corps of Engineers, 78 p.

FISK, H. N., 1947, Fine-Grained Alluvial Deposits and their Effects on Mississippi River Activity: Vicksburg, Mississippi River Commission, United States Army Corps of Engineers, 82 p.

GALLOWAY, W. E., 1981, Depositional architecture of Cenozoic Gulf Coastal Plain fluvial systems: in Ethridge, F. G. and Flores, R. M., eds., Recent and Ancient Non-Marine Depositional Environments: Society of Economic Paleontologists and Mineralogists Special Publication 31, p. 127–156

GIBLING, M. R. AND BIRD, D. J., 1994, Late Carboniferous cyclothems and alluvial paleovalleys in the Sydney Basin, Nova Scotia: Geological Society of America Bulletin, v. 106, p. 105–117.

HARLAND. W. B., ARMSTRONG, R. L., COX, A. V., CRAIG, L. E., SMITH, A. G., AND SMITH, D. G., 1989, A Geologic Time Scale 1989: Cambridge, Cambridge University Press, 263 p.

HAYES, C. W. AND KENNEDY, W., 1903, Oil Fields of the Texas-Louisiana Gulf Coastal Plain: Washington, United States Geological Survey Bulletin 213.

HELLAND-HANSEN, W. AND GJELBERG, J. G., 1994, Conceptual basis and variability in sequence stratigraphy: a different perspective: Sedimentary Geology, v. 92, p. 31–52.

HUNT, D. AND TUCKER, M. E., 1992, Stranded parasequences and the forced regressive systems tract: deposition during base-level fall: Sedimentary Geology, v. 81, p. 1–9.

IMBRIE, J. J., HAYS, J. D., MARTINSON, D. G., MCINTYRE, A., MIX, A. C., MORELEY, J. J., PISIAS, N. G, PRELL, W. L., AND SHACKLETON, N. J., 1984, The orbital theory of Pleistocene climate: support from a revised chronology of the marine $\delta^{18}O$ record, in Berger, A. L., Imbrie, J., Hays, J., Kukla, G. and Saltzman, B., eds., Milankovitch and Climate – Part 1: Dordecht, Reidel Publishing, p. 269–306.

JERVEY, M. T., 1988, Quantitative geological modelling of siliciclastic rock sequences and their seismic expression, in Wilgus, C. K., Hastings, B. S.,

Posamentier, H. S., Van Wagoner, J. C., Ross, C. A., and Kendall, C. G., eds., Sea-level Changes: An Integrated Approach: Tulsa, Society of Economic Paleontologists and Mineralogists Special Publication 42, p. 47–70.

KRAUS, M. J. AND T. M. BOWN, 1986, Paleosols and time resolution in alluvial stratigraphy, in Wright, V. P. ed., Paleosols: Their Recognition and Interpretation: Princeton, Princeton University Press, p. 180–206.

KRAUS, M. J. AND T. M. BOWN, 1988, Pedofacies analysis: a new approach to reconstructing ancient fluvial sequences, in Reinhardt, J. and Sigleo, W. R., eds., Paleosols and Weathering Through Geologic Time: Boulder, Geological Society of America Special Paper 216, p. 143–152.

KUKLA, G. J. AND OPDYKE, N. D., 1972, American glacial stages in paleomagnetic time scale: Boulder, Geological Society of America, Annual Meeting Programs with Abstracts, v. 4, p. 569–570.

MCGOWAN, J. H. AND L. E. GARNER, 1970, Physiographic features and stratification types of coarse-grained point bars, modern and ancient examples: Sedimentology, v. 14, p. 86–93.

MCGOWAN, J. H., BROWN, L. F., EVANS, T. J., FISHER, W. L., AND GROAT, C. G., 1975, Environmental Geologic Atlas of the Texas Coastal Zone: Bay City-Freeport Area: Austin, University of Texas at Austin Bureau of Economic Geology, 98 p.

MCGOWAN, J. H., PROCTOR, C. V., BROWN, L. F., EVANS, T. J., FISHER, W. L., AND GROAT, C. G., 1976, Environmental Geologic Atlas of the Texas Coastal Zone: Port Lavaca Area: Austin, University of Texas at Austin Bureau of Economic Geology, 107 p.

MIALL, A. D., 1985, Architectural element analysis: A new method of facies analysis applied to fluvial deposits: Earth Science Reviews, v. 22, p. 261–308.

MIALL, A. D., 1988, Facies architecture in clastic sedimentary basins, in Kleinspehn, K. L. and Paola, C., eds, New Perspectives in Basin Analysis: New York, Springer-Verlag, p. 67–81.

MIALL, A. D., 1991a, Heirarchies of architectural units in clastic rocks, and their relation to sedimentation rates, in Miall, A. D. and Tyler, N., eds., Three-Dimensional Facies Architecture of Terrigenous Clastic Sediments, and its Implication for Hydrocarbon Discovery and Recovery: Tulsa, Society of Economic Paleontologists and Mineralogists Concepts in Sedimentology and Paleontology 3, p. 6–12.

MIALL, A. D., 1991b, Stratigraphic sequences and their chronostratigraphic correlation. Journal of Sedimentary Petrology, v. 61, p. 497–505.

MORRISON, R. B., 1991, Introduction, in Morrison, R. B., ed., Quaternary Non-Glacial Geology of the Conterminous United States: Boulder, Geological Society of America, Geology of North America Volume K–2, p. 1–12.

MORTON, R. A. AND PRICE, W. A., 1987, Late Quaternary sea-level fluctuations and sedimentary phases of the Texas Coastal Plain and shelf, in Nummedal D. and Pilkey, O. H., eds., Sea-level Fluctuations and Coastal Evolution: Tulsa, Society of Economic Paleontologists and Mineralogists Special Publication 15, p.181–198.

MORTON, R. A. AND SUTER, J. R., 1996, Sequence stratigraphy and composition of Late Quaternary shelf margin deltas, northern Gulf of Mexico: American Association of Petroleum Geologists Bulletin, v. 80, p. 505–530.

NANSON, G. C., YOUNG, R. Y., PRICE, D. M., AND RUST, B. R., 1988, Stratigraphy, sedimentology, and Late Quaternary chronology of the Channel Country of Western Queensland, in R. F. Warner, ed., Fluvial Geomorphology of Australia: Sydney, Academic Press, p. 151–175.

NANSON, G. C., PRICE, D. M., AND SHORT, S. A., 1992, Wetting and drying of Australia over the past 300 Ka: Geology, v. 20, p. 791–794.

NANSON, G. C., CHEN, X. Y., AND PRICE, D. M., 1995, Aeolian and fluvial evidence of changing climate and wind patterns in the western Simpson Desert, Australia: Paleogeography, Paleoclimatology, and Paleoecology. v. 113, p. 87–102.

PAINE, J. G., 1991, Late Quaternary Depositional Units, Sea-level, and Vertical Movements Along the Central Texas Coast: Unpublished Ph.D. Dissertation, University of Texas at Austin, Austin, 261 p.

PATTON, P. C., PICKUP, G., AND PRICE, D. M., 1993, Holocene paleofloods of the Ross River, Central Australia: Quaternary Research, v. 40, p. 201–212.

POSAMENTIER, H. W., JERVEY, M. T., AND VAIL, P. R., 1988, Eustatic controls on clastic deposition I: conceptual framework, in Wilgus, C. K., Hastings, B. S., Posamentier, H. S., Van Wagoner, J. C., Ross, C. A., and Kendall, C. G., eds., Sea-level Changes: An Integrated Approach: Tulsa, Society of Economic Paleontologists and Mineralogists Special Publication 42, p. 110–124.

POSAMENTIER, H. W. AND VAIL, P. R., 1988, Eustatic controls on clastic deposition II: sequence and systems tract models, in Wilgus, C. K., Hastings, B. S., Posamentier, H. S., Van Wagoner, J. C., Ross, C. A., and Kendall,

C. G., eds., Sea-level Changes: An Integrated Approach: Tulsa, Society of Economic Paleontologists and Mineralogists Special Publication 42, p. 125–154.

POSAMENTIER, H. W., ALLEN, G. P., JAMES, D., AND TESSON, M, 1992, Forced regressions in a sequence stratigraphic framework: concepts, examples, and exploration significance: American Association of Petroleum Geologists Bulletin, v. 76, p. 1687–1709.

PORTER, S. C., 1989, Some geological implications of average Quaternary glacial conditions. Quaternary Research, v. 32, p. 245–262.

READHEAD, M. L., 1984, Thermoluminescence Dating of Some Australian Sedimentary Deposits: Unpublished Ph.D. Thesis, Australian National University, Canberra.

RETTALACK, G. J., 1986, Fossil soils as grounds for interpreting long-term controls on ancient rivers: Journal of Sedimentary Petrology, v. 56, p. 1–18.

SAUCIER, R. T. AND FLEETWOOD, A. R., 1970, Origin and chronologic significance of Late Quaternary terraces, Quachita River, Arkansas and Louisiana: Geological Society of America Bulletin, v. 81, p. 869–890.

SCHUMM, S. A., 1993, River response to baselevel change: implications for sequence stratigraphy: Journal of Geology, v. 101, p. 279–293.

SHANLEY, K. W. AND MCCABE, P. J., 1991, Predicting facies architecture through sequence stratigraphy: an example from the Kaiparowits Plateau, Utah: Geology, v. 19, p. 742–745.

SHANLEY, K. W. AND MCCABE, P. J., 1993, Alluvial architecture in a sequence stratigraphic framework: a case study from the Upper Cretaceous of southern Utah, U.S.A., in Flint, S. and Bryant, I., eds., Quantitative Modelling of Clastic Hydrocarbon Reservoirs and Outcrop Analogs: Blackwell Scientific, International Association of Sedimentologists Special Publication 15, p. 21–55.

SHANLEY, K. W. AND MCCABE, P. J., 1994, Perspectives on the sequence stratigraphy of continental strata: report of a working group at the 1991 Nuna Conference on High Resolution Sequence Stratigraphy: American Association of Petroleum Geologists Bulletin, v. 74, p. 544–568.

SUTER, J. R., 1987, Fluvial systems, in Berryhill, H. L., ed., Late Quaternary Facies and Structure, Northern Gulf of Mexico – Interpretations from Seismic Data: Tulsa, American Association of Petroleum Geologists Studies in Geology 23, p. 81–129.

SUTER, J. R. AND BERRYHILL, H. L., 1985, Late Quaternary shelf-margin deltas, Northwest Gulf of Mexico: American Association of Petroleum Geologists Bulletin, v. 69, p. 77–91.

THOMAS, M. A. AND ANDERSON, J. B., 1994, Sea-level controls on the facies architecture of the Trinity/Sabine incised-valley system, Texas continental shelf, in Dalrymple, R. W., Boyd, R., and Zaitlin, B. A., eds., Incised-Valley Systems: Origins and Sedimentary Sequences: Tulsa, Society of Economic Paleontologists and Mineralogists Special Publication 51, p. 63–82.

TOOMEY, R. S. III, BLUM, M. D., AND VALASTRO, S. JR., 1993, Late Quaternary climates and environments of the Edwards Plateau, Texas: Global and Planetary Change, v. 7, p. 299–320.

VAN SICLEN, D. C., 1985, Pleistocene meanderbelt ridge patterns in the vicinity of Houston, Texas: Gulf Coast Association of Geological Societies Transactions, v. 35, p. 525–532.

VAN SICLEN, D. C., 1991, Surficial geology of the Houston area: an offlapping series of Pleistocene (& Pliocene?) highest-sealevel fluviodeltaic sequences: Gulf Coast Association of Geological Societies Transactions, v. 41, p. 651–666.

VAN WAGONER, J. C., MITCHUM, R. M., CAMPION, K. M., AND RAHMANIAN, V. D., 1990, Siliciclastic sequence stratigraphy in well logs, cores, and outcrops: concepts for high-resolution correlation of time and facies: Tulsa, American Association of Petroleum Geologists Methods in Exploration 7. 55 p.

VERSTAPPEN, H. Th., 1980, Quaternary climate changes and natural environments of southeast Asia: Geojournal. v. 4, p. 45–54.

WESCOTT, W. A., 1993, Geomorphic thresholds and complex response of fluvial systems—some implications for sequence stratigraphy: American Association of Petroleum Geologists Bulletin, v. 77, p. 1208–1218.

WILLIAMS, D. F., THUNNELL, R. C., TAPPA, E., RIO, D., AND RAFFI, I., 1988, Chronology of the Pleistocene oxygen isotope record, 0–1.88 Ma before present: Paleogeography, Paleoclimatology, and Paleoecology, v. 64, p. 221–240.

WILLIS, B. J. AND BEHRENSMEYER, A. K., 1994, Architecture of Miocene overbank deposits in northern Pakistan: Journal of Sedimentary Research, v. b64, p. 60–67.

WINKER, C. D., 1979, Late Pleistocene Fluvial-Deltaic Deposition on the Texas Coastal Plain and Shelf: Austin, University of Texas at Austin Unpublished MA Thesis, 187 p.

WRIGHT, V. P., 1992, Paleopedology: stratigraphic relationships and empirical models: in Martini, I. P. and Chesworth, W., eds., Weathering, Soils and Paleosols: Elsevier, Amsterdam, p. 475–499.

WRIGHT, V. P. AND MARRIOTT, S. B., 1993, The sequence stratigraphy of fluvial depositional systems: role of flood plain sediment storage: Sedimentary Geology, v. 86, p. 203–210.

SEQUENCE STRATIGRAPHY, SUBSIDENCE RATES, AND ALLUVIAL FACIES, MANNVILLE GROUP, ALBERTA FORELAND BASIN

DOUGLAS J. CANT

Geological Survey of Canada, 3303 33rd St. N.W., Calgary, Alberta T2L 2A7, Canada

ABSTRACT: Within the Mannville Group (largely Aptian-Lower Albian) of the Alberta foreland basin, two distinct types of alluvial facies occur in different sequence stratigraphic contexts. One is relatively clean, braided-stream conglomerates and sandstones with prominent unconformities identifiable by abrupt lithologic changes; the other is comprised mainly of mudstones with thin coals and subsidiary (less than 25%) fine sandstones.

Basal Lower Mannville sediments were deposited during a period of very low (<10 m/my) subsidence rates or even during uplift of the basin and orogen, because overthrusting in the Cordillera had not been initiated. These relatively thin braided-river deposits fill paleotopographic depressions on a long-period (second order) tectonic unconformity. They consist of at least 90% sandstones and/or conglomerates, with one or more unconformities or disconformities within them. Internal unconformities are marked by channelled surfaces which separate non-marine (and some marginal marine) units with different lithologies, facies, and directions of transport.

The mudstone-dominated facies in the Upper Mannville was deposited during a period of high subsidence rates (>40 m/my) caused by active overthrusting in the associated Cordillera. Small-scale relative sea-level (RSL) fluctuations are documented in equivalent transgressive/regressive shoreline sequences. Deposition of the volumetrically dominant and widespread mudstones, thin (to 8 m) sandstones and coals, all interpreted as low-energy anastomosed stream and floodplain sediments, occurred during periods of coastal transgression. Each maximum flooding surface is correlative to a muddy and/or coaly zone immediately south of the shoreline. The allostratigraphic units defined by these surfaces contain thin sands, variable amounts of mudstone and thin coals, and can be correlated on well logs at least 350–km inland from the shorelines. They are generally sheet-like in geometry and slope seaward at a very low angle, with an offlapping stratal pattern. Episodes of falling sea-level caused incision of 20–m–deep valleys (fourth-order unconformities) through these deposits, however, rendering them difficult to correlate. In the early stages of the next RSL rise, the valleys were infilled by braided-river sands, with some estuarine sands in proximity to the shoreline. The major fluvial sandstones are therefore not contemporaneous with the mudstones, coals, and minor thin sandstones laterally adjacent to them. In interfleuve areas between valleys, the subaerial unconformities cannot be correlated on well logs or easily recognized visually because they lack lithologic expression.

The differences between the unconformity-related, sheet-like braided stream sandstones/conglomerates and the fine-grained meandering to anastomosing alluvial facies result from their sequence stratigraphic settings, controlled largely by the tectonic subsidence rates of the basin.

SEQUENCE STRATIGRAPHY OF NON-MARINE SEDIMENTS

Fluvial systems and the resulting non-marine deposits have long been known to respond to changes in base level (e.g., Fisk, 1944), and the growth of sequence stratigraphic ideas (Posamentier and Vail, 1988) has reemphasized this. Non-marine sediments have been interpreted in a sequence stratigraphic context, in most cases by integrating information on relative sea-level changes inferred from laterally equivalent marine sediments, as summarized by Shanley and McCabe (1994). The different styles of fluvial facies are a function of at least the following variables: amount and type of sediment supplied, the amount and variation of fluid discharge (Miall, 1977), as well as regional slope (Schumm and Khan, 1972). Several of these factors are influenced directly or indirectly by relative sea-level variation. Sea level functions in most cases as fluvial base level, thus affecting regional gradient and the caliber of sediment which can be moved. Discharge variations are controlled largely by climate which is itself affected by sea-level changes (Shanley and McCabe, 1994). Climate can also affect processes of chemical weathering and therefore the type of sediment supplied.

Falls in relative sea level are expressed in the ancient record by subaerial unconformities which extend across non-marine and shallow-marine facies (Posamentier and Vail, 1988). While a eustatic interpretation for base-level drops was emphasized by Posamentier and Vail (1988), they acknowledged some effects of subsidence. In contrast, in this paper, I will attempt to show that the main controls on the expression of unconformities and related fluvial facies is the variation in subsidence rates of the basin.

One of the fundamental problems in application of sequence-stratigraphic concepts in non-marine facies is the recognition of various types of surfaces. Subaerial unconformities are very difficult to distinguish from exposure surfaces due to river stage fluctuations or surfaces generated by lateral channel migration. Indeed, these exposure surfaces could be considered theoretically to be higher-order, less significant unconformities resulting from smaller-scale fluctuations in base level (Embry, 1995). This paper will document some criteria used for recognition of subaerial unconformities in non-marine facies of the Mannville Group.

Although Shanley et al. (1992) have documented tidal effects in sediments near a maximum flooding surface, by definition these do not extend inland beyond the range of marine influence. All marine surfaces important in sequence stratigraphy (regressive surface of erosion, ravinement surface, transgressive surface, slope onlap surface, and the maximum flooding surface) do not extend beyond the bayline, so recognition of subaerial unconformities must be extremely important for non-marine sequence stratigraphy. However, as noted above, changes of relative sea level should affect the lithologies, facies, and stratal geometries of equivalent alluvial deposits, suggesting that correlatable units might be recognizable in some examples. This report will document surfaces in non-marine rocks interpreted to be correlative to maximum flooding surfaces in equivalent marine sequences.

THE MANNVILLE GROUP

The Mannville Group is a clastic wedge of dominantly Aptian-Early Albian age (Fig. 1) deposited in the western Canda or Alberta foreland basin overlying tilted Jurassic to Devonian rocks (Fig. 2). The unit is penetrated by at least 70,000 logged petroleum bore holes and many thousands of cores are available. This paper compares and contrasts the basal non-marine deposits of the Lower Mannville Group, the Cadomin and basal Gething Formations in western Alberta and British Columbia

Relative Role of Eustasy, Climate, and Tectonism in Continental Rocks, SEPM Special Publication No. 59
Copyright © 1998, SEPM (Society for Sedimentary Geology), ISBN 1-56576-042-5

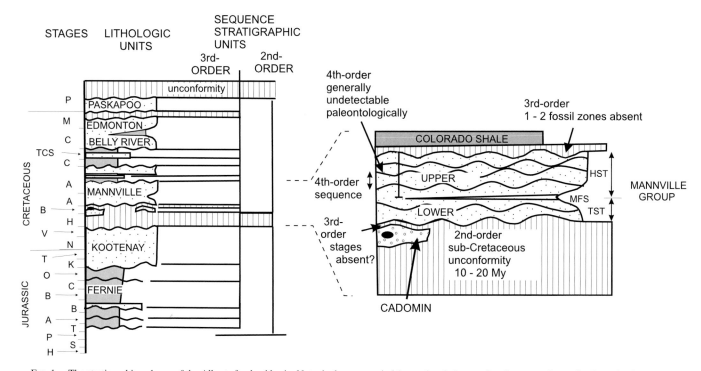

FIG. 1.—The stratigraphic column of the Alberta foreland basin. Note the longest-period (second-order) unconformity occurs immediately under the Mannville Group. The inset shows more detailed stratigraphy of this unit, suggesting that the unconformity between the Cadomin Conglomerate (including basal Gething) may be more substantial (third order) than the internal unconformities (fourth order) within the Mannville Group.

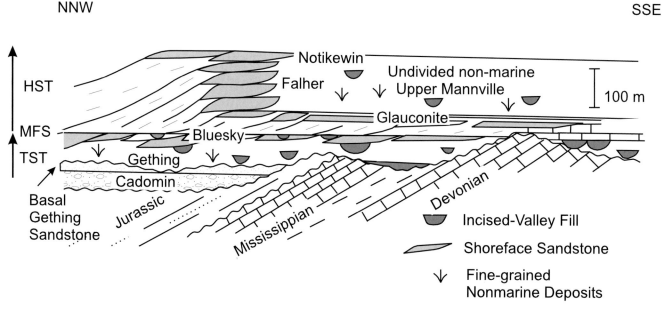

FIG. 2.—A longitudinal cross section showing the stratigraphy of the Mannville Group clastic wedge in the western part of the foreland basin. This paper deals with the basal part of the Lower Mannville Group, the Cadomin and Gething Formations as well as the Falher Member coastal sandstones and the undivided non-marine deposits within the Upper Mannville Group. The maximum flooding surface (Mfs) separates the transgressive systems tract (Tst) below from the highstand systems tract (Hst) above.

(Fig. 3) with fluvial (and related shoreface deposits) in the Upper Mannville Group from the western part of the basin (Fig. 4).

The non-marine to estuarine Lower Mannville Group and the overlying backstepping shorefaces comprise the basal trans- gressive systems tract (TST); the entire progradational Upper Mannville Group is the highstand systems tract (HST) of this unit (Cant and Stockmal, 1993). As documented by Chamberlin et al. (1989) the subsidence rates of the Alberta foreland basin increased markedly from Lower to the Upper Mannville times,

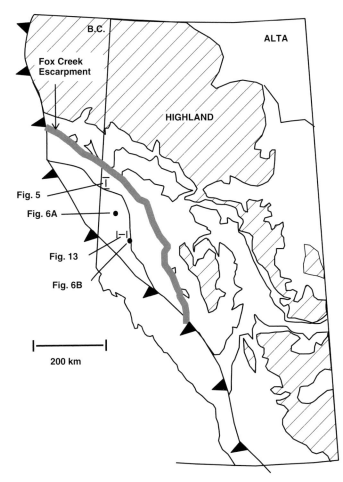

FIG. 3.—A simplified map of the Lower Mannville Group showing areas with no accumulation, valleys cut on the unconformity, and the Fox Creek Escarpment. The locations of other Lower Mannville figures are also shown. The thickness of the Lower Mannville deposit is largely controlled by the paleotopography on the basal unconformity and perhaps some flexural subsidence in the west.

from 10–20 m/my to 50–130 m/my. The range reflects that the northern part of the basin was subsiding faster than the south, consistent with paleoflow directions. Their estimates are approximate as no unconformities were recognized within the Mannville Group, and age dating is difficult, particularly in Lower Mannville non-marine deposits. Note that they have overestimated the Lower Mannville rate in central Alberta by apparently using a thrust-repeated thickness. As indicated on Figure 3, large parts of the basin did not subside substantially during Lower Mannville deposition. However, notwithstanding these problems, Chamberlin et al. (1989) have compiled sufficient data to conclude that the subsidence rate in the western part of the basin near the overthrust belt did increase by a factor of about five times from Lower to the Upper Mannville times. Using the time divisions of Chamberlin et al. (1989) for the area studied in western Alberta yields average rates for Lower Mannville accommodation formation of 5 m/my and the Upper Mannville of 38 m/my. Note these rates do not take into account internal unconformities or variations. As will be discussed below, it is interpreted here that the basal Lower Mannville Group

was deposited during periods of extremely low rates, before foreland subsidence was initiated. The higher part of the Lower Mannville was deposited during a period of higher overthrust-induced subsidence rates.

RECOGNITION OF STRATIGRAPHIC SURFACES IN NONMARINE FACIES

As was discussed above, most stratigraphically important surfaces are marine (Embry, 1995), so are unavailable for use in nonmarine facies. This section will review the criteria used to recognize two types of surfaces in the Mannville Group.

Subaerial Unconformities

The criteria used are suggestive but not definitive of an erosional/incision event. These are: (1) good correlations across a channellized sand body suggesting a continuous regional lithologic succession, subsequently eroded, (2) channel fills cutting down from the same stratigraphic level, (3) abrupt changes in inferred stream type and scale (e.g., from low-energy anastomosing rivers with channels less than 8 m deep to significantly higher-energy braided rivers with channel complexes over 20

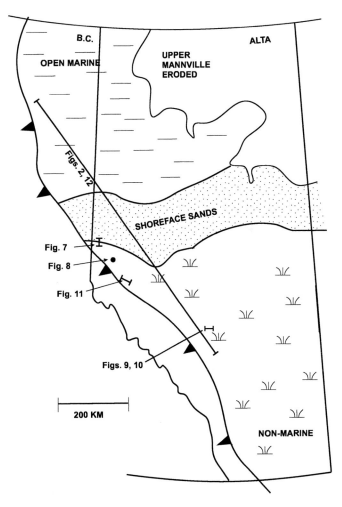

FIG. 4.—A map of generalized facies trends for the Upper Mannville Falher Member and equivalent, indicating longitudinal progradation parallel to the orogenic front. Locations of other Upper Mannville figures are indicated.

m thick), (4) marked changes in sandstone petrography (e.g., from quartzarenites to lithic sandstones), (5) abrupt vertical changes in lithologies (e.g., conglomerate or pure quartzose sandstone sharply overlain by mudstone-dominated facies) and (6) abrupt changes in *overall* directions of sediment supply in a basin (not on the basis of a few paleocurrent measurements or even the different paleoflow direction of a single channel). These criteria are suggestive of an important erosion surface but do not prove one exists; all could be duplicated by contemporaneous channels or abrupt but aggradational changes in fluvial regime because of factors such as climate or tectonic gradient change. Criteria 3 to 6 are applicable to unchannelized sheets of non-marine sediment rather than incised valleys.

On a sedimentological scale, few facies are unique to valley fills as compared to unconfined alluvial plains; however, the topography of valley walls may allow development of mass-flow deposits (Morison and Hein, 1987) supplied transversely to the stream, in that case as massive and normally graded, very poorly sorted gravels. Other types of features perhaps not unique to valleys but certainly characteristic of many modern ones are large-scale (tens of meters high with displacements of several meters) syn-sedimentary rotational slumps and normal faults affecting the wall material. Where active channels impinge against valley walls, scree deposits or coarser lag deposits derived from the material in the walls are common in modern rivers. However, all these facies are volumetrically minor and may be reworked by the stream.

Many examples of fluvial channels have been described in the literature which are reinterpretable as valley-fill deposits. Particularly in subsurface studies, many sinuous but essentially non-laterally migrating channels have been mapped (e.g., Berg, 1968) or imaged from seismic data. Unconfined active meander belts tend to be at least 2–3 times the width of the channel meander amplitude, thus depositing sand bodies from several hundred meters to a few kilometers in lateral extent. Significant lateral accretion of point bars or lateral movement of channels therefore creates an irregularly shaped sand body, with a high width/thickness ratio in many cases with inclined lateral-accretion stratification, lacking the very sinuous shape of a single channel form. Preservation of the shape of the channel at one instant implies that constraints to lateral migration existed, likely the walls of the valley. In a paper ahead of its time, Friend et al. (1979) compared fluvial sheet sands showing lateral migration to ribbon-shaped channel sand bodies. They concluded that incision resulting from vertical movement of the area was one possible factor preventing lateral migration. High-quality coals deposited in peat swamps very near to major active channels is a contradiction in terms of environmental conditions (i.e., overbank flooding and sediment transport vs. low-ash content) as discussed by McCabe (1984). If the sandstone occurs as a later incised-valley fill, the peat swamps are effectively separated (temporally and topographically) from the large channels. Although not specifically mentioned by McCabe (1984), base-level changes resulting in incised valleys should be a most effective mechanism for isolating peat-forming or other low-energy environments from active sediment transport.

In interfluve areas between incised valleys, distinguishing exposure surfaces due to base-level falls from those caused by climatic and seasonal river-stage fluctuations is, almost by definition, extremely difficult. Some parts of long-period base-level-fall exposure surfaces may be marked by very mature paleosols (including coals), but these can also occur on contemporaneous floodplains (Bernard et al., 1970). In general, criteria are lacking to distinguish between the surfaces of intervalley upland areas and floodplain surfaces exposed by normal river stage or climatic fluctuations.

Non-marine Equivalents to Transgressive Surfaces

Theoretically, a transgression can occur during steady sea-level conditions if enough sediment is removed by shoreline erosion which would induce river downcutting. However, it is unlikely this is an important process geologically. Most transgressions involve rising relative sea level which causes a backwater effect in fluvial systems and reduction of non-marine gradients. On floodplains, this would be reflected by deposition of finer sediment. Depending on climate, this zone could be host to caliches or coals. Because of the backwater effect, the gradients of the channels are also reduced and more fine sediment will be deposited at least in the topographically higher parts of the fluvial system, although sands will still occur in the deeper channels. An important flooding or transgressive surface therefore should correlate to a non-marine surface capping a zone of generally finer deposits. However, because of the presence of the active channels, this surface may be difficult to identify in all occurrences. In some occurrences, the invasion of marine water up channels may provide direct evidence for the transgression in the form of tidal sedimentary structures or body and/or trace fossils (Shanley et al., 1992).

In the Upper Mannville example to be discussed, zones of non-marine mudstones and coals could be correlated with variable difficulty. These zones were related to transgressive surfaces above the related shoreface sandstones.

LOWER MANNVILLE CONGLOMERATES AND SHEET SANDSTONES

The Lower Mannville Group rests on a long-period unconformity which truncates tilted Jurassic to Devonian rocks (Figs. 1, 2), related to reorganization of the orogen-foreland basin system (Cant and Stockmal, 1993). The unconformity was deeply incised (Cant and Abrahamson, 1995) with major valleys in some places (Fig. 3) and extensive soil and residual deposits in others. The Cadomin Conglomerate extends eastward from the overthrust belt and onlaps a northwest-oriented, westward-facing erosional slope (Fig. 3) termed the Fox Creek Escarpment (McLean, 1977). The Cadomin Conglomerate is dominantly a sand-supported, crudely stratified conglomerate interpreted as a gravelly braided-river deposit (Varley, 1984) which grades finer and sandier to the northeast. Because of its stratigraphic position on the sub-Cretaceous unconformity and its grain size, it was once interpreted to mark the initiation of overthrusting (McLean, 1977) but has been reinterpreted to reflect general uplift of the basin and the overthrust belt (Heller and Paola, 1989; Cant and Abrahamson, 1996). Preservation of the unit is due to the paleotopography on the unconformity which trapped the gravels between the Fox Creek Escarpment and the overthrust belt (as well as in some reactivated grabens). In the north-central part of the basin, facies and grain-size trends in the subsurface and paleocurrent measurements in outcrop (Varley, 1984) all indicate northeastward transport; this is also consistent with the pre-thrusting age of the Cadomin, before the northwestward longitudinal slope of the basin was established by differential loading along its margin (Cant and Stockmal, 1993).

The Cadomin Conglomerate is overlain by two distinctly different units; one consists of fluvial to estuarine sandstones up to about 30 m thick (Figs. 5, 6A); the other is the volumetrically more typical fine-grained Gething facies (Figs. 5, 6B). The contact between these two units is channelized (Fig. 5); in many places the sandstone is absent, interpretable as a result of complete erosion (Cant and Abrahamson,1996). The contact between the conglomerate and the fine-grained Gething facies is abrupt in cores and outcrops with no gradation (Fig. 6B). The base of the fine-grained facies is therefore interpreted as an unconformity. Limited palynological analysis has shown the Cadomin to be Barremian age, while the fine-grained Gething is Aptian age (E. Davies, pers. commun., 1994), suggestive but not conclusive of an unconformity. The nature of the contact between the conglomerate and the overlying sandstone is less certain. It is marked in some occurrences by a thin (2–3 m) shale bed (Fig. 6A) but in others, lacks this unit with the result

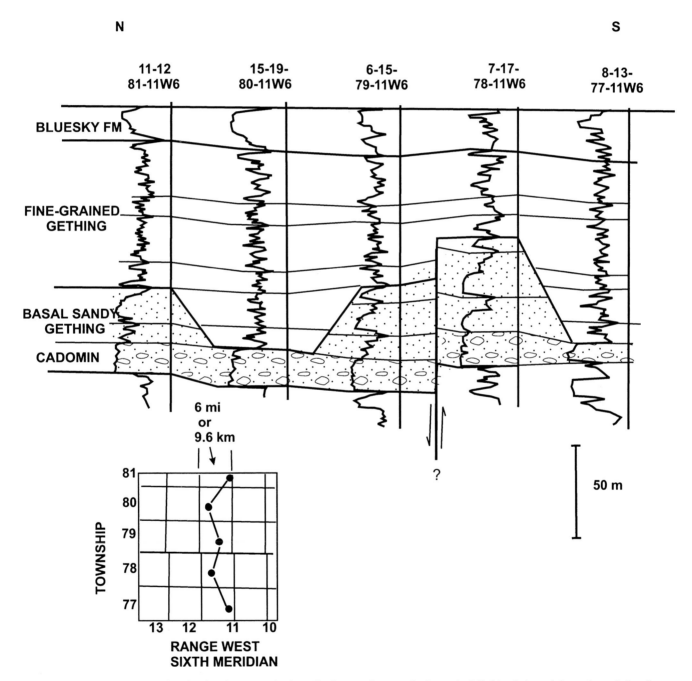

FIG. 5.—A gamma-ray cross section showing the apparently channelized contact between the fine-grained Gething facies and the sandstone below. In many places, the sandstone is removed entirely, and the mudstones rest directly on the Cadomin Conglomerate. This section is 5 km long, and is located on Figure 3. Inset shows detailed well locations. The fault is inferred from the offset correlations in the basal sandy Gething Formation.

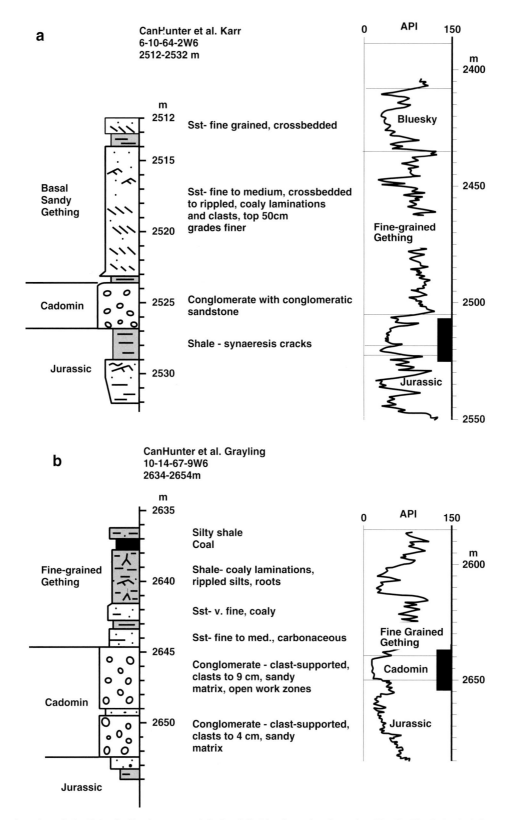

FIG. 6.—Two core logs through the Cadomin Conglomerate and the basal Gething formation (located on Fig. 3). The Cadomin is interpretable as a braided-river conglomerate transported northeastward from the overthrust belt (Varley, 1984). Core A shows the Cadomin overlain by about 15 m of sandstone. This basal Gething sandstone is interpreted to be dominantly fluvial, although some cores show *Planolites* burrows, possibly suggesting an estuarine influence. Core B shows the Cadomin overlain abruptly by mudstones, coals, and thin sandstones of the fine-grained Gething facies. This fine-grained facies is very similar to the Upper Mannville non-marine deposits. Cored intervals shown on gamma-ray logs by black bars.

that the sandstones and conglomerates are difficult to separate on logs. In some outcrops west of the study area (Stott, 1973) and in southern Alberta, the sandstone lies directly above the unconformity, interpretable as if the conglomerate had been completely eroded. Some gravel occurs in the sandstone, also suggesting the possibility of erosional recycling, but the significance of this cannot be estimated. Several cores show the sandstone to be fluvial to estuarine in origin (Cant and Abrahamson, 1996). Tentatively, this surface is interpreted to be a subaerial unconformity as well because of its abrupt nature and the sharp change in facies and lithology.

The rate of sedimentation in the Cadomin-basal Gething interval was very low (Chamberlin et al., 1989), although a real estimate is difficult because of problems in dating the conglomerate and determining the periods of the unconformities. The rates of 10–20 m/my for the Lower Mannville Group from Chamberlin et al. (1989) appear to be derived mainly from biostratigraphic dating of the fine-grained facies. Except in areas where the unit thickens over reactivated grabens of the Peace River Arch, a major Paleozoic structure, the conglomerate is generally less than 20 m thick. Palynological dating of some shaly interbeds has yielded Barremian ages (E. Davies, pers. commun., 1994). The Barremian stage was about 6 my long, yielding a minimum accumulation rate of about 3 m/my. However, precise dating of the top and base of the preserved unit has not been achieved, and actual rates during deposition may have been significantly greater. However, in this pre-overthrusting period, subsidence rates were likely even to be negative, with general uplift occurring (Heller and Paola, 1989).

Over much of the remainder of the foreland basin, the basal Lower Mannville consists of thin sheets, erosional remnants, and valley fills in areas of paleotopography on the basal unconformity, with variable proportions of sandstones (Cant and Abrahamson,1996). Individual unconformity-bounded units may be as thin as 15–20 m in the southern headwater areas. The basal sandstones were not supplied from the Cordillera; their mineralogies are dominated by quartz, over 90% in places, ultimately supplied from the Canadian Shield, with small proportions of chert grains supplied from then-outcropping Mississippian carbonates. These basal deposits are also believed to pre-date foreland subsidence and sediment supply. In southern and central Alberta, the valley fills consist of two or three unconformity-bounded sandstone- or mudstone-dominated units, as documented by truncation of units on cross sections (Cant and Abrahamson, 1996), and abrupt facies and petrographic changes. The abrupt changes were documented from almost pure quartz braided-river sandstones to a unit of with equal amounts of lithic sandstones and mudstones (Hayes, 1986), similar to the fine-grained Gething facies or Upper Mannville Group. At least some of these sandstones are considerably older than the finer-grained Lower Mannville sediments above. Some may be contemporaneous with the Cadomin Conglomerate.

UPPER MANNVILLE FINE-GRAINED ALLUVIAL FACIES

The unit exhibits longitudinal paleoflow to the north-northwest (Fig. 4) along the basin axis as established by many authors (Eisbacher et al.,1974; Jackson, 1984; Cant and Stockmal, 1989). Cross sections approximately normal to the shorelines and regional facies belts extend in this direction, with non-ma-

rine deposits to the south. The unit overall is strongly progradational, with some 200 km of shoreline migration along the basin axis before the shorefaces stacked vertically (Fig. 2) at the margin of the Peace River Arch, which was subsiding more rapidly than the remainder of the basin. This portion of the paper discusses the facies and sequence stratigraphy of the Falher Member shorefaces and equivalent non-marine deposits to the south (Figs. 2, 4).

Shoreface-Non-marine Relationships

The 20- to 30-m-thick Falher B shoreface sandstone and conglomerate body shows a very abrupt (on a scale related to the 2 to 5 km well spacing) landward change into mudstones, coals, and minor sandstones (within 3 km between the wells 6–19–69–10W6 and 6–7–69–10W6, Fig. 7). A few well logs and cores show that the southern margin of the shoreface sandstone is underlain by non-marine deposits (Fig. 7) and essentially onlaps them. The beds of mudstone and coal bounding the Falher B at the top of the Falher C cycle and the base of the Falher A cycle are continuous across the facies boundary into the non-marine area, thus constraining the lateral equivalence of the coarse shore-zone deposits to the mudstones and coals at the southern end of the section. The basal part of the shoreface rests sharply on a regressive surface of erosion, and is capped by a subaerial unconformity (Cant, 1995) resulting from a drop in relative sea level. Falling and lowstand sea-level shoreface and coaly deposits occur some 30 km north of the main shoreline, unattached directly to it. As sea level rose, transgressive conglomerates with linear, barrier-island morphologies were laid down on top of the unconformity. Offshore, the top surface of the transgressive sandstones and conglomerates can be traced to the flooding surface on the top of the equivalent coarsening-upward succession. The Falher B shoreface unit therefore consists of two successions, the basal regressive shoreface sandstone and the overlying transgressive barrier conglomerate (Cant, 1995). The much finer non-marine deposits immediately behind also consist of two successions (Fig. 7), each composed of mixed sandstone and mudstone, capped by a thin coal. It seems reasonable to correlate the two phases of development of the Falher B shore-zone complex to the two dominantly fine-grained, coal-capped non-marine units although only the upper coaly unit connects physically. The upper coal has been interpreted as part of the transgressive barrier-island system (Demarest and Kraft, 1987) but alternatively could be interpeted as resulting from another transgression which occurred after a minor sea-level fall (Cant, 1995). A similar relationship between coals and transgressive-regressive parasequences in equivalent marine strata has been documented by Kamola and Van Wagoner (1995). The Upper Mannville Falher B coal-capped non-marine units are absolutely continuous for about 30 km. More regional correlations of less-than-continuous zones will be discussed below.

Behind both the Falher A and B shoreface sandstones (within 20 km), marine water intermittently flooded the low-energy area. Core logs show bioturbated intervals, and brackish to marine dinoflagellates occur in some muddy beds. Both the A and B units also show estuarine sandstones extending some 30 km inland (Cant, 1995), but no estuary exists behind the B shoreface on the cross section illustrated (Fig. 7); one has been found

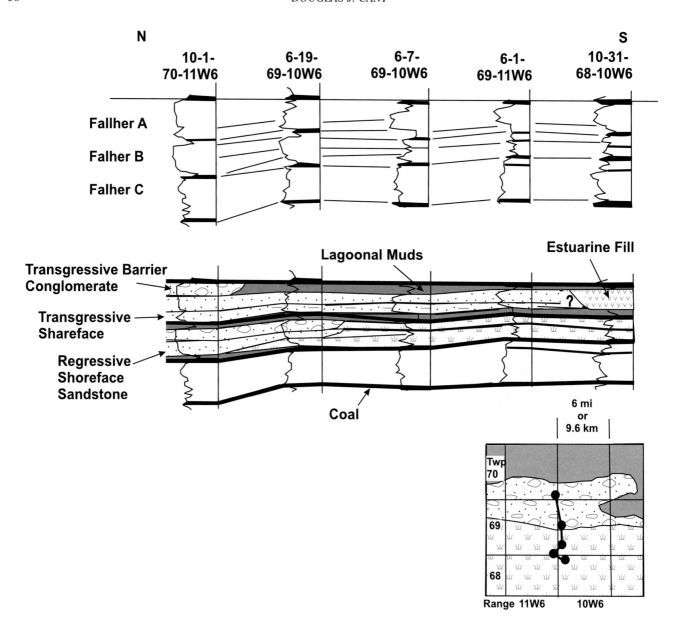

FIG. 7.—A duplicated gamma-ray log cross section showing the landward termination of Falher A and B shoreface sandstones. The upper section is correlated but uninterpreted, with the interpreted section below. Coals are documented from sonic and density logs, and conglomerates and sandstones are distinguished in cores. The Falher B shoreface sandstones and conglomerates onlap against fine-grained non-marine to marginal marine deposits. The regressive shoreface sandstones and transgressive barrier conglomerates each appear to correlate to a sandstone and mudstone unit capped by a thin coal bed. Immediately behind the shoreline, several mudstone beds show bioturbation and contain marine dinoflagellates, indicating that the area was intermittently flooded. The sandstones at the south end of the Falher A shoreface deposit are incised by estuarine sandstones (Cant, 1995); the contact between them cannot be identified on logs. The inset shows a detailed map with the overall context in Figure 4.

some 50 km east. Cores show the upper parts of the sandstones are pebbly and cross-bedded, with coaly and shaly interbeds and laminations. The lower parts have more hummocky cross-stratification and bioturbated zones. In-situ roots, pelecypod shell molds, and burrows such as *Planolites* and *Palaeophycus* occur in different interbedded shales and sandstones. The sand bodies are narrow channels some distance landward from the coastline but widen and coalesce toward it (Cant, 1995). The upper, coarser-grained and current-dominated portion is interpreted as almost entirely fluvial in origin, the final highstand fill of the estuary. The presence of these large estuarine com-

plexes is a direct result of the relative sea-level falls which formed the lowstand shorelines to the north, analogous to the process which formed modern estuaries. Lateral to the estuarine sands, a fine-grained coaly unit can also be correlated southward from the top of the Falher A shoreface sandstone and conglomerate body. The relative sea-level drops in both the Falher A and B units are inferred to have occurred between deposition of the regressive shoreface sandstone and the transgressive barrier conglomerates (Cant, 1995). Thus, in nonmarine sediments close to the shoreline, fine-grained and coaly zones correlating to the major transgressive surfaces can be identified. The subaerial unconformities can be identified on logs only at the bases of estuarine channels. They are not recognizable where only fine-grained lagoonal/floodplain deposits occur.

Non-marine Deposits

In general, farther south away from the shore-zone, Upper Mannville deposits are dominated by mudstones and thin coals, with about 25% fine-grained sandstones (Fig. 8). The sandstones are sharp-based and fine upward, with numerous silty and shaly interbeds, generally showing small crossbeds and current ripples, but also rooting and minor bioturbation. This assemblage of sediments could be interpreted as the deposits of low-energy meandering channels on the order of 2- to 8-m depth, with contemporaneous overbank floodplains accumulating muds and peat. This conventional interpretation can be made because of their fining-upward character, the upward reduction of scale of cross-bedding and ripples, and the low proportion of sandstone (Allen, 1965). Alternatively, because of the presence of apparently contemporaneous low-energy suspended-load channels set in a matrix of fine-grained overbank material, a model of anastomosed channels can be proposed (e.g., Nadon, 1994). This model is favored because no evidence is available for lateral accretion of the channels. However, few criteria exist to clearly separate low-energy meandering stream from similar anastomosed stream deposits in the subsurface because of limited well control and the inability to be certain of the contemporany of channels. In either case, because of the intrinsically laterally restricted nature of the channel sandstone bodies and the lack of major cycles or sequences, correlation through this facies is extremely difficult, and the Upper Mannville Group (ranging up to 200 m in thickness) here has been undivided stratigraphically.

These deposits are interrupted by narrow fluvial channel sandstones, commonly 15 to 20 m thick but ranging to 30 m in some cases (Fig. 9). These much larger channels have also been interpreted conventionally in the past as major trunk river channels in the area, surrounded by contemporaneous smaller tributary channels along with overbank fines and peat swamps. These larger channel deposits occur at different stratigraphic levels throughout the Upper Mannville Group, laterally equivalent to all the Falher shorelines. The channel sands are almost entirely cross-bedded (Fig. 10), and show little grain-size variation in their lower parts. Their upper parts grade finer, with a higher proportion of shale. In many cases, coal beds lie immediately (as far as can be determined from local well spacing of a few kilometers) beside them. As shown in the example in Figure 9, correlations can be established from the finer sediments on one side across to the other, suggesting that the mudstone and coal beds were once continuous. Although not absolute proof, the correlation is suggestive that the sandstones may be interpreted as filling incised valleys resulting from the relative sea-level falls discussed previously and more completely elsewhere (Cant, 1995). From the vertical succession of sedimentary structures (Fig. 10) and lack of pronounced grain-size trends, at least the lower portions of the valley fills are interpreted to be sandy braided-river deposits, although the pos-

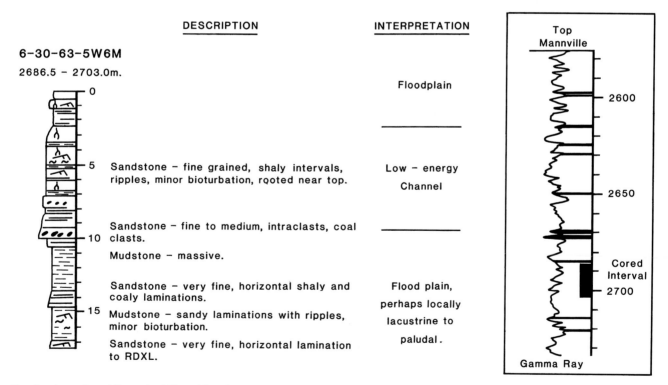

FIG. 8.—A core log of fine-grained Upper Mannville non-marine sediments at a distance back from the shoreline (see Fig. 4 for location). These consist dominantly of mudstones with minor rippled sandstones, interpreted as floodplain deposits, succeeded by low-energy channel sandstones about 8 m thick. These sandstones are generally uncorrelatable between wells, suggesting that they are narrow and did not meander laterally. They are interpreted as anastomosed channel deposits set in the volumetrically dominant floodplain mudstones and minor coals. The Upper Mannville Group is comprised of about 80% of this facies.

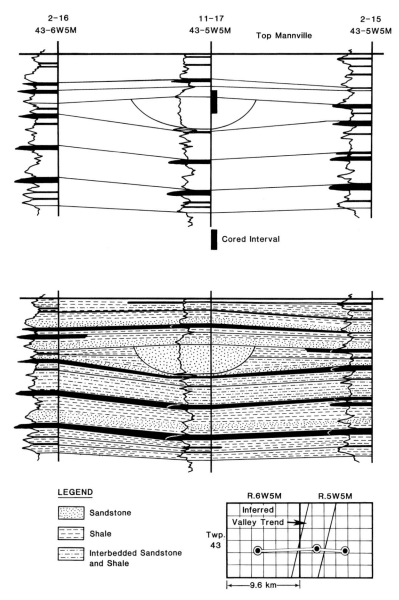

FIG. 9.—A correlated gamma-ray log cross section at the top and its lithologic interpretation below, with location map, showing the context of the core in Figure 10. Overall location is shown on Figure 4. The good correlations between the two outer wells (about 12 km apart) suggest the finer sediments were once continuous. The coal (in black) on the east (equivalent to the top of the sandstone) is the only unit which is not correlatable. The thick sandstone is therefore interpreted as a valley fill rather than a channel contemporaneous with the finer sediments on either side of it. The inferred valley trend is determined from surrounding wells and is consistent with the regional paleogeography (Fig. 4).

sibility of laterally amalgamated sinuous channel fills cannot be ruled out. The valleys were backfilled in the early stages of relative sea-level rise when large volumes of coarse sediment were available. The sandy estuarine complexes extending inland from the shorelines are believed to be the downstream ends of some of the valley fills; the cored valley fills shown here probably intersected the shoreline much farther east than the estuary fill shown in Figure 7. Valley fills cannot be traced far back from the shoreline in westen Alberta because of truncation by the Tertiary overthrust belt, and they are not cored farther east. The finer sediment in the higher parts of the valleys suggest that the type of stream changed, perhaps in response to diminishing gradients as relative sea-level rise quickened.

Regional Correlations

As noted by Cross (1988), coastal-plain coals are influenced, if not controlled, by relative sea-level changes, and they occur preferentially at the tops of shoreface sands where they are stacked vertically. Kamola and Van Wagoner (1995) believed these coals to be regressive, following the shoreface. The Upper Mannville coals on the Falher shoreface deposits are interpreted as transgressive, part of the barrier succession, as interpreted by Demarest and Kraft (1987) or as a result of transgression after a minor sea-level fall (Cant, 1995). A regressive interpretation seems less likely because several of them extend some 30–km inland beyond the shorefaces. In a regressive situation,

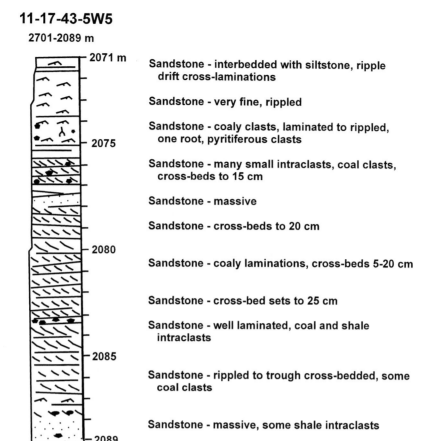

11-17-43-5W5

2701-2089 m

2071 m — Sandstone - interbedded with siltstone, ripple drift cross-laminations

Sandstone - very fine, rippled

2075 — Sandstone - coaly clasts, laminated to rippled, one root, pyritiferous clasts

Sandstone - many small intraclasts, coal clasts, cross-beds to 15 cm

Sandstone - massive

Sandstone - cross-beds to 20 cm

2080 — Sandstone - coaly laminations, cross-beds 5-20 cm

Sandstone - cross-bed sets to 25 cm

Sandstone - well laminated, coal and shale intraclasts

2085 —

Sandstone - rippled to trough cross-bedded, some coal clasts

Sandstone - massive, some shale intraclasts

2089 —

M F

FIG. 10.—A core log of the anomalously thick sandstone on the cross section in Figure 9 from central Alberta (see Fig. 4 for location). The basal 14 m are almost entirely cross-bedded, and can be interpreted as the deposits of a bedload-dominated braided river. The finer grained, lower energy, rippled and rooted deposits at the top of the core are believed to comprise a separate unit as shown in Figure 9. M and F are medium and fine-grained sandstone.

the coals on top of the shorefaces would either be eroded, or no deposition would occur because of water-table fall. The coals are therefore inferred to be related to rises in relative sea level.

Virtually all of the thin sequences in the Upper Mannville have coals developed on top of the shoreface sandstones. Some extend continuously into the non-marine area to the south, but others are cut away by channels in places (e.g., Falher A, Fig. 7) immediately behind the shoreline (Cant, 1995) or farther south as described previously. A number of the sequences have multiple coals at their tops, forming zones up to 8 m thick with 3 or 4 individual seams comprising about 50% of the zone. One of these on top of the Falher E sequence has been recognized as a consistent marker (informally termed the "Fourth Coal") and has been used in establishing the stratigraphy of the major gas field in the shoreface sandstones (Wyman, 1984).

Correlations through the Upper Mannville nonmarine deposits are very difficult, partly because fine-grained nonmarine facies are intrinsically laterally variable, with thin channel (meandering or anastomosing) sands in close juxtaposition to contemporaneous floodplains. In addition, the thick sandy valley fills cut through the succession at numerous stratigraphic levels, further compounding the difficulty of correlation. However, (1) the recognition that coals may be related to fluctuations in relative sea level, (2) the presence of some coals extending continuously for 30 km landward from the flooding surfaces at the tops of shoreface sandstones, (3) the widespread occurrence of zones of fine sediment with one or more coals and (4) the observation that coaly zones some 350 km to the south showed about the same regularity of spacing as those near the shoreline, all suggested that it might be possible to correlate through the non-marine facies. In addition, upward-fining and upward-coarsening successions, both sandstone and mudstone-dominated, expressed on the gamma-ray logs, could be correlated with some degree of confidence over limited areas (Fig. 11). It should be noted however, that other coals, generally thinner and appearing between the coals at the tops of the shoreline successions, do occur in the non-marine facies, sometimes rendering correlations difficult. The coaly surfaces correlated in the non-marine facies are believed to be the expression of the maximum transgressions, when rates of RSL rise were greatest. This hypothesis is consistent with the conclusions of Cross (1988) about the times of maximum accumulation of coastal-plain coals. In the non-marine facies, the surfaces cannot be called maximum flooding surfaces because they were not flooded. They are here termed flooding surface equivalents (FSE) and the units between them are termed base-level rise (BLR) units.

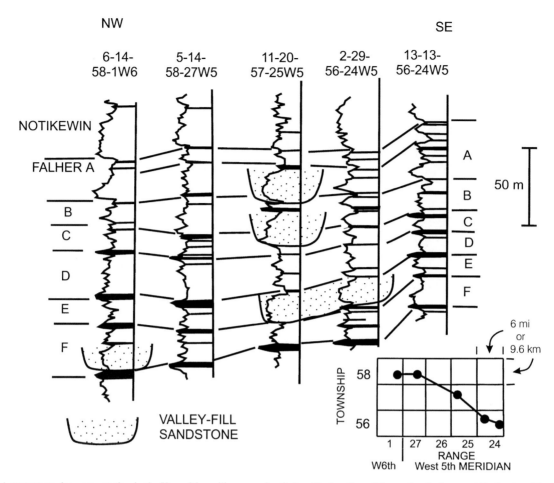

FIG. 11.—A gamma-ray log cross section in the Upper Mannville nonmarine facies. The location of the section is shown on the inset and located regionally on Figure 4. The correlated surfaces correspond to flooding surfaces of marine sequences to the north. The section illustrates the difficulties in correlation in this facies because of channelling and inherent lateral variability.

The correlations were made in the context of this laterally variable facies assemblage with numerous incised-valley fills at different stratigraphic levels. In marine parts of the Mannville Group, correlations were made through virtually every well on the cross section. In this case, because of the factors mentioned, a less rigorous procedure was designed. Where correlations broke down because of channeling or lateral variability, wells dominated by the finer facies were preferentially used to establish correlations. Because some proportion of wells in different areas of the sections could not be linked at each stratigraphic level, the correlations are not as sure as those in which every well can be linked. However, in spite of those caveats, the recognizability of the coaly zones allows some conclusions, however tentative, to be drawn (Fig. 12).

The correlation lines on the coal zones rise landward at a very low angle toward the top of the Mannville Group (Fig. 12). The entire non-marine facies of the Upper Mannville Group is comprised of these units. The overall configuration is one of offlap, where the surfaces originate at the top of the Mannville unit and slope downward and seaward through it (Fig. 12). Because of the difficulties in correlation, it is not certain that all units extend proximally to the top of the Mannville; the thinnest ones at the shoreline may essentially terminate landward by onlap. Traced northward to their distal ends, the equivalent marine sequences downlap onto the maximum flooding surface of the Mannville Group (Fig. 12).

Some surfaces are easier to correlate because the sediments associated with them are more distinctive and can be traced with a greater degree of confidence. For example, the base of the Falher D and top of the Falher E units was more recognizable because the Falher D showed slightly thicker clastics, and the intervening coal was slightly more continuous. Northward in the shore zone, this surface is equivalent to the top of the "Fourth Coal" zone of Wyman (1984). It rests on the top of the Falher E shoreface sandstone, but differs from the other Falher coaly intervals on the shorefaces in that it is thicker (up to 5 m) with three individual coals developed. Presumably, it represents a greater rise in relative sea level than the others. The surface equivalent to the fourth coal zone was traced from the shore in northwest Alberta some 350 km to the southeast.

A lower coal zone behind one of the progradational Glauconite Formation shorefaces (Fig. 2) was traced about the same 350 km. Its slope (corrected for differential subsidence to the north) was about 0.0002 (Fig. 12). The direction of the cross section was constrained by the preserved shape of the basin and well control; the section is probably not a true dip section, so

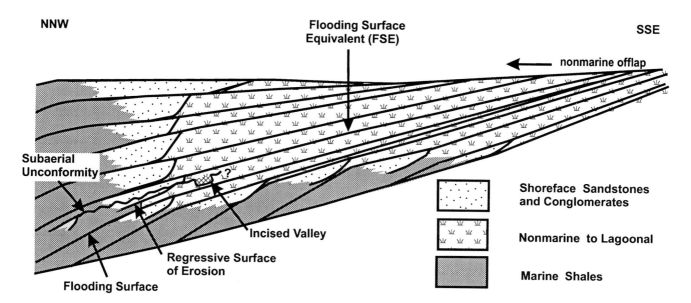

FIG. 12.—A diagrammatic cross section of the Upper Mannville Group showing the offlap of the flooding surface equivalents. They pass onto the tops of the shoreface sandstones and offshore marine coarsening-upward successions which clinoform and downlap against the top of the Lower Mannville Group. In one flooding-surface-bounded unit, the regressive surface of erosion, subaerial unconformity with lowstand sandstone, and incised valley are shown. The impossibility of recognizing this unconformity in non-marine facies is indicated by the question mark. The line of section is the same as that of Figure 2 (located on Fig. 4) but cut off to the north.

the real slope may be fractionally higher. The implication of these correlations is that a relative sea-level rise caused non marine deposition some 350 km from the shoreline. Using the estimated slope, the coal zone would have been some 75 m above sea level at the time.

SEQUENCE STRATIGRAPHY, FACIES, AND SUBSIDENCE

The contrast between the braided-river sandstones and conglomerates of the basal Lower Mannville and the anastomosing-river mudstones and coals of the Upper Mannville Group is striking. The basal Lower Mannville sediments show multiple, recognizable, relatively closely spaced unconformities. Where any palynological data is available, rates of sedimentation and subsidence were slow, with many hiatuses. These low rates are a fundamental control on the facies and stratigraphy of the non-marine sediments; because of high fluvial gradients, they allow transport of very coarse sediment far into a basin, in contrast to periods of more rapid subsidence with lower fluvial gradients when coarse deposits are trapped near their sources (Blair and Bilodeau, 1988). The style of fluvial sedimentation, (i.e., coarse braided-river sandstones and conglomerates (Figs. 5, 6), as relatively thin sheets in unconfined areas or as valley fills in incised areas) results because of high rates of supply and transport of sand and gravel compared to rates of subsidence.

The Upper Mannville sediments consist of dominantly muddy and coaly, anastomosed to meandering stream deposits (Figs. 8, 9, 11), cut by multiple incised valleys. The high rates of subsidence and supply of large volumes of fine-grained sediment result because of active overthrusting.

Other rock units in this foreland basin also display similar relationships between unconformities, non-marine facies and overall subsidence rates. As overthrusting began in the Aptian time (indicated by lithic sandstones) (Stott, 1973), the coarse

sandstones and conglomerates were succeeded by fine-grained facies (Figs. 5, 13) almost identical to those described from the Upper Mannville, again cut by sandstone-filled valleys (Fig.13). Thick Late Cretaceous (Edmonton Group) and earliest Tertiary (Paskapoo Formation) rocks (Fig. 1) are dominantly non-marine as sediment supply rates had overwhelmed subsidence rates (Cant and Stockmal, 1993) of 50 to 80 m/my (Chamberlin et al., 1989). Most of this clastic wedge is fine grained, similar to the Upper Mannville Group. Near the Cretaceous-Tertiary boundary, the basal Paskapoo Formation is a widespread sheet of medium-grained sandstone averaging about 50 m in thickness, with thick (800 m) fine-grained deposits below it. The Paleocene sediments above have been truncated by erosion, but up to 1000 m have been preserved in some places, and these are dominantly mudstones (Lerbekmo et al., 1992). By means of magnetostratigraphy, Lerbekmo et al. (1992) showed that within this apparently continuous section, a 2– to 3–my hiatus exists at the base of the sandstone (Fig. 1). The basal Paskapoo Formation is similar to the basal Lower Mannville examples in that (1) it was deposited immediately above an unconformity, therefore in a period of slow *average* subsidence or even general uplift, (2) it is relatively shale-free and (3) it is succeeded by finer, anastomosed to meandering stream deposits. Internal unconformities within the basal Paskapoo Sandstone have not been identified.

CONCLUSIONS

1. Basal Lower Mannville non-marine sediments are sandy to conglomeratic, fill erosional topography on the basal unconformity, are braided-river deposits and are bounded by important (second- and third-order) unconformities.

2. In the Upper Mannville Group, volumetrically dominant fine-grained non-marine sediments were deposited in low-

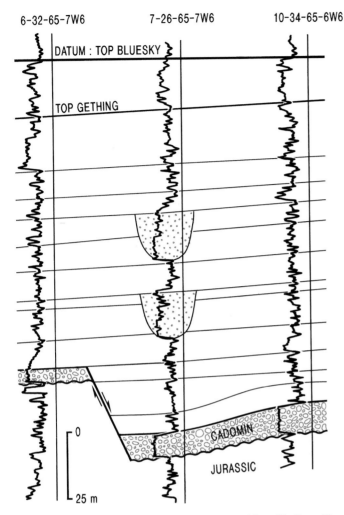

6-32-65-7W6   7-26-65-7W6   10-34-65-6W6

DATUM : TOP BLUESKY

TOP GETHING

0

25 m

CADOMIN

JURASSIC

FIG. 13.—A gamma-ray cross section in the Lower Mannville Group illustrating correlation in the fine-grained Gething facies interrupted by incised valley sandstones. These occur over the hanging wall of an inferred normal fault (because of offset of the Cadomin Conglomerate) with some syn-sedimentary growth at the base of the fine-grained Gething facies. Location on Figure 3.

energy floodplains and channels. Valleys were incised into these finer-grained sediments as relative sea level fell (creating small fourth-order unconformities) and the shoreline regressed. The fills of these valleys are sandy braided-river (with some estuarine) sandstones deposited in the early phase of the next rise in relative sea level. Most thick Upper Mannville fluvial sandstones are these braided-river valley fills, separated physically and temporally from the adjacent mudstones and coals.

3. The fine-grained non-marine facies in the Upper Mannville Group can be divided into correlatable allostratigraphic units, commonly delineated at their tops by surfaces capping shalier and coalier zones generated during relative sea-level fluctuations. The surfaces correlate to maximum flooding surfaces in marine sequences. The units defined by these surfaces offlap, sloping gradually seaward.

4. In the Alberta foreland basin, periods of low subsidence rates (without active thrusting) are associated with thin

sheets or infills of erosional paleotopographic lows of braided-river sandstones and conglomerates, commonly with internal unconformities. Periods of active thrusting and high subsidence rates are associated with thick clastic wedges of dominantly fine-grained anastomosed to meandering stream deposits with isolated sandy incised-valley fills.

5. In non-marine sediments deposited during periods of low accommodation formation, unconformities are the most recognizable stratigraphic surface. In non-marine sediments deposited during periods of high accommodation formation, surfaces equivalent to maximum flooding surfaces are most recognizable, but still difficult correlate.

ACKNOWLEDGMENTS

Earlier versions of this paper benefitted substantially from careful reviews by T. A. Cross, G. Plint, and K. W. Shanley. A. P. Hamblin pointed out the significance of the Tertiary magnetostratigraphic example. A. Embry shared his sequence-stratigraphic experience on similar deposits.

REFERENCES

ALLEN, J. R. L., 1965, A review of the origin and characteristics of recent alluvial sediments: Sedimentology, v. 3, p. 89–108.
BERNARD, H. A., MAJOR, C. F. JR., PARROT, B. S., AND LEBLANC, R. J., 1970, Recent Sediments of Southeast Texas. A Field Guide to the Brazos Alluvial and Deltaic Plains and the Galveston Barrier Island Complex: Austin, Bureau of Economic Geology, University of Texas, Guidebook No. 11.
BERG, R. R., 1968, Point-bar origin of Fall River sandstone reservoirs, northeastern Wyoming: American Association of Petroleum Geologists Bulletin, v. 52, p. 2116–2112.
BLAIR, T. C. AND BILODEAU, W. L., 1988, Development of tectonic cycles in rift, pull-apart, and foreland basins: response to episodic tectonism: Sedimentary Geology, v. 16, p. 517–520.
CANT, D. J., 1995, Sequence stratigraphic analysis of individual depositional successions: Effects of marine/non-marine sediment partitioning and longitudinal sediment transport, Mannville Group, Alberta foreland basin: American Association of Petroleum Geologists Bulletin, v. 79, p. 749–762.
CANT, D. J. AND ABRAHAMSON, B., 1995, Isopach of transgressive system–Mannville Group: Calgary, Geological Survey of Canada, Open File 3090.
CANT, D. J. AND ABRAHAMSON, B., 1996, Regional distribution and internal stratigraphy of the Lower Mannville: Bulletin of Canadian Petroleum Geology, v. 44, p. 508–529.
CANT, D. J. AND STOCKMAL, G. S., 1989, The Alberta foreland basin: relationship between stratigraphy and terrane-accretion events: Canadian Journal of the Earth Sciences, v. 26, p. 1964–1975.
CANT, D. J. AND STOCKMAL, G. S., 1993, Some controls on sedimentary sequences in foreland basins: examples from the Alberta Basin, in Frostick, L. E. and Steel, R. J., eds., Tectonic Controls and Signatures in Sedimentary Successions: International Association of Sedimentologists Special Publication 20, p. 49–66.
CHAMBERLIN, V. E., LAMBERT, R. ST. J., AND MCKERROW, W. S., 1989, Mesozoic sedimentation rates in the Western Canada Basin as indicators of the time and place of tectonic activity: Basin Research, v. 2, p. 189–202.
CROSS, T. A., 1988, Controls on coal distribution in transgressive – regressive cycles, Upper Cretaceous, Western Interior, U.S.A., in Wilgus, C. K., Hastings, B. S., Kendall, C. G. St. C., Posamentier, H. W., Ross, C. A., and Van Wagoner, J. C., eds., Sea-Level Changes: An Integrated Approach: Tulsa, Society of Economic Paleontologists and Mineralogists Special Publication 42, p. 371–380.
DEMAREST, J. M. II AND KRAFT, J. C., 1987, Stratigraphic record of Quaternary sea levels: implications for more ancient strata, in Nummedal, D., Pilkey, O. H., and Howard, J. D., eds., Sea-Level Fluctuation and Coastal Evolution: Tulsa, Society of Economic Paleontologists and Mineralogists Special Publication 41, p. 223–239.
EISBACHER, G. H., CARRIGY, M. A., AND CAMPBELL, R. B., 1974, Paleodrainage pattern and late-orogenic basins of the Canadian Cordillera, in Dickin-

son, W. R., ed., Tectonics and Sedimentation: Tulsa, Society of Economic Paleontologists and Mineralogists Special Publication 22, p. 143–166.

EMBRY, A. F., 1995, Sequence boundaries and sequence hierarchies: problems and proposals, *in* Steel, R. J., Felt, V. L., Johanneson, E. D. and Mathieu, C., eds., Sequence Stratigraphy on the Northwest European Margin: Bergen, Norwegian Petroleum Foundation Special Publication 5, p. 1–11.

FISK, H. N., 1944, Geological investigation of the alluvial valley of the lower Mississippi River: Vicksburg, United States Army Corps of Engineers Mississippi River Commission, 78 p.

FRIEND, P. F., SLATER, M. J., AND WILLIAMS, R. C., 1979, Vertical and lateral building of river sandstone bodies, Ebro Basin, Spain: Journal of the Geological Society of London, v. 136, p. 39–46.

HAYES, B. J., 1986, Stratigraphy of the basal Cretaceous Lower Mannville Formation, southern Alberta and north-central Montana: Bulletin of Canadian Petroleum Geology, v. 34, p. 30–48.

JACKSON, P. C., 1984, Paleogeography of the Lower Cretaceous Mannville Group of western Canada, *in* Masters, J. A., ed., Elmworth – Case Study of a Deep Basin Gas Field: Tulsa, American Association of Petroleum Geologists Memoir 38, p. 49–78.

HELLER, P. L. AND PAOLA, C., 1989, The paradox of Lower Cretaceous gravels and the initiation of thrusting in the Sevier orogenic belt, United States Western Interior: Geological Society of America Bulletin, v. 101, p. 864–875.

KAMOLA, D. L. AND VAN WAGONER, J. C., 1995, Stratigraphy and facies architecture of parasequences with examples from the Spring Canyon Member, Blackhawk Formation, Utah, *in* Van Wagoner, J. C. and Bertram, G. T., eds., Sequence Stratigraphy of Foreland Basin Deposits: Tulsa, American Association of Petroleum Geologists Memoir 64, p. 27–54.

LERBEKMO, J. F., DEMCHUK, T. D., EVANS, M. E., AND HOYE, G. S., 1992, Magnetostratigraphy and biostratigraphy of the continental Paleocene of the Red Deer Valley, Alberta, Canada: Bulletin of Canadian Petroleum Geology, v. 40, p. 24–35.

McCABE, P. J., 1984, Depositional environments of coal and coal-bearing strata, *in* Rahmani, R. A. and Flores, R. M., eds., Sedimentology of Coal and Coal-bearing Sequences: Oxford, International Association of Sedimentologists, Special Publication 7, p. 13–42.

McLEAN, J. R., 1977, The Cadomin Formation, stratigraphy, sedimentology and tectonic implications: Bulletin of Canadian Petroleum Geology, v. 25, p. 792–827.

MIALL, A. D., 1977, A review of the braided river depositional environment: Earth Science Reviews, v. 13, p. 1–62.

MORISON, S. R. AND HEIN, F. J., 1987, Sedimentology of the White Channel gravels, Klondike area, Yukon Territory: Fluvial deposits of a confined valley, *in* Etheridge, F. C., Flores, R. M., and Harvey, M. D., eds., Recent Developments in Fluvial Sedimentology: Tulsa, Society of Economic Paleontologists and Mineralogists Special Publication 39, p. 205–216.

NADON, G. C., 1994, The genesis and recognition of anastomosed fluvial deposits: data from the St. Mary River Formation, southwestern Alberta, Canada: Journal of Sedimentary Research, v. B64, p. 451–463.

POSAMENTIER, H. W. AND VAIL, P. R., 1988, Eustatic controls on clastic deposition II—sequence and systems tracts models, *in* Wilgus, C. K., Hastings, B. S., Kendall, C. G. St. C., Posamentier, H. W., Ross, C. A., and Van Wagoner, J. C., eds., Sea-Level Changes—An Integrated Approach: Tulsa, Society of Economic Paleontologists and Mineralogists Special Publication 42, p. 125–154.

SCHUMM, S. A. AND KHAN, H. R., 1972, Experimental study of channel patterns: Geological Society of America Bulletin, v. 83, p. 1755–1770.

SHANLEY, K. W. AND McCABE, P. J. 1994, Perspectives on the sequence stratigraphy of continental strata: American Association of Petroleum Geologists Bulletin, v. 78, p. 544–568.

SHANLEY, K. W., McCABE, P. J., AND HETTINGER, R. D., 1992, Significance of tidal influence in fluvial deposits for interpreting sequence stratigraphy: Sedimentology, v. 39, p. 905–930.

STOTT, D. F., 1973, Lower Cretaceous Bullhead Group between Bullmoose Mountain and Tetsa River, Rocky Mountain Foothills, northeastern British Columbia: Ottawa, Geological Survey of Canada Bulletin 219, 228p.

VARLEY, C. J., 1984, Sedimentology and hydrocarbon distribution of the Lower Cretaceous Cadomin Formation, northwest Alberta, *in* Koster, E. H. and Steel, R. J., eds., Sedimentology of Gravels and Conglomerates: Calgary, Canadian Society of Petroleum Geologists Memoir 10, p. 175–187.

WYMAN, R. E., 1984, Gas resources in Elmworth coal seams, *in* Masters, J. A., ed., Elmworth – Case Study of a Deep Basin Gas Field: Tulsa, American Association of Petroleum Geologists, Memoir 38, p. 173–187

A DYNAMIC SYSTEMS APPROACH TO THE REGIONAL CONTROLS ON DEPOSITION AND ARCHITECTURE OF ALLUVIAL SEQUENCES, ILLUSTRATED IN THE STATFJORD FORMATION (UNITED KINGDOM, NORTHERN NORTH SEA)

MARK DALRYMPLE, JEREMY PROSSER AND BRIAN WILLIAMS

Dept. of Petroleum Geology, University of Aberdeen, Aberdeen, Scotland

ABSTRACT: In recent years much emphasis has been placed upon unravelling the sequence evolution of coastal and marine deposystems whereas continental depositional environments have received little attention. Sequence stratigraphic models developed for marine and coastal environments have delineated the regional controls which influence the deposition and preservation of sediments, and the spatial distribution of the different facies within the basin. These regional controls are tectonics, climate and eustasy.

This single suite of allocyclic, or regional, variables also controls the deposition of facies within the upstream alluvial realm, but their relative importance differs greatly from downstream coastal and marine depositional systems. For example, the relative influence of eustasy will progressively decrease landward of the coastline. A 'lag time' exists between the actual change of a controlling variable (e.g., sea level) and the response and attempted equilibration of the alluvial deposystem to that change. In the case of eustasy, this lag time will show progressive increase landward of the coastline.

The variables which influence the sedimentology and geomorphology of alluvial depositional environments are nested within complex feedback loops, in which driving mechanism and effected change is re-iterated, so that the effected change in the deposystem becomes the driving mechanism for subsequent change. A paradigm for the regional evolution of different fluvial architectural styles has been developed utilizing the concept of the *stream equilibrium profile*. The stream equilibrium profile is the hypothetical instantaneous stream profile resulting from amalgamation of all variables controlling the system and is analogous to the 'graded stream' concept of Leopold and Bull (1979). Amalgamation is essential due to the complexity of interaction between numerous controlling variables in particularly dynamic systems. The stream equilibrium profile acts as a 'pseudo-base level' in the alluvial environment, base level *sensu stricto* in the continental environment being the regional water table in the hinterland and relative sea level at the coastline.

The Statfjord Formation in the Brent, Statfjord and Snorre fields of the northern North Sea are used to illustrate the principles of 'pseudo-base level' variation as a result of changes in variables controlling regional scale development of fluvial systems, and the ensuing architectural differences are discussed in the proposed model.

THE STREAM EQUILIBRIUM PROFILE AND ALLUVIAL BASE LEVEL

Semantic differences associated with the definition of base level still incite enthusiastic debate. At its simplest, "base level is effectively sea level" (Schumm, 1993), and as a consequence is effectively the surface to which, given a long enough period of time, subaerial erosion and submarine deposition will try to proceed. It is the surface above which the process of erosion is dominant and below which the process of deposition is dominant (both processes act at the same time, but at different rates). Marine base level thus controls the creation and destruction of accommodation space (*sensu* Jervey, 1988) which is needed for sediments to be deposited and subsequently preserved. Depositional surfaces within the subaerial alluvial basin, however, are rarely coincident with sea level, and the 'base-level' surface, as defined above, is best expressed as the regional water table (cf. Schumm, 1993) which may correlate with sea level at the coastline (Fig. 1). Thus base level variation in the subaerial alluvial realm is not the dominant control on accommodation space. For this we must look to the stream equilibrium profile.

Shanley and McCabe (1991) and Wright and Marriott (1993) use a 'base-level concept' to delineate systems tracts in the continental realm. Wright and Marriott (Fig. 1, 1993) identify Lowstand, Transgressive and Highstand Systems Tracts for the alluvial realm, but do not acknowledge that it is the 'pseudo base level' of the stream equilibrium profile and not 'ultimate base level' which controls regionally developed fluvial architectures. A consequence of this is that the conceptual model of Wright and Marriott (1993) develops systems tracts (LST, TST and HST, which imply a dominant relative sea-level control) at the 'wrong time' in the 'pseudo base-level' fluctuation cycle. The conceptual model presented here indicates that the pseudo base-level fluctuation cycle may be completely out of phase with the relative sea-level fluctuation curve in the upstream al-luvial environment. Shanley and McCabe (1991) also recognize LST, TST and HST deposits within the Straight Cliffs Formation of southern Utah and pick their Calico and Drip Tank Sequence Boundaries where successions of alluvial plain braided river deposits unconformably overly either coastal plain sediments, or sediments which can be correlated to a coeval coastline. However the identification of sequence boundaries within wholly alluvial successions which cannot be correlated to contemporaneous coastline deposits becomes somewhat more complex.

Schumm (Fig. 1, 1993) and Wheeler (1964) while recognising the importance of the stream equilibrium profile in terms of sequence development, inaccurately referred to it as base level. The stream equilibrium profile can act only as a 'pseudo base level', existing over a shorter time and smaller spatial scale than 'ultimate base level' (i.e., sea-level). Ultimate base level has been discussed under the title of 'geomorphic base level' by Shanley and McCabe (1994). This title can be misleading, however, as the dynamic nature of the alluvial system does not allow its geomorphology to equilibriate with this fundamental surface. The geomorphological principles controlling sedimentation and erosion are more evident in the dynamic stream equilibrium profile (here referred to as 'pseudo base level') which Shanley and McCabe (1994) discussed under the title of 'stratigraphic base level', based on the work of Barrell (1917), Sloss (1962) and Wheeler (1964). Shanley and McCabe (1994) alluded to various 'equilibrium surfaces' which control the creation and destruction of accommodation space and it is the nature of these hypothetical surfaces which this conceptual model attempts to illustrate.

The stream equilibrium profile controls the creation and destruction of accommodation space and thus deposition and erosion of sediments within the alluvial realm. Hypothetically, however, given a long enough period of time without change

Relative Role of Eustasy, Climate, and Tectonism in Continental Rocks, SEPM Special Publication No. 59
Copyright © 1998, SEPM (Society for Sedimentary Geology), ISBN 1-56576-042-5

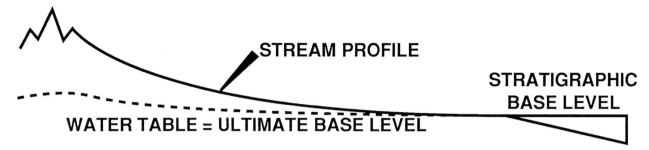

FIG. 1.—Relationship between the stream profile (or topography), continental base level (regional water table) and coastal/marine base level.

in the system, fluvial erosion could proceed to ultimate base level (*sensu* Schumm, 1993), negating the influence of any change in the stream equilibrium profile. In reality, the alluvial system (including the stream equilibrium profile) does not remain static, or without perturbation, for any considerable length of time (if at all), on any but the largest of scales. This may be comparable to the development of hierarchical sequences, (as discussed in Mitchum and Van Wagoner, 1991), over differing temporal and spatial scales, except that sequence development is not generated primarily by changes in relative sea-level.

Figure 2 illustrates (i) how the 'ultimate base level' surface is physically manifested as the regional water table and sea level, and (ii) its relationship with the stream equilibrium profile which is, essentially, a hypothetical surface having no physical expression in the real world. The stream equilibrium profile is controlled by the dynamics of the unstable alluvial depositional environment. The *stream equilibrium profile* is the hypothetical stream profile (in two dimensions) and the depositional surface (in three dimensions) upon which the drainage system of the basin is most efficient. This stream equilibrium profile is somewhat similar to the 'graded stream profile'. It is not the same as the *stream profile*, which is the elevation profile, or topography, of the stream in two dimensions at any particular moment in time.

All dynamic systems attempt to attain equilibrium, and as such the fluvial system is always 'seeking to become more efficient', aggrading or degrading to attain the hypothetical status of its equilibrium profile. Ouchi (1985) noted in flume experiments that subsidence causes aggradation and uplift causes degradation in the fluvial environment. Aggradation can be thought of simply as a relative rise in elevation of the stream equilibrium profile with respect to the stream profile or topography (*pseudo base level high or Stream Equilibrium Profile High*). Degradation may be thought of simply as representing a relative fall (*pseudo base level low or Stream Equilibrium Profile Low*) in elevation of the stream equilibrium profile. When the stream profile, or topography, lies above the stream equilibrium profile then the dominant sedimentary process acting at that point in the alluvial basin will be erosion (degradation), as the fluvial system attempts to lower its elevation to equilibrate. When the stream profile is below the stream equilibrium profile the dominant sedimentary process acting in the system is deposition (aggradation). The creation and destruction of accommodation space is thus controlled by the stream equilibrium profile which acts in the short term as a 'pseudo base level' surface (not ultimate base level). Both processes of deposition and erosion act contemporaneously within the alluvial basin, although their

rates may be low enough at times, or in places, to be negligible. One process will however dominate at any specific point within a basin. The hypothetical stream equilibrium profile thus represents the surface at which the rate of erosion is balanced by the rate of deposition, a concept suggested by Wheeler (1964), but inaccurately referred to as base level. The process which is visibly dominant in the natural system is decided by the relative positions of the stream profile itself and the hypothetical stream equilibrium profile. Figure 2 illustrates the situation where accommodation space has been created on the alluvial plain and deposition is dominant, allowing the development of an alluvial clastic wedge. Subsequent alteration of the stream equilibrium profile could result in creation of more accommodation space if the stream equilibrium profile is elevated, or in erosion of the previously deposited sediment, or underlying country rock, if the stream equilibrium profile is lowered.

An alluvial pseudo base-level curve for the generation of the unconformity bounded sequences of Mitchum et al. (1977) can be easily derived from the marine relative sea-level curve (Fig. 3). The resulting pseudo base-level fluctuation curve need not be related solely to the effects of variation in the marine relative sea-level curve. However, the pseudo base-level concept is best derived and illustrated in near coastal settings, where the effects of base-level variations are best understood. During a type I lowstand event (Fig. 3A), given enough time, an erosional unconformity, and incised valley system will develop through headward erosion. Since, for alluvial degradation (i.e., incision) to occur, the stream profile must lie above the stream equilibrium profile, an idealized pseudo base-level fluctuation curve (Fig. 3B) can be generated. Where accommodation space is unavailable, the rates of aggradation are lowest causing lateral switching of fluvial channels and reworking of the floodplain, resulting in sheet sand development as shown in flume experiments by Wood et al. (1990, 1993). Amalgamated channel sands sometimes overly regionally extensive sequence bounding unconformities (Shanley and McCabe, 1991) caused by a lowering of the stream equilibrium profile to a level at or below its geomorphological profile. Incision and incised valley development can occur if the elevation of the stream equilibrium profile (pseudo base level) is varied as a result of changes in the allocyclic controlling variables, and these effect changes in the rates of aggradation or degradation. Figure 4 illustrates the elevation of the stream equilibrium profile fluctuating with time. Following this curve, after a certain lag time, is the stream profile curve. The curve in Figure 4A is constructed for a point near the coastline, where the potential for elevation change is high because the lag time for the relative sea-level control to effect a response in the coastal plain fluvial system is small. Figure 4B illustrates a curve constructed for a point inland from

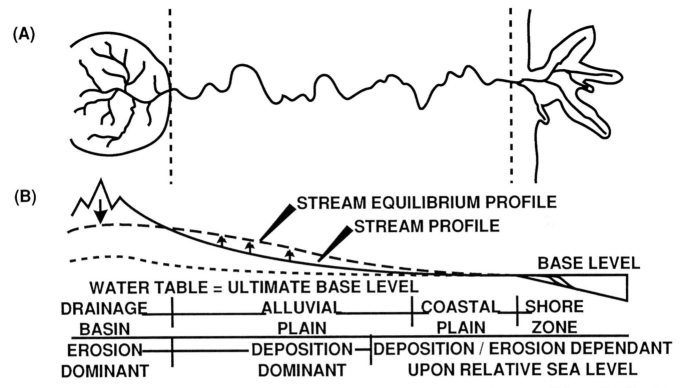

FIG. 2.—(A) Areal distribution of the components of the alluvial environment (from Schumm, 1977) and (B) the stream equilibrium profile and its role in defining these sub-environments.

a coastline, where the time lag for the effects of sea-level variation to reach this inland location will be greater. In this model, zones of amalgamated channel sandstones will develop where the rates of relative elevation change of the actual stream profile are lowest (ie: resulting in lateral switching of channels as opposed to vertical channel movement). Figure 5 illustrates how the difference in lag times at proximal (Fig. 4A) and distal (Fig. 4B) points within a basin results in decreasing influence of the controlling variable away from the source of variation (e.g., an upstream decrease in depth of incision during a relative sea-level lowstand). The knickpoint of an incised valley system represents the limit of influence of that allocyclic perturbation. Areas upstream from knickpoints that are generated by relative sea-level fall will be controlled primarily by the climatic and tectonic regimes prevailing at that time.

If the stream profile becomes more unstable as a result of changes in elevation of the stream equilibrium profile, it will attempt to equilibrate itself by either aggrading up to the stream equilibrium profile or by degrading down to it. However the drainage system, or stream profile, can never reach its most efficient or equilibrated geomorphology. Koss et al. (1990, 1994) and Shanley and McCabe (1994) noted a lag time between a causal change (or perturbation in the systems relative stability) and an effectual response, and this lag time is proposed here to be a major driving mechanism behind the systems dynamics.

Lag time is here suggested to be made up from two separate *lag components*. The first lag component is the *'causal' lag time*, **lt(c)**. This component represents the time difference between introduction of change to a system and the instant a *stream profile* begins to respond to that change. The change in the stream equilibrium profile, because it is a hypothetical surface, is instantaneous. The causal lag time appears to be controlled by allocyclic variables. In the case of relative sea level, the further inland from the coast we travel, the greater is the causal lag time until the effects of a relative sea-level fall are felt by the stream system. If changes occur in the drainage basin, brought about by tectonic or climatic perturbations, there will be a lag time before the effects of these changes are felt downstream. Increasing distance from the source of allocyclic change has the effect of decreasing the gradient of the curve describing changes in the elevation of the stream profile, so that the amount of elevation change possible per unit time is diminished. The second lag time component is the *'effectual' lag time*, **lt(e)**. This represents the time interval between the instant a stream system begins to respond to a change in its equilibrium status and adjustment of the stream profile to the new conditions of equilibrium generated by change. Effectual lag time is controlled by the autocyclic mechanisms which operate in the alluvial realm, such as the nature of the sediment load, local water tables, local lake levels, vegetation cover and fluvial morphology.

CONTROLS ON STREAM EQUILIBRIUM PROFILE FLUCTUATION

Allocyclic controls and their causal lag times control the fluvial systems regional development. The allocyclic controls upon the elevation of the stream equilibrium profile are tectonics, climate (including climatically induced changes in vegetation cover) and eustasy. Interaction between these controlling

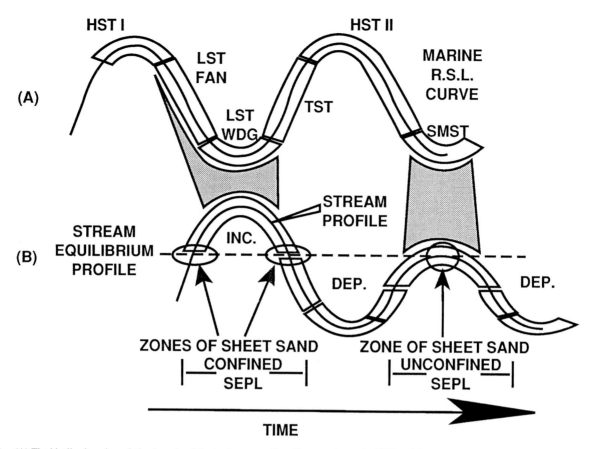

FIG. 3.—(A) The idealized marine relative base-level fluctuation curve (from Posamentier et al., 1988) and the systems tracts which develop with time through the cycle at that point in the basin. (B) During continental lowstand events which fall below the shelf slope break, incision develops at which point in the basin the stream profile is above the stream equilibrium profile. During continental highstand, the stream profile is below the stream equilibrium profile and so aggradation occurs.

factors alter the stream equilibrium profile, or pseudo base level, instantaneously. With time, the system may approximately equilibrate itself regionally to large scale allocyclic changes, at which time the autocyclic variables and the effectual lag time become the dominant controls on alluvial morphology. This view contradicts Knox (1976) who suggested that preserved alluvial deposits are representative of periods of allocyclic stability, but most aggradation and degradation of sediments occurs when the system has been forced out of relative equilibrium (i.e., during a period of allocyclic instability). Most workers recognize the importance of the main allocyclic controls, which are tectonics, climate and eustasy. Wood et al. (1990, 1993) also indicated that the rate of base level change (here stream equilibrium profile change) is important. However, when the lag time is examined, the rate of stream equilibrium profile change becomes important only when it is slow enough to allow the system to effectively respond immediately, ie: relatively equivalent to the lag time.

Figure 6 illustrates how alteration of these allocyclic controls affects the relative positions of the stream equilibrium profile and the stream profile. Geometrically there are three regional control points affecting the sedimentological dynamics of the basin. Point A is the highest point in the drainage basin and is controlled by climate and tectonics. Point B is defined as the landward limit of subaerial accommodation space and is con-

trolled by tectonics and sediment supply (which is a function of climate). Point B is also the point of intersection of the stream equilibrium profile (pseudo base level) and the topography. Point C is defined as the basinward limit of subaerial accommodation space and is controlled by the interaction between eustasy and tectonics (relative sea level) and sediment supply, determined dominantly by climate. It is important to remember that it is the system in disequilibrium which drives aggradational or incisional events and that perturbation, or the introduction of change into the system, causes the development of the differing fluvial architectures which are seen in the rock record.

Climate

Regional climatic variations arise as a function of plate latitudes. Climate controls the humidity or aridity of the environment and as such has a 'great impact' on the fluvial depositional environment (Knox, 1976). It has a strong influence on the rates of erosion and sediment transport within the drainage and depositional basins. The rate of erosion of control point A will increase with increasing humidity, and decrease with increasing aridity (Fig. 7). If climatic change results in increased humidity the amount of sediment being carried through the system will increase with increased fluid flux and control point B will move

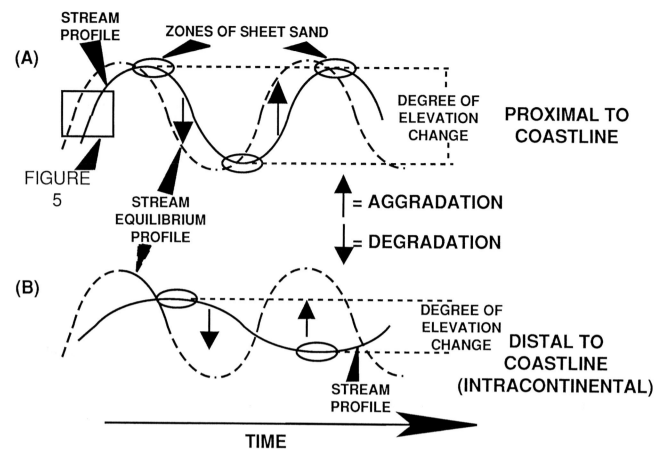

FIG. 4.—As the elevation of the stream equilibrium profile fluctuates, the elevation of the stream profile follows the same trends after a lag time. (A) Large elevation changes (incisional events) are possible in coastal area. (B) As the lag time increases inland, the possibility of high degrees of incision decreases. These curves can be superimposed on tectonic subsidence and/or water table elevation fluctuation curves to more accurately define periods of creation or destruction of continental accommodation space.

towards the sediment source. This increased sediment load implies an increased capacity for deposition during low stages of flow and so must be stored during long term discharge fluctuations from the drainage basin. Thus the threshold at which the rate of sedimentation equals erosion is reached earlier in the drainage basin (area A–B in Fig. 6). During changes from a humid to a more arid environment, control point B will move towards the coastline.

Sediment flux through the drainage and depositional basins to the coast also controls the position of point C, but in this case there will also be a strong influence exerted by relative sea level. Depending upon the rates of creation of accommodation space (controlled by relative sea level) and the rates of accommodation space infill (controlled by sediment supply) depositional systems at point C can aggrade, prograde, retrograde and incise, as discussed by Posamentier and Vail (1988), Posamentier et al. (1988) and Miall (1991). Sediment and fluid flux when combined with local and regional tectonic movement will control the convexity of the stream equilibrium profile between points B and C. Stable climate (ie: little tectonic activity and static global plate position) is dominated by autocyclic, small scale geomorphological processes in a situation analogous to the 'quasi-equilibrium' of Knox (1976). However regionally it is climatic change which alters the development of the alluvial

environment by shifting the system out of 'quasi-equilibrium' and into dis-equilibrium.

Tectonics

Tectonism plays an important role in the position of all main control points illustrated in Figure 6. Regional aseismic warping may uplift or subside the whole, or large areas of, the land mass. If uplifted the continental depositional basin will extend in a marine or basinward direction, with point C following the coastline (Fig. 8). However, the continental depositional basin (B–C) will decrease in size if regional subsidence occurs and a low sediment supply cannot keep pace with the transgression of relative sea level into the previously continental realm. Control point A can be thrust up and towards the marine basin in a foreland setting, or may be lowered by extensional faulting caused by hinterland loading or rifting. Local areas of increased accommodation space may be created by extensional faulting on the continental landmass.

Eustasy

In the continental realm eustatic variation exerts a control only on the areas of the alluvial plain and the coastal plain which are exposed to a perturbation for a duration longer than

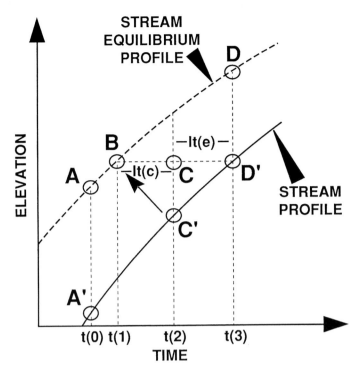

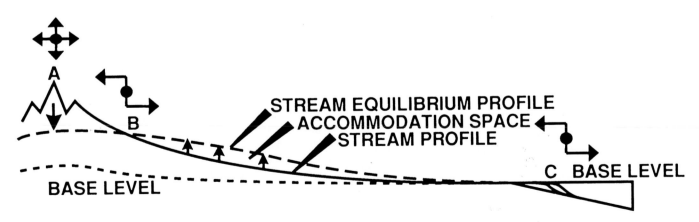

lt(c)=CAUSAL LAG TIME
lt(e)=EFFECTUAL LAG TIME

FIG. 5.—At t(0), the stream profile is at elevation A', and the stream equilibrium profile is at elevation A. Changes in controlling variables move the equilibrium profile to elevation B. However the stream profile does not begin to respond to any change in equilibrium until after the 'causal lag time' (lt(c)) at which time the stream profile, now at elevation C' due to previous alterations in controlling variables, attempts to reach the equilibrium conditions of point B. The amount of time taken to equilibriate to the conditions of point B is here referred to as the 'effectual lag time' (lt(e)). Meanwhile the equilibrium profile has moved to a new status, position D, which the stream profile must now try to equilibrate with. Increasing the causal or effectual lag times simply decreases the rate of change of elevation of the stream profile. Generally, 'causal lag time' is controlled by allocyclic mechanisms and 'effectual lag time' is controlled by autocyclic controls.

the causal lag time for that particular point in the basin. This allows a 'pulse' of equilibration to move upstream until is it absorbed by the systems allocyclic and then autocyclic mechanisms (decreasing in scale upstream). These are most commonly the areas proximal to the coast or low lying areas more inland. The eustatic control is incorporated into relative sea level, but with tectonic quiescence, sea-level rise will move control point C up and landward (Fig. 9), decreasing subaerial accommodation space if sediment supply is relatively low and thus cannot keep pace with the transgression (Van Wagoner et al., 1988). Likewise, control point C will move out and downslope during sea-level fall. Control point C is the point to which the stream system grades itself, as discussed by Miall (1991).

SEQUENCE ARCHITECTURE AND STREAM EQUILIBRIUM PROFILE FLUCTUATION

Wescott (1993) emphasizes the importance of the study of modern geomorphology and applying it to the rock record. The bounding surfaces which define a 'sequence' *sensu* Mitchum et al. (1977) are illustrated in Figure 10 (cf. Fig. 11, Shanley and McCabe, 1994, by Uliana and Legarreta). In Figure 10 alluvial sequences are bounded both above and below by regionally extensive unconformity surfaces and their correlative conformities. Deposition of the alluvial suite commences with amalgamation of channel sand bodies due to a lack of accommodation space causing local lateral migration or avulsion of the channel units, floodplain reworking, abandonment and preservation of channel sands before lateral avulsion or migration of the channel system re-occurs.

Sequence boundaries in the non-marine realm are caused by a lowering of the alluvial systems Stream Equilibrium Profile (SEP) with respect to the actual stream profile. The extent to which the SEP is lowered relative to the actual stream profile, and the lag time until a response occurs governs whether a confined or unconfined sequence will develop. As in coastal and marine sequence stratigraphy, two types of sequence boundary can be differentiated (Fig. 10). The 'confined amalgamated channel unit' is the geomorphological equivalent of the incised valley system and Type I sequence boundary of Van Wagoner et al. (1988). The 'unconfined amalgamated channel unit' is the geomorphological equivalent of the Type II se-

FIG. 6.—The three main geometric control points involved in the creation of subaerial accommodation space are A and B, which are primarily controlled by climate and tectonic activity, and C which has the additional influence of eustasy. Control point A is the highest point in the drainage basin. Control point B is the landward limit of subaerial accommodation space, where the topography is intersected by the stream equilibrium profile (i.e., where net deposition is equal to net erosion). Control point C is the basinward limit of subaerial accommodation space.

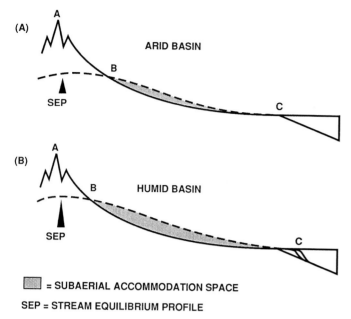

= SUBAERIAL ACCOMMODATION SPACE

SEP = STREAM EQUILIBRIUM PROFILE

FIG. 7.—As the climate changes from a relatively arid environment to a more humid one, erosion in the drainage basin increases, so providing a greater sediment flux through the fluvial system. This greater volume of sediment must be stored during its flux through the subaerial basin. As a result of this, control point B must move upward, toward the hinterland, thus raising the stream equilibrium profile and creating accommodation space. Control point C, with increasing sediment throughput to the coastline, will prograde if relative sea level allows.

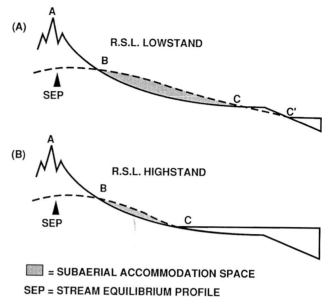

= SUBAERIAL ACCOMMODATION SPACE

SEP = STREAM EQUILIBRIUM PROFILE

FIG. 9.—As relative sea level floods the coastal plain during rise, control point C will move towards the hinterland (if the rate of sediment supply is too low to keep pace with the relative sea-level rise), so destroying subaerial accommodation space. Sediment supply to the coast is controlled by the prevailing climatic and tectonic regime in that area.

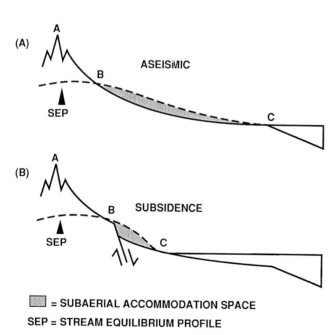

= SUBAERIAL ACCOMMODATION SPACE

SEP = STREAM EQUILIBRIUM PROFILE

FIG. 8.—With subsidence, accommodation space is created in the hinterland areas and, if sediment supply is high enough, infilled. Note that relative sea level rises at the coast, thus destroying subaerial accommodation space. If sediment is trapped upstream, the amount of sediment reaching the coastline will decrease, thus affecting the position of control point C, depending upon the status of relative sea level at that time. If uplift occurs, then subaerial accommodation space is destroyed and erosion and sediment bypass will occur.

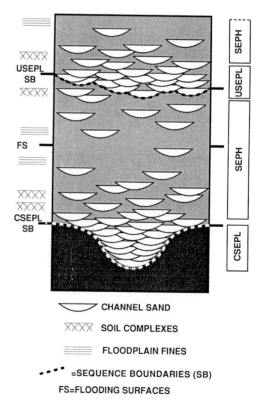

⌣ CHANNEL SAND

XXXX SOIL COMPLEXES

≡ FLOODPLAIN FINES

- - - - =SEQUENCE BOUNDARIES (SB)

FS=FLOODING SURFACES

FIG. 10.—Illustration of gross sequence architecture and continental systems tracts which develop with stream equilibrium profile elevation fluctuation, resulting from alteration of allocyclic controlling mechanisms.

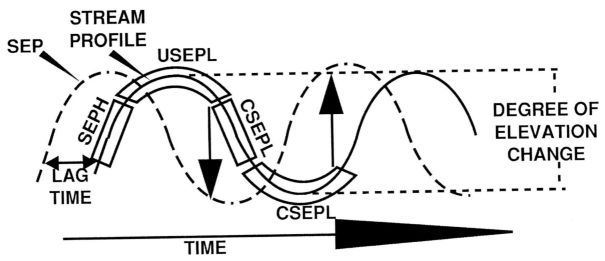

Fig. 11.—Continental systems tracts and their time of evolution with respect to fluctuation of the stream profile and the stream equilibrium profile.

quence boundary. The terms 'Type I' and 'type II' sequence boundaries are not used here in this model of continental sequence stratigraphy as they imply a dominant relative sea-level control on the depositional system.

Similar types of stratal bounding surfaces occur within sequences in 'hinterland' regions (Mitchum, 1977). Confined continental sequences are bounded below by an erosional valley surface (caused by an 'SEP Low') and above by a confined or unconfined sequence boundary. Unconfined continental sequences are bounded below by an erosional surface (again caused by an 'SEP Low') which follows the bases of the lowermost erosive channel sand units which constitute the amalgamated channel sand sheet resulting from a stream equilibrium profile relative lowstand. An unconfined sequence is bounded above by either a confined or unconfined continental sequence boundary. Unless a confining incised valley can be demonstrated, amalgamated channel sands should be described as being unconfined. SEP Lows can occur without the development of incised valleys.

Wright and Marriott (1993) and Shanley and McCabe (1991, 1994) have subdivided continental sequences into lowstand (LST), transgressive (TST), and highstand (HST) systems tracts to temporally equate alluvial plain sediments with sequences in coastal and marine environments. However use of terminology such as transgressive systems tract in environments where influence of relative transgression on the coastal plain has a low probability of being physically expressed further upstream (i.e., over most of the alluvial plain), can create confusion in correlation of stratal surfaces (e.g., the coastal type I; Van Wagoner et al., 1988) and regional erosional surfaces that develop entirely within the subaerial realm. Due to the differing importances of allocyclic controls in different reaches of the fluvial system, a sequence boundary (in terms of a chronostratigraphic surface) in 'upstream' regions need not be contemporaneous with a sequence boundary developed as a result of a relative sea-level fall at the coast. However, the two different sequence boundaries may tie together at some point in the subaerial depositional basin, even though they are not chronostratigraphically equivalent.

Figure 11 illustrates the continental phases of deposition and erosion which generate the subaerial equivalent of systems tracts, developing as a result of fluctuation of the stream equilibrium profile. The amalgamated channel sands within an incised valley constitute the initial deposits of the Confined Stream Equilibrium Profile Low (CSEPL) of Figure 10. The fine grained sediments of this unit have poor preservation potential due to lateral switching of, reworking by, and preservation of channel sands. A CSEPL essentially generates a continental sequence boundary, or incised valley surface. On the alluvial plain it is unlikely that this will be caused by a drop in relative sea level. Soil horizons may develop (Bown and Kraus, 1987; Wright and Marriott, 1993) on well drained river terraces produced by a lowstand event, and these could represent some of the correlative conformities corresponding to the hiatal sequence boundary. The sediments of the CSEPL are comprised mainly of amalgamated channel sand bodies deposited within the confining incised valley system. These sandy units are frequently exploited for mineral and hydrocarbon resources. As aggradation continues, so the incised valley fills.

As the Stream Equilibrium Profile fluctuation cycle moves on into the Stream Equilibrium Profile High (SEPH) the rate of aggradation increases, causing channel sand bodies to become more isolated within better preserved fine grained sediments of the floodplain (Figs. 10, 11). Very early and very late SEPH floodplain deposits, where rates of sedimentation (aggradation) are quite low, may be dominated by soil development. However, when aggradation rates increase, development of thick, mature soil horizons will be inhibited, and floodplain fines will be preserved without significant pedification. Posamentier and Vail (1988), Posamentier et al. (1988), Shanley and McCabe (1994) and the present conceptual model all suggest that alluvial flooding surfaces reflect rapidly rising base level, a consequence of which would be rapid aggradation. Alluvial flooding surfaces are regionally extensive subaqueously deposited complexes of fine grained floodplain deposits, which exhibit a higher degree of preservation of original sedimentary structure (i.e., have a low degree of pedification). This contradicts the model of Wright and Marriott (1993, Fig. 1) which suggests

that amalgamated channel sheet sands are generated by a Stream Equilibrium Profile High.

Unconfined Stream Equilibrium Profile Lows will develop upon lowering of the rate of aggradation after the relative SEPH, causing a higher degree of channel amalgamation and reworking of any floodplain fines or soil complexes that have developed. Given enough time a regionally extensive amalgamated channel sheet sand will develop due to low rates of elevation change. If the stream profile remains above the stream equilibrium profile then the system will continue to degrade, incising into its previous deposits, possibly generating the deposits of the Confined Stream Equilibrium Profile Low as previously discussed.

This conceptual model suggests that a distinctive lithostratigraphic architecture develops which closely corresponds to its chronostratigraphic equivalent. Chronostratigraphic surface dating can be difficult in the continental environment due to the low preservability of biological samples in higher energy environments, and their degree of oxidation in the subaerial realm. However, a variety of other correlation methods are available (heavy mineral, geochemical and wireline correlation techniques) for delineation of sequence defining surfaces in continental successions.

STATFJORD FORMATION, NORTHERN NORTH SEA

Brent, Statfjord and Snorre Fields

The alluvial plain sediments of the Statfjord Formation were deposited at the end Triassic-Jurassic transition. The medium to coarse grained fluvial channel sandstones form prolific reservoir horizons in several fields in the Viking Graben area, among them the Brent Field (block 211/29), the Statfjord Field (blocks 33/9 and 33/12) and the Snorre Field (blocks 34/4 and 34/7) (Fig. 12). Hydrocarbons were first discovered in the Statfjord Formation of the Brent Field in 1974 (Struijk and Green, 1991). A depositional period of *c.* 12 my between late Rheatian and mid-Sinemurian has been assigned to these deposits by Steel (1993), who refers to the alluvial clastic wedge of the Statfjord Formation as the Statfjord megasequence. Figure 13 shows both the regional stratigraphy in the Viking Graben and the internal stratigraphy of the Statfjord Formation. The Statfjord megasequence (Steel, 1993) incorporates the lowermost sediments of the Dunlin Group and the uppermost Cormorant Formation (equivalent to the Lunde Formation in the Norwegian sector, northern North Sea). The Statfjord Formation was initially defined as a lithostratigraphic unit by Bowen (1975) and was subsequently subdivided into three formal members termed the Raude, Eiriksson and Nansen by Deegan and Scull (1977), although the Nansen has recently been upgraded to formation status in its own right by the British Geological Survey (1993). Tectonism, climate change and eustasy all played major roles in the development of the Statfjord Formation. The continental landmass during Triassic times was as shown in Figure 14.

Tectonism

Two phases of Kimmerian rifting occurred in the northern North Sea. The initial rifting event occurred between the Permian and early Triassic (Roe and Steel, 1985; Badley et al.,

1988; Yielding et al., 1992; Steel, 1993). By middle Triassic times a post-rift thermal subsidence basin had evolved (Yielding et al., 1992). The second rift event peaked in the late Jurassic creating many of the structural traps seen in the Brent province. Triassic to middle Jurassic thermal subsidence of the Viking Graben was accommodated on steep north-south basin bounding faults (Badley et al., 1988). Figure 15 illustrates the initial syn-rift and post rift sedimentation trends across the basin.

The basin's dominantly axial drainage system is interpreted as having flowed essentially from south to north during deposition of the Statfjord Formation (Roe and Steel, 1985; Struijk and Green, 1991; Steel, 1993). However, Dalland (1994) suggested that, in places, the lower Statfjord Formation drained from north to south on the basis of Samarium-Neodymium isotope data, invoking early Kimmerian tectonism as the primary cause of this provenance change. However, whilst Sm-Nd model ages can be used to detect a provenance change, they cannot always assign a specific geographical provenance to any dataset. As a result, caution must be used when assigning specific provenances, or provenance changes within a drainage system. Steel (1993) suggests that the sand rich core of the 'Statfjord megasequence' results from uplift of basin margin areas.

Regional subsidence during the thermal sag phase of rift evolution would have decreased slowly (Badley et al., 1988). With decreasing regional tectonic activity, major climatic change becomes a more important allocyclic mechanism for the creation and destruction of subaerial accommodation space in upstream, coastline distal locations. Decreases in the subsidence rates of the thermal sag basin will have also contributed to the overall coarsening upwards profile of the Statfjord Formation, as the rate of creation of subaerial accommodation space, due to a lack of tectonic extension, decreases with time (cf. Steel, 1993; Steel and Ryseth, 1990).

Climate

A major climate change is associated with the Triassic-Jurassic transition which is evident in the sediments of the Statfjord Formation in the Viking Graben area. As the Laurasian-Gondwanan plate system drifted slowly northwards, it crossed the climatic zones illustrated in Figure 16. By the late Triassic the Viking Graben area had reached approximately 40°N (Hay et al., 1982) and was continuing to drift northward into the temperate subhumid zone (Perlmutter and Matthews, 1989). The mainly red bed sediments of the Lunde Formation underlying the Statfjord Formation are interpreted to have been deposited in an arid to semi-arid alluvial plain setting (Nystuen et al., 1990, Steel and Ryseth, 1990). The increased humidity associated with climatic transition from the temperate arid zone to the temperate subhumid zone will have caused the relative elevation of the alluvial systems equilibrium profile due to the increased fluid and sediment flux. This, combined with continuing but decreasing subsidence rates, created the accommodation space necessary for deposition and preservation of the Statfjord Formation. The climate change is reflected in the alluvial sediments. The lower floodplain sediments are characterised by red, highly oxidized soil complexes with extensive calcrete development (typical of arid zone climates), and thin isolated channel sands (4–8 m). Higher in the stratigraphy, floodplain

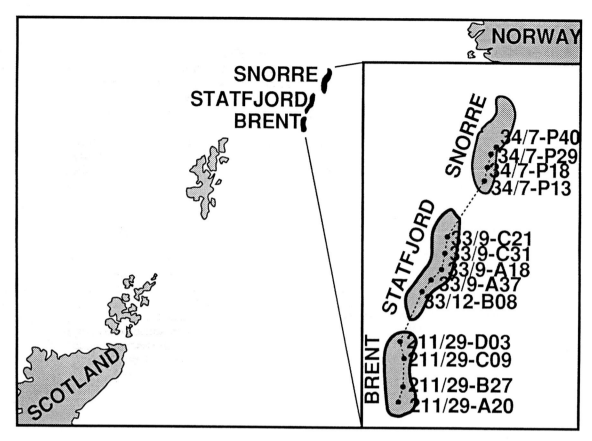

FIG. 12.—Location map showing the relevant field locations in the northern North Sea and the well locations for Figure 19.

suites are less oxidized and less pedified, and more extensive channel sands were deposited under conditions of perennial flow. The climate change towards more humid conditions will have contributed to a Stream Equilibrium Profile Highstand (i.e., creation of subaerial accommodation space, within which sediments were deposited, although quantitatively the amount of space created is still unclear). As the accommodation space was infilled, the channel systems began to amalgamate and so sheet sands developed near the top of the sequence during the ensuing Unconfined Stream Equilibrium Lowstand.

Eustasy

Figure 17 illustrates coastal onlap curves, from which eustatic fluctuations can be inferred, both for the North Viking Graben and globally. The Statfjord Formation sediments were deposited during cycles TR3.3 and J1.1 which mark the onset of the long term Jurassic transgression. The relative base level (sea level) lowstand at end Hettengian times possibly contributed to the incisional event in the coastal plain sediments of the upper Statfjord Formation in Snorre field. The lowstand will also have contributed to the generation of alluvial accommodation space by moving control point C in a basinward direction (Fig. 6), although its effect would have been concentrated at the downstream end of the fluvial system (present only near the very top of the stratigraphic column in the Brent, Statfjord and Snorre Fields). However the early Jurassic transgression quickly outpaced this minor incisional event.

SEQUENCE BOUNDARY IDENTIFICATION

In the absence of any biostratigraphically controlled surfaces within the Statfjord Formation, various other methods have had to be utilized in order that sequence defining surfaces can be identified. Although not all techniques could be applied to every well, in those which allowed a full suite of analyses the same surfaces were identified by the different methods used. These surfaces could then be correlated beyond sections which had cored and full wireline suite intervals. Figure 18A shows a gamma log curve for well 211/29-6 in the Brent Field.

Morton (pers. commun.) has identified several discrete heavy mineral zones which correlate with the isotope stratigraphy of Mearns et al. (1989) and Dalland (1994). Using the garnet, zircon, rutile, monazite ratios, the surfaces across which provenance changes occur are identified (Fig. 18B). Again, in terms of the conceptual sequence stratigraphic model, SEP lows and highs produce the same effects as described above, as these analytical techniques are dealing directly with the nature of the sediments preserved. Heavy mineral grains show differing weathering characteristics, indicative of exposure time to early diagenetic fluid flux (Bouch, pers. commun.).

Figure 18C illustrates trace element analysis for Strontium and Rubidium for well 211/29-C16 (Prosser et al., 1995). The differing concentrations of the trace elements are thought to be related to residence time on the floodplain during early fluxing with meteoric waters (Morton and Berge, pers. commun.).

Mearns et al. (1989) identified discontinuities in the Samarium-Neodymium isotope stratigraphy of the Snorre Field (Fig.

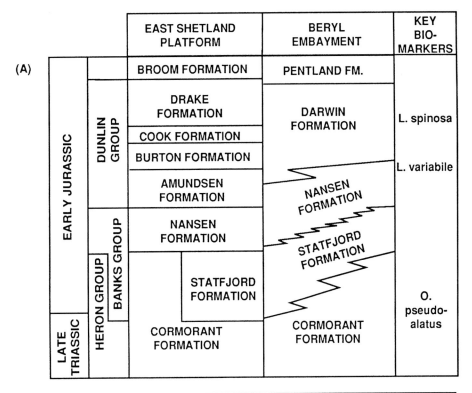

(A)

	EAST SHETLAND PLATFORM	BERYL EMBAYMENT	KEY BIO-MARKERS

FIG. 13.—(A) Stratigraphy of the Viking Graben and Beryl Embayment area and (B) stratigraphy of the Statfjord Formation (redrawn from Knox and Cordey, 1993)

18D). Samarium-Neodymium isotope stratigraphy reflects the model ages of the provenance of sediments from which the sample was taken. Dalland (1994) also noted these, but invoked different source areas to explain the varying provenance ages, which results in his interpretation of switching of the direction of drainage. Regional variations in provenance can be explained by examination of the states of the allocyclic mechanism at the time of deposition (taking lag times into account). SEP low-stands will encourage erosion, allowing bypass of sediment and introduction of new mineralogical suites which were sourced further into the hinterland. SEP highstands will develop aggradationally and so are more likely to trap sediment sourced relatively proximal to the site of deposition.

Several other wireline and geochemical correlation techniques were employed by Prosser et al. (1995), including wireline log crossplots, soil correlations from core and wireline, geochemical analysis and simple lithostratigraphic correlation

(Fig. 18), in the Brent Field. All the major sequence defining surfaces identified by analytical techniques were independently picked by application of the conceptual model presented here.

SEQUENCE DEVELOPMENT

The deposits of the alluvial plain are extremely heterogeneous and their distribution and geometries are difficult to predict, to any degree of high resolution. However general trends in the nature of sediments usually occur and can be explained in terms of the 'pseudo base level', or stream equilibrium profile, concepts discussed previously. Figure 19 illustrates gamma log correlations of the Statfjord Formation across the Brent, Statfjord and Snorre fields of the northern North Sea. The approximately 70–km–long section runs from SSW to NNE along the proposed fluvial drainage system which drained the hinterland basins of the late Triassic-early Jurassic Viking Graben.

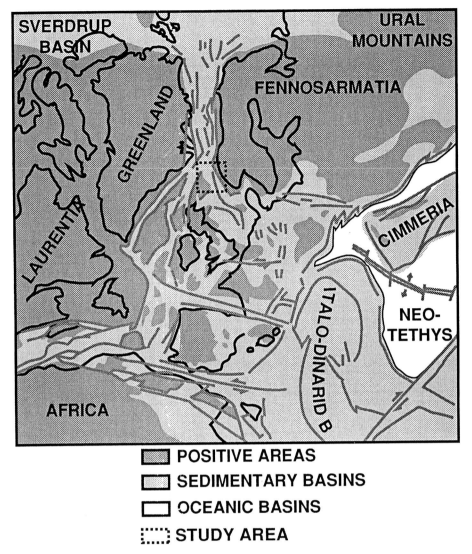

FIG. 14.—Palaeogeographic reconstruction of Triassic rift systems in the Arctic-North Atlantic realm (from Ziegler, 1982).

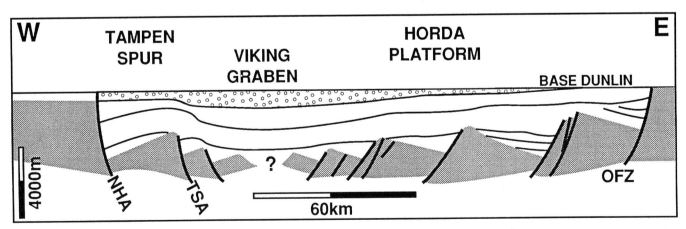

FIG. 15.—Schematic cross section showing the structural setting at late Triassic times across the Viking Graben. Nha-Ninian/Hutton Alignment, Tsa-Tampen Spur Alignment, Ofz–Oygarden fault zone (from Steel and Ryseth, 1990).

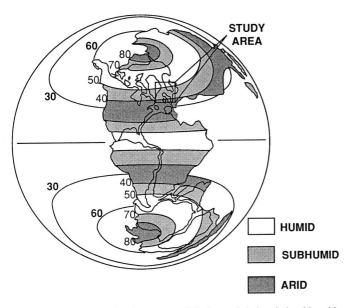

FIG. 16.—Late Triassic plate system latitudes and their relationship with climatic zones (from Hay et al., 1982).

The Upper Statfjord Sequence

The Statfjord Formation consists of two sequences, the upper lying entirely within the Statfjord Formation, and the lower one straddling the Cormorant/Lunde and Statfjord Formation lith-ostratigraphic boundary (Fig. 19), which is gradational (Struijk and Green 1991). Due to the change in the depositional environment from the upper Statfjord Formation coastal plain sediments into the marine transgressive sands of the Nansen Formation at the most basinward locality, the upper boundary of the upper Statfjord Formation is relatively easy to identify when heavy mineral, geochemical and wireline correlation techniques are used. As a result the Statfjord Formations upper sequence boundary can be picked at the base of the Nansen Formation at its most basinward locality. This sequence boundary can then be correlated into the coastal plain sediments of the uppermost Statfjord Formation.

The lower boundary of the upper sequence also becomes apparent, when heavy mineral, geochemical and reservoir production data are combined, and lies below the amalgamated channel sheet sand complex which is so prominent in the Brent Field. Unless evidence for a regionally confining valley system can be demonstrated then the amalgamated channel sands which are preserved at the base of the upper Statfjord sequence should be assigned to an Unconfined Stream Equilibrium Profile Low (USEPL) systems tract. The upper Statfjord Formation is overlain by the Nansen Formation, a transgressive marine sandstone, the base of which represents a regionally erosive ravinement surface within the Brent and Statfjord fields, generated by sediment reworking during the lowstand and transgressive phases of base-level fluctuation in the early Jurassic. No Nansen Formation is present in the Snorre Field, to the north of Brent and Statfjord, as a result of its location upon a low

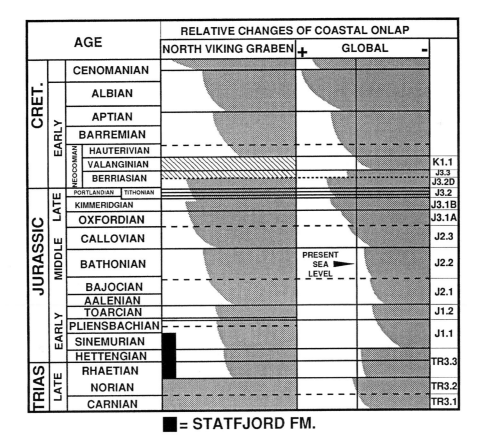

FIG. 17.—Global and North Viking Graben coastal onlap curve (after Vail and Todd, 1981), from which eustatic fluctuation can possibly be inferred.

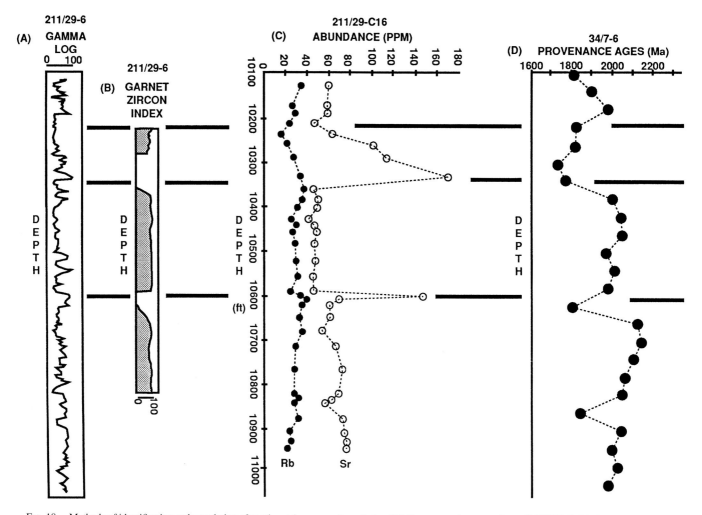

FIG. 18.—Methods of identification and correlation of continental sequence boundaries: (A) Gamma-ray log curve for well 211/29-6, (B) Garnet-zircon indices for well 211/29-6 (A. Morton, pers. commun., 1994), (C) Rubidium-strontium trace element analysis for well 211/29-C16 (Prosser et al., 1995), (D) Samarium-neodymium isotope analysis of provenance age from well 34/7-6 (Mearns et al., 1989). All techniques identify identical key surfaces.

lying coastal plain at the time of Nansen Formation deposition ie: as transgression commenced, the coastline moved rapidly southwards across the low gradient coastal plain of the Snorre Field area and began reworking sediments in the Statfjord Field area where either the gradient of the palaeoslope increased or the relative rate of transgression decreased. A minor coastal plain incisional event can be seen in the upper Statfjord sequence in the Snorre Field which marks the relative lowstand event preceding the transgression. In the Stream Equilibrium Profile High (SEPH) systems tract of the upper Statfjord sequence shale horizons thin northwards (i.e., towards the palaeo-coastline), perhaps indicating increased degrees of headward incision and erosion of floodplain shales due to eustatic fluctuations.

The Lower Statfjord Sequence

The lower Statfjord Formation is a floodplain dominated alluvial suite, which generally coarsens up to some degree, with highly pedified and oxidized soil complexes typical of arid zone climates. Small single storey fluvial sandstones are preserved in the lower sections, but exhibit poor lateral continuity (ribbon

sands) and are here interpreted to be the deposits of a Stream Equilibrium Profile High systems tract. A major, in places highly pedified, shale horizon at the top of this SEPH forms an extensive barrier to fluid flow during production in the Brent Field (Struijk and Green, 1991). However this reservoir barrier breaks down when traced northward due to increased incision by the overlying sands of the upper Statfjord sequence. This generally pedified horizon has been termed the 'D' shale in the Statfjord Field (MacDonald and Halland, 1993). The surface between this shale and an overlying amalgamated channel sandstone is interpreted as the lower sequence boundary of the upper Statfjord Sequence. Across this surface provenance related changes in the detrital constituents of sandstones occur (Mearns et al., 1989; Dalland, 1994; Morton, pers. commun.).

The SEPH systems tract in which the lower Statfjord sequence was deposited by changes in the states of the allocyclic controlling factors (climatic change, tectonics and eustasy), creates accommodation space. As this accommodation space was filled, a relative 'pseudo base level' lowstand (USEPL) resulted, during which the amalgamated channel sheet sand unit was deposited at the base of the upper Statfjord sequence. This amalgamated channel sheet sand is underlain by an unconfining

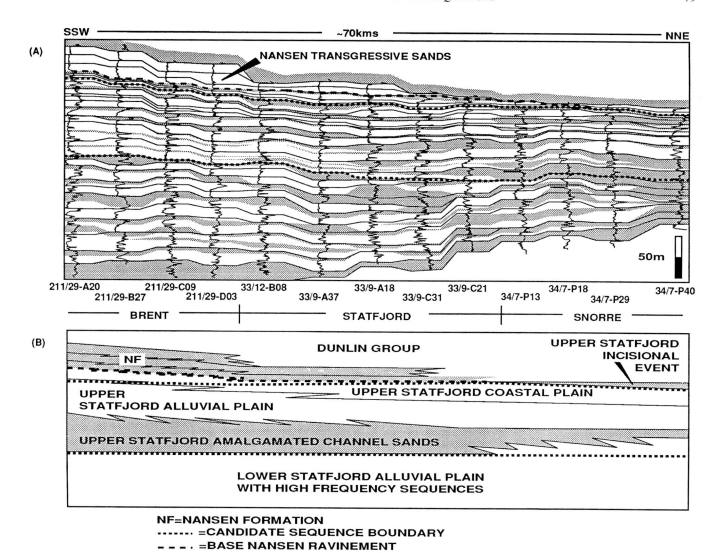

FIG. 19.—(A) Sequence analysis of the Statfjord Formation. The Lower Statfjord consists of many high frequency sequences. The Upper Statfjord contains a type II continental lowstand amalgamated channel sand sheet. Main sequence boundaries are shown as fine dashed lines, The base Nansen Formation ravinement surface is shown as a coarse dashed line. (B) Schematic interpretative diagram of Figure 19A (NF-Nansen Formation).

sequence boundary (due to a lack of sufficient west-east lateral well coverage it is difficult to demonstrate whether a confining incised valley system exists).

The Lower Statfjord sequence seems to be made up of many higher frequency, smaller scale sequences (Van Wagoner, 1994), with minor channel sand amalgamation, topped by abandonments and then minor pedified complexes.

CONCLUSIONS

Differentiation must be made between ultimate base level and the pseudo base level (stream equilibrium profile) which controls the creation and destruction of subaerial accommodation space. In the Statfjord Formation the relative roles of allocyclic controlling mechanisms changes throughout the time of deposition of alluvial sediments, and with palaeogeographic position within the depositional system. Deposition in the lower Statfjord Stream Equilibrium Profile High systems tract is controlled by the creation of accommodation space by climatic

change and tectonic subsidence. As this accommodation space was infilled, a relative lowstand of the Stream Equilibrium Profile occurred, generating the extensive amalgamated channel sheet sand unit which is underlain by an Unconfining sequence boundary (USEPL). These amalgamated sheet sandstones form prolific hydrocarbon reservoirs in the Brent, Statfjord and Snorre Fields of the northern North Sea. A period of increasing aggradation then occurred, resulting in less amalgamation of the channel sands and greater preservation of the floodplain units in the succession (with a marked decrease in the amount of pedification compared to the lower Statfjord). During early Jurassic transgression the sediments of the upper Statfjord Formation coastal plain and shoreline deposits of the Nansen Formation were deposited.

ACKNOWLEDGMENTS

The authors gratefully acknowledge the support of Esso Exploration and Production, U.K., for funding this research. Cato

Berge at Shell Exploration and Production, U.K., and Oivind Hostad at Esso Norge provided much logistical support in accessing wireline and core data, as well as providing some fruitful discussions. Statoil Oil Company, Saga Petroleum and Norsk Hydro AS are also thanked for providing access to data and permission to publish the results of this work. Thanks also go to Leslie Wood and Ron Steel for there comments in reviewing this paper.

REFERENCES

BADLEY, M. E., PRICE, J. D., RAMBECH DAHL, C., AND AGDESTEIN, T., 1988, The structural evolution of the northern Viking Graben and its bearing upon extensional modes of basin formation: Journal of the Geological Society of London, v. 145, p. 455–472.

BARRELL, J., 1917, Rhythms and the measurement of geologic time: Geological Society of America Bulletin, v. 28, p.745–904.

BOWN, T. M. AND KRAUS, M. J., 1987, Integration of channel and floodplain suites.I Developmental sequence and lateral relations of alluvial paleosols: Journal of Sedimentary Petrology, v. 57, p. 587–601.

DALLAND, A., 1994, Sm-Nd provenance age data as a tool in sequence stratigraphic interpretation and detailed reservoir correlation of the Statfjord Formation, Gullfaks Field: Conference on High Resolution Sequence Stratigraphy (abs): Liverpool, Innovations and Applications, Abstract Volume, p. 116.

DEEGAN, C. E. AND SCULL, B. J., 1977, A proposed standard lithostratigraphic nomenclature for the Central and Northern North Sea: Report of the Institute of Geological Sciences, 77/25.

HAY, W. W., BEHENSKY JR., J. F., BARRON, E. J., AND SLOAN, J. L., 1982, Late Triassic-Liassic palaeoclimatology of the proto-central North Atlantic rift system: Palaeogeography, Palaeoclimatology, Palaeoecology, v. 40, p. 13–30.

JERVEY, M. T., 1988, Quantitative geological modelling of siliciclastic rock sequences and their seismic expressions, in Wilgus, C. K., Hastings, B. S., Kendall, C. G. St. C., Posamentier, H. W., Ross, C. A. and Van Wagoner, J. C., eds., Sea-level Changes: An Integrated Approach: Tulsa, Society of Economic Paleontologists and Mineralogists Special Publication 42, p. 47–69.

KNOX, J. C., 1976, Concept of the graded stream, in W. N. Melhorn and R. C. Flemal, eds., Theories of Landform Development: Proceedings of the 6th Annual Geomorphology Symposia Series, p. 169–198.

KNOX, R. W. O'B. AND CORDEY, W. G., 1993, A Lithostratigraphic Nomenclature of the U.K. North Sea, V. 3, Jurassic (Central and Northern North Sea): British Geological Survey, p. 165–179.

KOSS, J. E., ETHERIDGE, F. G., AND SCHUMM, S. A., 1990, Effects of base-level change on coastal plain and shelf systems— an experimental approach (abs.): American Association of Petroleum Geologists Bulletin, v. 74, p. 697.

KOSS, J. E., ETHERIDGE, F. G., AND SCHUMM, S. A., 1994, An experimental study of the effects of base-level change on fluvial, coastal plain and shelf systems: Journal of Sedimentary Research, v. B64, p. 90–99.

LEOPOLD, L. B. AND BULL, W. B., 1979, Base level, aggradation and grade: Proceedings of the American Philosophical Society, v. 123, p. 168–202.

MACDONALD, A. C. AND HALLAND, E. K., 1993, Sedimentology and shale modelling of a sandstone-rich fluvial reservoir: Upper Statfjord Formation, Statfjord Field, Northern North Sea: American Association of Petroleum Geologists Bulletin, v. 77, p. 1016–1040.

MEARNS, E. W., KNARUD, R., RAESTAD, N., STANLEY, K. O., AND STOCKBRIDGE, C. P., 1989, Samarium-neodymium isotope stratigraphy of the Lunde and Statfjord Formations of Snorre oil field, northern North Sea: Journal of the Geological Society of London, v. 146, p. 217–228.

MIALL, A. D., 1991, Stratigraphic sequences and their chronostratigraphic correlation: Journal of Sedimentary Petrology, v. 61, p. 497–505.

MITCHUM, JR., R. M., VAIL, P. R., AND THOMPSON III, S., 1977, Seismic stratigraphy and global changes of sea-level, part 2: the depositional sequence as a basic unit for stratigraphic analysis, in Payton, C. E., ed., Seismic Stratigraphy—Applications to Hydrocarbon Exploration: Tulsa, American Association of Petroleum Geologists Memoir 26, p. 53–62.

MITCHUM, JR., R. M. AND VAN WAGONER, J. C., 1991, High frequency sequences and their stacking patterns: sequence stratigraphic evidence of high frequency eustatic cycles: Sedimentary Geology, v. 70, p. 131–160.

NYSTUEN, J. P., KNARUD, R., JORDE, K., AND STANLEY, K. O., 1990, Correlation of Triassic to Lower Jurassic sequences, Snorre Field and adjacent areas, northern North Sea, in Collinson, J. D., ed., Correlation in Hydrocarbon Exploration: Norwegian Petroleum Society, Graham and Trotman, p. 273–289.

OUCHI, S., 1985, Response of alluvial rivers to slow active tectonic movement: Geological Society of America Bulletin, v. 96, p. 504–515.

PERLMUTTER, M. A., AND MATTHEWS, M. D., 1989, Global Cyclostratigraphy—A Model, in Cross, T. A., ed., Quantitative Dynamic Stratigraphy: p. 233–260.

POSAMENTIER, H. W. AND VAIL, P. R., 1988, Eustatic controls on clastic deposition II—sequence and systems tract models, in Wilgus, C. K., Hastings, B. S., Kendall, C. G. St. C., Posamentier, H. W., Ross, C. A. and Van Wagoner, J. C., eds., Sea-level Changes: An Integrated Approach: Tulsa, Society of Economic Paleontologists and Mineralogists Special Publication 42, p. 125–154.

POSAMENTIER, H. W., JERVEY, M. T., AND VAIL, P. R., 1988, Eustatic controls on clastic deposition I —conceptual framework, in Wilgus, C. K., Hastings, B. S., Kendall, C. G. St. C., Posamentier, H. W., Ross, C. A. and Van Wagoner, J. C., eds., Sea-level Changes: An Integrated Approach: Tulsa, Society of Economic Paleontologists and Mineralogists Special Publication 42, p. 109–124.

PROSSER, D. J., BINGJIAN, L., HOLE, M. J., WILLIAMS, B. P. J. AND MORTON, A. C., 1995, Correlation of fluvial reservoir sequences: a case study using the Brent Field Statfjord Formation, northern North Sea (abs.): Houston, Proceedings of American Association of Petroleum Geologists Conference, p. 78A.

ROE, S. L., AND STEEL, R. J., 1985, Sedimentation, sea-level rise and tectonics at the Triassic-Jurassic boundary (Statfjord Formation), Tampen Spur, Northern North Sea: Journal of Petroleum Geology, v. 8, p. 163–186.

SCHUMM, S. A., 1977, The Fluvial System: New York, John Wiley and Sons, 338 p.

SCHUMM, S. A., 1993, River response to base level change: Implications for sequence stratigraphy: Journal of Geology, v. 101, p. 279–294.

SHANLEY, K. W. AND MCCABE, P. J., 1991, Predicting facies architecture through sequence stratigraphy—an example from the Kaiparowits Plateau, Utah: Geology, v. 19, p. 742–745.

SHANLEY, K. W. AND MCCABE, P. J., 1994, Perspectives on the sequence stratigraphy of continental strata: American Association of Petroleum Geologists Bulletin, v. 78, p. 544–568.

SLOSS, L. L., 1962, Stratigraphic models in exploration: American Association of Petroleum Geologists Bulletin, v. 46, p. 1050–1057.

STEEL, R. J., 1993, Triassic-Jurassic megasequence stratigraphy in the Northern North Sea: rift to post-rift evolution, in Parker, J. R., ed., Petroleum Geology of Northwest Europe: Proceedings of the 4th Conference, p. 299–315.

STEEL, R. J. AND RYSETH, A., 1990, The Triassic-early Jurassic succession in the northern North Sea: megasequence stratigraphy and intra-Triassic tectonics, in Hardman, R. F. P and Brooks, J., eds., Tectonic Events Responsible for Britain's Oil and Gas Reserves: Geological Society Special Publication 55, p. 139–168.

STRUIJK, A. P. AND GREEN, R. T., 1991, The Brent Field, Block 211/29, U.K. North Sea, in Abbotts, I. L., ed., United Kingdom Oil and Gas Fields, 25 Years Commemorative Volume, London Geological Society Memoir 14, p. 63–72.

VAIL, P. R. AND TODD, R. G., 1981, Northern North Sea Jurassic unconformities, chronostratigraphy and sea-level changes from seismic stratigraphy, in Illing, L. V. and Hobson, G. D., eds., Petroleum Geology of the Continental Shelf of Northwest Europe: London, Heyden and Son, p. 216–235.

VAN WAGONER, J. C., 1994, Models for fluvial architecture and non-marine sequence stratigraphy, in Johnson, S. D., ed., High Resolution Sequence Stratigraphy (abs): Liverpool, Innovations and Applications, Conference Abstract Volume, Liverpool University, p. 17–18.

VAN WAGONER, J. C., POSAMENTIER, H. W., MITCHUM, R. M., VAIL, P. R., SARG, J. F., LOUTIT, T. S., AND HARDENBOL, J., 1988, An overview of the fundamentals of sequence stratigraphy, in Wilgus, C. K., Hastings, B. S., Kendall, C. G. St. C., Posamentier, H. W., Ross, C. A. and Van Wagoner, J. C., eds., Sea-level Changes: An Integrated Approach: Tulsa, Society of Economic Paleontologists and Mineralogists Special Publication 42, p. 39–45.

WESCOTT, W. A., 1993, Geomorphic thresholds and complex response of fluvial systems— some implications for sequence stratigraphy: American Association of Petroleum Geologists Bulletin, v. 77, p. 1208–1218.

WHEELER, H. E., 1964, Base level, lithostratigraphic surface and time stratigraphy: Geological Society of America Bulletin, v. 75, p. 599–610.

WOOD, L. J., ETHERIDGE, F. G., AND SCHUMM, S. A., 1990, Effects of base-level change on coastal plain-shelf slope systems: an experimental approach (abs.): American Association of Petroleum Geologists Bulletin, v. 74, p. 1349.

WOOD, L. J., ETHERIDGE, F. G., AND SCHUMM, S. A., 1993, Effects of base-level fluctuation on coastal-plain, shelf and slope depositional systems: an experimental approach, *in* Posamentier, H. W., Summerhayes, C. P., Haq, B. U., and Allen, G. P., eds., Sequence Stratigraphy and Facies Associations,

International Association of Sedimentologists Special Publication 18, p. 43–53.

WRIGHT, V. P. AND MARRIOTT, S. B., 1993, The sequence stratigraphy of fluvial depositional systems: the role of floodplain sediment storage: Sedimentary Geology, v. 86, p. 203–210.

ZIEGLER, P. A., 1982, Geological Atlas of Western Europe and Central Europe: Shell Internationale Petroleum Maatschappij.

ANATOMY OF HINTERLAND DEPOSITIONAL SEQUENCES: UPPER CRETACEOUS FLUVIAL STRATA, NEUQUEN BASIN, WEST–CENTRAL ARGENTINA

LEONARDO LEGARRETA AND MIGUEL ANGEL ULIANA

ASTRA C.A.P.S.A. Exploration and Production Division, Tucumán 744 (Piso 10), 1049 – Buenos Aires, Argentina

ABSTRACT: Modern terminal fans, that is lobate-shaped wedges formed at the downstream end of endorheic drainage systems, develop a distinctive facies signature derived from rapid basinward decay of stream flow and lateral merger into the background hinterland deposition. This type of alluvial pattern has been recognized in the Upper Cretaceous redbed successions of the Neuquén basin. Paleogeographic reconstructions suggest the former presence of ephemeral streams that episodically discharged into a mud-dominated alluvial plain-alluvial basin setting and constructed 30– to 50–m–thick and 20– to 30–km–wide sand-rich fans. Stratigraphic analysis shows that the terminal fan accumulations are contained within 100– to 150–m–thick alluvial packages bounded by stratigraphic discontinuities ("hinterland sequences"). These packages record the repetition of a definite internal architecture. When fully developed, the stratigraphic motif consists of sandy forward-stepping stacks, followed by sandy to sandy-shaly backward stepping complexes, finally capped by mudrock-rich aggradational stacks punctuated by paleosols. Conditions forcing these stratal patterns produced a basin-wide stratigraphic overprint and were iterated nine times, suggesting a 1– to 3–my episodicity. Stratigraphic sections and paleogeographic reconstructions based on the recognition and mapping of the Upper Cretaceous hinterland sequences have been used to support exploration for hydrocarbons, underground water resources and strata-bound mineral concentrations.

INTRODUCTION

The core of the sequence stratigraphy approach that has become popular for analyzing sedimentary successions (among others Vail, 1987; Posamentier and Vail, 1988; Posamentier at al., 1988; Van Wagoner et al., 1990) developed mainly from the study of coastal and marine deposits (Payton, 1977). The application of sequence stratigraphic principles to terrestrial rock successions ("hinterland" sequences) has made much more restricted progress and is considered to be still in its infancy (Shanley and McCabe, 1994). Areas of uncertainty include fundamental aspects such as the relative importance of allocyclic forcing on the development of sequence boundaries and stratigraphic architecture (i.e., incidence of tectonic, climatic and eustatic drive). Perhaps the most serious limitation for short-term progress in this area is the relative scarcity of well-documented actual examples tackled from a sequence-stratigraphic perspective.

The presence in the geologic record of unconformity-bounded packages made up of terrestrial strata is well established (e.g., Simpson, 1940; Retallack, 1983). Also, some of these genetic intervals display a definite organization (Jol and Smith, 1991; López-Gómez and Arche, 1993; Wright and Marriot, 1993), contradicting the notion that non-marine successions are internally random and difficult to predict. Following this line, we examine, from a sequence stratigraphy perspective, the stratal patterns and facies architecture of the Upper Cretaceous alluvial deposits preserved in the Neuquén basin of west-central Argentina.

Hinterland Setting: The Rock Record of Intra–plate Alluviation Removed from Direct Oceanic Influence

For the purpose of the present study, we use the notion of hinterland sequences advanced by the Exxon group (Vail et al., 1977). Thus, a hinterland depositional sequence is understood as a relatively conformable succession of genetically related strata bounded by unconformities or their relative conformities, that "consists entirely of nonmarine deposits laid down at a site interior to the coastal area, where depositional mechanisms are controlled only indirectly or not at all by the position of sea level" (Mitchum, 1977).

Many hinterland deposits such as the widespread Mesozoic and Cenozoic epicratonic accumulations preserved across South America, consist of lithologically simple alluvial successions formed by variable proportions of sandstone-conglomerate and mudrock. When attempting to break these thick and relatively homogeneous successions into lower-hierarchy mappable genetic units, one of the most obvious alternatives is to look for abrupt vertical changes in the dominant grain size or in the net-to-gross sandstone ratio (e.g., Cazau and Uliana, 1973; Legarreta and Gulisano, 1989). Contrasting facies juxtapositions record the occurrence of pronounced displacements of alluvial depositional tracts. In addition, many hinterland series are punctuated by regionally extensive dissected horizons associated with scours and intra-basinal gullying (e.g., Retallack, 1983; Kraus and Middleton, 1987) that are generally interpreted to reflect time intervals lacking physical representation (Bown and Kraus, 1993). Away from the regional scours, the stratal discontinuities may be recorded as cryptic interfaces that separate relatively similar mudrock intervals. In some instances, the breaks may be highlighted by the presence of mature soil profiles (Bown and Larriestra, 1990). Paleosol concentrations in the vicinity of these hinterland stratal discontinuities have been used in regional correlation and are given chronostratigraphic significance (Hanneman and Wideman, 1991; Dreyer, 1993).

Basin-wide observations on the internal architecture recorded in hinterland packages bounded by stratal discontinuities, demonstrate the repeated occurrence of changes in the style of lithofacies stacking (Cazau and Uliana, 1973; Uliana, 1979; Legarreta and Gulisano, 1989). The repeated appearance of sandy forward-stepping stacks, followed by sandy-shaly backstepping complexes, and mudstone-rich aggradational stacks has been found to be a common depositional motif in some hinterland basins dominated by alluvial infilling (Legarreta et al., 1993b). These organized facies stacks have been tentatively compared (Legarreta et al., 1993b) with systems tracts recognized in marginal marine sequences (see Van Wagoner et al., 1988) and are considered as a record of baselevel transit cycles (cf. Wheeler, 1964).

A number of the Cretaceous and Triassic successions of Argentina that we are familiar with show specific facies assemblages that may be best attributed to depositional settings supplied by terminal fans of endorheic systems. Because these ephemeral stream systems (Friend, 1978) are poorly known when compared to perennial stream settings (Olsen, 1987), it seems pertinent to include a brief review of two presently active terminal clastic lobes that provide useful analogues for the sequence model discussed below.

The "Ephemeral Stream" Depositional Context: Present-day Examples

Some of the controls on sand and mud accumulation at the distal ends of large endorheic river systems can be assessed by analyzing the present depositional dynamics of the African Okavango delta, a "land of disappearing rivers" (Lee, 1990) and the Indo–Gangetic terminal fans such as the Markanda (Parkash et al., 1983).

Apart from the obvious size differences (Figs. 1, 2), these fan-shaped inland features have several characteristics in com-

mon. In both cases, deposition relates to far-travelled rivers that depend on a heavy but seasonal rainfall pattern, at variance with the rather arid climate acting in the depositional tract. Both streams have a highly variable discharge regime and dump most of their coarse sediment load through a multi-branch distributive system that feeds several depositional lobes. The topography is relatively subdued in the two areas and, reportedly, rapid infiltration is one of the main causes behind rapid downfan stream-decay below the apex zone. Because of conditions in the source area, gravel-sized material is scarce. Sandy facies decrease in proportion from the apex to the fan toe. The facies spectrum across the depositional tract has been described only for the Markanda (Parkash et al., 1983) and includes: multi-storey channelized sands with trough and planar cross-bedding (distributary channels); and tabular beds made up of laminated and climbing rippled sands, associated with siltstones and mud-

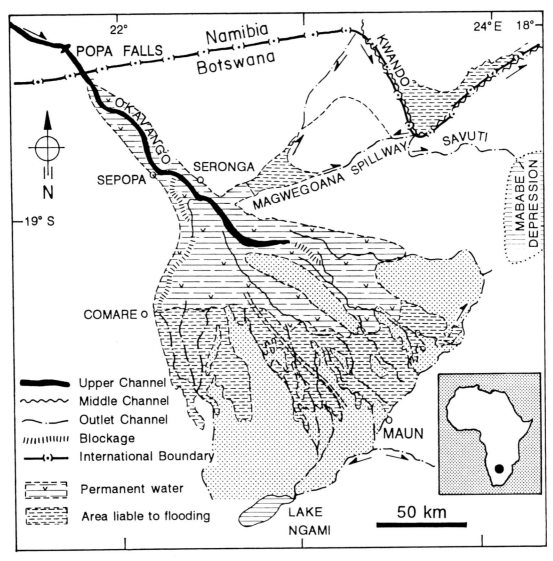

FIG. 1.—Map view of the Okavango "delta" in NW Botswana (modified after Johnson and Bannister, 1977). The water derives from 2500–m–high mountains located in eastern Angola. After crossing the down to the SE Gomare fault, the river splits into a number of branches that spread out from the lobe axis and form a labyrinth of shallow waterways. They come back together along a curving front controlled by the down to the NW Kunyere–Thamalakane fault system (McCarthy, 1993; Stanistreet and McCarthy, 1993). Due to extremely flat physiography the waters reaching the peripheral drainage system may reverse flow direction as the depressions fill and dry up.

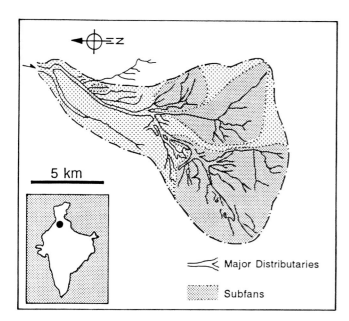

FIG. 2.—Map view of the Markanda terminal fan in north-central India (modified from Parkash et al., 1983). Water catchment over 120 km away in the high Siwalik Mountains. Notice the network of channels following a multi-branch distributive pattern feeding several active lobes. In spite of minimal surface relief (5–m) flood flow deceleration relates to rapid water loss by lateral infiltration.

stones with rootlets and desiccation cracks (overbank deposition).

Some of the contrasts between these two examples provide ideas on the local variability to be expected in this type of depositional system. The Markanda is a relatively unconfined "dry" system, where the river bed is empty most of the time, and the fan area can lose most of the monsoonal discharge by rapid infiltration. Conversely, the Okavango is a comparatively "wet" setting featuring perennial swamps, where the flood water becomes temporarily ponded and vanishes gradually through evaporation. In addition, the Okavango can be considered as a tectonically-constrained depositional site. The fan location is determined by the subtle but clear morphostructural control provided by a presently active cross-trending graben structure (Mccarthy, 1993; Stanistreet and McCarthy,1993).

MODELLING ALLUVIAL SEQUENCE DEVELOPMENT IN AN "EPHEMERAL
STREAM" HINTERLAND SCENARIO

The generic block diagrams in Figure 3 and the idealized sketch in Figure 4 summarize what we perceive as critical spatial and genetic relationships to be expected in an inland terminal-fan depositional context. These figures are an attempt to assemble a working scheme that takes into consideration the stratigraphic elements listed above, the knowledge derived from presently active terminal systems and field and subsurface observations on the stratal patterns recorded in ephemeral-stream deposits preserved in Argentina (Legarreta et al., 1993b) and in the basins of Greenland and Europe (Friend, 1978; Olsen, 1987; López-Gómez and Arche, 1993).

The depositional context is assumed to be a broad hinterland basin, where background sedimentation is represented by mixed-load and suspended-load deposits which accumulated in

alluvial plain and alluvial basin environments. The diagrams show the expected sedimentary and stratigraphic response to a regional event triggering the accumulation of a sandy ephemeral-stream complex, and the subsequent return to the fine-grained alluvial background. The relative curve at the bottom-right corner of Figure 4 suggests how different baselevel stages might be related to the successive facies stacking modes. These successive changes might be attributed to many alternative factors (e.g., climatic and/or tectonic drive, changing discharge, water table and vegetation, rejuvenation of the drainage network, changes in the morphology of the fluvial system, and relocation of the preferred site for sand sedimentation). From a stratigraphic standpoint, however, the patterns in Figure 4 may be seen to reflect a variable relationship between sediment supply and accommodation potential ("baselevel transit cycle").

Depositional Patterns and Stratal Architecture

Because of the characteristic occurrence of an abrupt change in alluvial style, granulometric contrast and association with regional scouring, the initial sequence development often coincides with a break in the depositional record. Moreover, terminal fan activity in itself can be seen to reflect episodic and relatively abrupt introduction of coarse-grained material, not

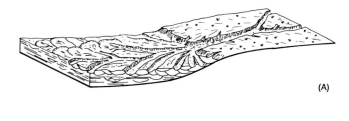

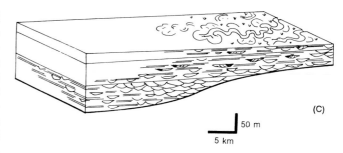

FIG. 3.—Conceptual diagram showing fluvial-style evolution and changes in stratal patterns during the accumulation of a hinterland deposit related to ephemeral stream activity. (A) Basin shrinkage, marginal incision and fore-stepping deposition; (B) basin expansion and backstepping deposition; and (C) limited aggradation and depositional stasis.

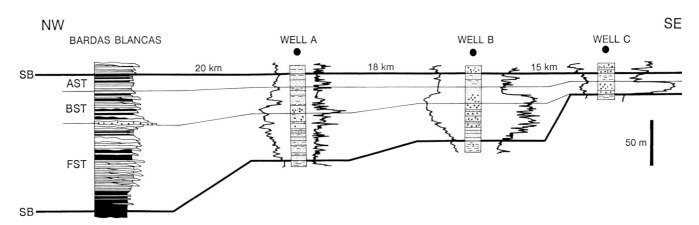

FIG. 8.—NW–SE oriented section of the lower Neuquén Group (Río Limay Formation, Cenomanian), based on facies and stacking patterns observed in outcrops and lateral tracing through wireline log (SP/RT) correlation. Systems tracts codes as in Figure 4. See Figure 7 for location.

sandstones feature lenticular shape and festoon cross-stratification, usually obliterated by bioturbation. Associated mudrock interbeds show carbonate concretionary horizons and modification by soil processes and suggest low sedimentation rates and hiatuses. These characteristics and the association of mixed and suspended load deposited under traction-plus-fallout conditions point out a depositional setting dominated by sheet floods and subordinate low-relief channels, feeding depositional lobes at the distal end of a distributary fluvial system.

Electric logs (spontaneous potential and resistivity) from wells A and B (Fig. 8), supplemented by observations at nearby outcrops in Pampa Amarilla (PA in Fig. 7), were used to identify rock types, facies, and stacking patterns closer to the head of the dispersal system. Outstanding upslope changes involve overall sequence thinning, a sharper basal contact and increase of the average grain size. Maximum grain size still occurs in strata in the middle part of the succession. Also, a stacking pattern beginning with an upward coarsening and thickening followed by upward fining and thinning, can be recognized in both wells. The top part of the upward-coarsening interval is made up of coarse-grained pebbly sandstones and fine-to coarse-grained gravel conglomerates. Conglomerates are crudely stratified, but may show inclined bedding and reverse grading. The sandstones are organized in crude thick to medium strata, bounded by dominantly erosive surfaces. They are internally massive or show thick parallel-lamination, carpet traction features, and parting lineation. Mudrocks are reduced to scattered erosional relics. Facies development at wells A and B are thus interpreted to have been deposited under high-energy, upper flow regime, in a context of highly concentrated flows. Pebbly and conglomerate facies are understood to represent the presence of channels active at the upper reaches of the distributary system.

At even more proximal sites such as well C (Figs. 7, 8) the basal upward-coarsening member disappears through progressive onlap onto an erosional surface. Near Río Diamante, at the northern end of the basin, outcropping strata equivalent to the lower Río Limay beds are also found to be areally restricted. They are lost by onlap toward the hinterland, while as in well C, the upward-fining stack at the top, expands the basin fringe (Cruz, 1993, Fig. 4).

In terms of the two-dimensional diagram in Figure 4, the areally confined member featuring the upward-coarsening organization is referred to the forestepping systems tract. Alluvial expansion at the base of the upward fining package is seen to represent the backstepping tract, followed by the finer-grained aggradational stack just before the sequence boundary. The long distance persistence of these stratal patterns suggests an effect of, at least, regional control on alluvial development.

Possible Controls on Sequence Development

The mechanisms forcing waxing and waning of the Late Cretaceous Neuquén basin, where stratigraphic discontinuities and fairly organized aggregations of coarser and finer grained alluvial strata were formed, can be rationalized in terms of changing depositional baselevel and variable supply to accommodation rates (Legarreta et al., 1993). The forces driving these changes are less obvious, but basin-wide extent of the sequence boundaries and some of the contrasted facies juxtapositions suggest an allogenic or at least regional control on sediment flux and sedimentation.

Regional paleogeographic and paleotectonic reconstructions (Uliana and Biddle, 1988) point out cratonal derivation during a period of regional tectonic quiescence and subdued hinterland relief, while South America was drifting away from Africa. Backstripping attempts on the Neuquén basin fill (Legarreta and Uliana, 1991) also suggest that Late Cretaceous time was a period dominated by negative exponential subsidence (sag), associated with lithospheric cooling. These indicators thus hint that tectonic activity allowed a gently sloped watershed relief, provided a topographic sink, and influenced orientation and gradient of the regional paleoslope. However, subsidence was likely progressive rather than episodic, failing to induce tectonic landforms suitable to trigger pulses of clastic-shedding into the basinal site.

When the Upper Cretaceous hinterland sequences are confronted with older or younger Cretaceous depositional sequences showing oceanic (eustatic) influence (Legarreta and Gulisano, 1989; Legarreta et al., 1993), they are found to record a similar frequency (1–2 my) and to involve comparable volumes of sediment. In spite of these resemblances, the Upper

Cretaceous genetic packages lack evidences of marine imprint and we interpret them as being deposited in an ephemeral alluvial setting, in consequence stratigraphic forcing related to eustatic oscillations is considered an unlikely possibility.

Alternatively, stratigraphic packaging observed in the Neuquén stata might be explained as a fluvial response to climatically-driven changes in discharge regime. River systems draining into a hinterland site featuring low relief and stable upland slopes, and being subject to climatically-induced increase in fluvial discharge, should be expected to react by widespread morphological and sedimentary adjustments such as proximal bypass and remobilization of older alluvium (Heward, 1978; Blum et al., 1994). At stages when stream power was raised above the level needed to displace the background sediment load, likely adjustments would include channel entrenchment, disection, and truncation along the proximal reaches of the distributary system. The sediment incorporated through reworking across the proximal tract would become displaced basinward, until progressive decrease in competence and transport capacity would promote the down dip development of depositional lobes (the areally restricted FST). Eventually, progress of the climatic cycle driving systematic decrease of the stream power, would ensue a regime favoring increased net accumulation (Blum et al., 1994), updip migration of the sediment storage/bypass interface, and regional backfill (the expanding BST). Progressive shrinkage of the area providing reworked alluvium would be associated with average grain-size decrease, with extensive paleosol development and ultimately with depositional stasis (the widespread AST). Along the basin fringe, the succeeding stratigraphic discontinuity (SB) would be enhanced sysnchronously with a new phase of regional degradation and fluvial adjustment driven by the next long-term climatic perturbation (cycle ?).

The large-scale incidence of long-term (10^5- to 10^6-yr time framework) climatic events on sediment flux has been recently addressed by Cecil (1990), suggesting that the magnitude of siliciclastic shedding is closely related to climatic seasonality. According to Cecil (1990), coarsening upward Middle Pennsylvanian sequences in the Appalachian basin can be correlated with the increased frequency of seasonal (wet-dry) periods. In this context, arrival of heavier clastic flux and coarser siliciclastics might be considered as the result of periods of strongly seasonal rainfall. Minimum siliciclastic arrival and extensive development of paleosols should be expected under dominant non-seasonal climates. Tropical-rainy regimes would promote leached soils, and arid regimes would favor aridisols and pedogenic carbonates. Whether or not these climate to sediment-flux relationships developed for glacial periods can be confidently applied to the warm-equable global regimes prevailing during the Late Cretaceous (Frakes et al., 1992) is a matter of speculation.

CONCLUDING REMARKS

The persistence of terminal systems like the present Okavango, feeding an inland depositional basin capable of accommodating a yearly sand input of some 660,000 tons, should be able to leave a distinctive signature in the geologic record.

Consideration of several basin types across South America (Kokogián et al., 1989; Fitzgerald et al., 1990; Fernández Seveso et al., 1993), suggests that mega-continental aggregations like the Late Paleozoic–Early Mesozoic Gondwanaland are likely to produce climatic and runoff patterns leading to internally drained depositional sites. Stratigraphic descriptions depicting endorheic alluvial situations are, however, relatively uncommon in the specialized literature.

Whether hinterland (alluvial) depositional sequences are to be interpreted as responses to tectonic events, to base-level fluctuations, to climatic cycles or to combinations of these processes, it is presently difficult to assess (Shanley and McCabe, 1994). In our opinion, this ambiguity should not prevent stratigraphers from using the depositional sequence approach to unravel the stratigraphic architecture and the depositional history of non-marine successions deposited away from direct oceanic influence. Sequence discrimination should still be possible by integrated analysis of physical relationships such as those provided by bedding terminations, regional erosion and paleosols, stratal patterns and facies stacking trends.

The "ephemeral stream" sequence model discussed in this paper has been used to split a Cretaceous hinterland accumulation into mappable genetic units, and it may represent an effective tool when building up the stratigraphic studies and paleogeographic reconstructions required to support resource exploration in other sedimentary basins.

ACKNOWLEDGMENTS

We gratefully acknowledge the management of Astra C.A.P.S.A. for support and permission to publish this paper. Tony Tankard graciously provided literature on the Okavangom system. We would like also to thank the reviewers Kevin Biddle and Peter Friend and the editors Keith Shanley and Peter McCabe for their constructive criticism and helpful comments.

REFERENCES

BLUM, M. D., TOOMSEY, R. S., III, AND VALASTRO, S., JR., 1994, Fluvial response to Late Quaternary climatic and environmental change, Edwards Palteau, Texas: Paleogeography, Paleoclimatology, Paleoecology, v. 108, p. 1–21.

BOWN, T. M. and KRAUS, M. J., 1993, Time-stratigraphic reconstruction and integration of paleopedologic, sedimentologic, and biotic events (Willwood Formation, Lower Eocene, Northwestern Wyoming, U.S.A.): SEPM Research Reports 1, p. 68–80.

BOWN, T. M. AND LARRIESTRA, C. N., 1990, Sedimentary paleoenvironments of fossil platyrrhine localities, Miocene Pinturas Formation, Santa Cruz Province, Argentina: Journal of Human Evolution, v. 19, p. 87–119.

CAMPBELL, C. V., 1976, Reservoir geometry of a fluvial sheet sandstone: American Association of Petroleum Geologists Bulletin, v. 60, p. 1009–1020.

CAZAU, L. B. AND ULIANA, M. A., 1973, El Cretácico Superior Continental de la Cuenca Neuquina: Quinto Congreso Geológico Argentino Actas III, p. 131–163.

CECIL, C. B., 1990, Paleoclimate controls on stratigraphic repetition of chemical and siliciclastic rocks: Geology, v. 18, p. 533–536.

CONDAT, P., CRUZ, C. E., KOZLOWSKI, E., AND MANCEDA, R., 1990, Ambiente deposicional de las sedimentitas del Grupo Neuquén Inferior en el suroeste de Mendoza, Argentina: Undécimo Congreso Geológico Argentino Actas II, p. 65–68.

CRUZ, C. E., 1993, Facies y estratigrafía secuencial del Cretácico Superior en la zona Río Diamante, Provincia de Mendoza, Argentina: XII Congreso Geológico Argentino Actas I, p. 46–54.

CRUZ, C. E., CONDAT, P., KOZLOWSKI, E., AND MANCEDA, R., 1989, Análisis estratigráfico secuencial del Grupo Neuquén (Cretácico Superior) en el valle del Río Grande, Provincia de Mendoza: Primer Congreso Nacional de Exploración de Hidrocarburos Actas II, p. 689–714.

DIGREGORIO, J. H. AND ULIANA, M. A., 1980, Cuenca Neuquina, in Turner, J. C. M., ed., Segundo Simposio de Geología Regional Argentina: Academia Nacional de Ciencias II, p. 985–1032.

DREYER, T., 1993, Quantified fluvial architecture in ephemeral stream deposits of the Esplugafreda Formation (Paleocene), Tremp-Graus Basin, northern Spain, in Marzo, M. and Puigdefábregas, C., eds., Alluvial Sedimentation: Oxford, International Association of Sedimentology Special Publication 17, p. 337–362.

FERNANDEZ-SEVESO, F., PÉREZ, M. A., BRISSON, I. E., AND ALVAREZ, L., 1993, Sequence Stratigraphy and tectonic analysis of the Paganzo Basin, Western Argentina: XII°ICC–P Comptes Rendus, v. 2, p. 223–260.

FITZGERALD, M. G., MITCHUM, R. M., JR., ULIANA, M. A., AND BIDDLE, K. T., 1990, Evolution of the San Jorge Basin, Argentina: American Association of Petroleum Geologists Bulletin, v. 74, p. 879–920.

FRAKES, L. A., FRANCIS, J. E., AND SYKTUS, J. I., 1992, Climate Models of the Phanerozoic: the History of the Earth's Climate over the Past 600 Millions Years: Cambridge University Press, 274 p.

FRIEND, P. F., 1978, Distinctive features of some ancient river systems, in Miall, A. D., ed., Fluvial Sedimentology: Calgary, Canadian Society of Petroleum Geologists Memoir 5, p. 531–542.

FRIEND, P. F., 1983, Towards the field classification of alluvial architecture or sequence, in Collinson, J. D. and Lewin, J., eds., Modern and Ancient Fluvial Systems: Oxford, International Association of Sedimentology Special Publication 6, p. 345–354.

FROSTICK, L. E. AND REID, I., 1987, Tectonic control of desert sediments in rift basins ancient and modern, in Frostick, L. and Reid, I., eds., Desert Sediments: Ancient and Modern: Geological Society Special Publication 35, p. 53–68.

HANNEMAN, D. L. AND WIDEMAN, CH. J., 1991, Sequence stratigraphy of Cenozoic continental rocks, southwestern Montana: Geological Society of America Bulletin, v. 103, p. 1335–1345.

HEWARD, A. P., 1978, Alluvial fan sequence and megasequence models: with examples from Westphalian D–Stephanian B coalfields, Northern Spain, in Miall, A. D., ed., Fluvial Sedimentology: Calgary, Canadian Society of Petroleum Geologists Memoir 5, p. 669–702.

JOHNSON, P. AND BANNISTER, A., 1977, Okavango sea of land of water: Cape Town, C. Struik Publishers, 201 p.

JOL, H. M. AND SMITH, D. G., 1991, Ground penetrating radar of northern lacustrine deltas: Canadian Journal of Earth Sciences, v. 28, p. 1939–1947.

KOKOGIAN, D. A., BOGGETTI, D. A., AND REBAY, G. A., 1989, Cuenca Cuyana. El análisis estratigráfico secuencial en la identificación de entrampamientos estratigráficos sutiles: Primer Congreso Nacional de Exploración de Hidrocarburos, Actas 2, p. 649–674.

KRAUS, J. L. AND MIDDLETON, L. T., 1987, Dissected paleotopography and baselevel changes in a Triassic fluvial sequence: Geology, v. 15, p. 18–21.

LEE, D. B, 1990, Okavango Delta: National Geographic Magazine, v.178, p. 38–68.

LEGARRETA, L. AND GULISANO, C., 1989, Análisis estratigráfico secuencial de la Cuenca Neuquina (Triásico Superior–Terciario Inferior), in Chebli, G. and Spalletti, eds., Cuencas Sedimentarias Argentinas. Facultad de Ciencias Naturales, Universidad Nacional de Tucumán, Serie Correlación Geológica 6, p. 221–243.

LEGARRETA, L., GULISANO, C., AND ULIANA, M. A., 1993a, Las secuencias sedimentarias jurásico-cretácicas, in Ramos, V. A., ed., Geología y Recursos Naturales de Mendoza, XII° Congreso Geológico Argentino Relatorio, p. 87–114.

LEGARRETA, L. AND ULIANA, M. A., 1991, Jurassic–Cretaceous marine oscillations and geometry of back-arc basin fill, central Argentine Andes, in McDonald, D. I. M., ed, Sedimentation, Tectonics and Eustasy: Oxford, International Association of Sedimentologists Special Publication 12, p. 429–450.

LEGARRETA, L., ULIANA, M. A., LAROTONDA, C. A., AND MECONI, G., 1993b, Approaches to nonmarine sequence stratigraphy — Theoretical models and examples from Argentine basins, in Eschard, R., and Doligez, B., eds., Subsurface Reservoir Characterization from Outcrop Observations: Editions Technip, p. 125–143.

LÓPEZ-GOMEZ, J. AND ARCHE, A., 1993, Architecture of the Cañizar fluvial sheet sandstones, Early Triassic, Iberian Ranges, eastern Spain, in Marzo, M. and Puigdefábregas, C., eds., Alluvial Sedimentation: Oxford, International Association of Sedimentology Special Publication 17, p. 363–381.

MCCARTHY, T. S., 1993, Physical and biological processes controlling the Okavango Delta — A review of recent research: Botswana Notes and Records, v. 24, p. 57–86.

MITCHUM, R. M., JR., 1977, Seismic stratigraphy and global changes of sea level, Part 11: Glossary of term used in seismic stratigraphy, in Payton, Ch. E., ed., Seismic Stratigraphy—Applications to Hydrocarbon Exploration: Tulsa, American Association of Petroleum Geologists Memoir 26, p. 205–212.

OLSEN, H., 1987, Ancient ephemeral stream deposits: a local terminal fan model from the Bunter Sandstone Formation (L.Triassic) in the Tonder–3, –4 and –5 wells, Denmark, in Frostick, L. and Reid, I., eds., Desert Sediments: Ancient and Modern: Geological Society Special Publication 35, p. 69–86.

PARKASH, B., AWASTHI, K., AND GOHAIN, K., 1983, Lithofacies of the Markanda terminal fan, Kurukshetra district, Haryana, India, in Collinson, J. D. and Lewin, J., eds., Modern and Ancient Fluvial Ssystems: Oxford, International Association of Sedimentology Special Publication 6, p. 337–344.

PAYTON, CH. E., 1977, Seismic Stratigraphy—Applications to Hydrocarbon Exploration: Tulsa, American Association of Petroleum Geologists Memoir 26, 516 p.

PETRI, S., 1987, Cretaceous paleogeographic maps of Brazil: Palaeogeography, Palaeoclimatology, Palaeoecology, v. 59, p. 117–168.

POSAMENTIER, H. W. AND VAIL, P. R., 1988, Eustatic controls on clastic deposition II–sequence and systems tract models, in Wilgus, C. K., Hastings, B. S., Kendall, C. G. St. C., Posamentier, H. W., Ross, C. A., and Van Wagoner, J. C., eds., Sea-level Change: An Integrated Approach: Tulsa, Society of Economic Paleontologists and Mineralogists Special Publication 42, p. 125–154.

POSAMENTIER, H. W., JERVEY, M. T., AND VAIL, P. R., 1988, Eustatic controls on clastic deposition I–conceptual framework, in Wilgus, C. K., Hastings, B. S., Kendall, C. G. St. C., Posamentier, H. W., Ross, C. A., and Van Wagoner, J. C., eds., Sea-level Change: An Integrated Approach: Tulsa, Society of Economic Paleontologists and Mineralogists Special Publication 42, p. 110–124.

RETALLACK, G. J., 1983, Late Eocene and Oligocene paleosoils from Badlands National Park, South Dakota: Boulder, Geological Society of America Special Paper 193, 82 p.

SHANLEY, K. W. AND MCCABE, P. J., 1994, Perspectives on the sequence stratigraphy of continental strata: American Association of Petroleum Geologists Bulletin, v. 78, p. 544–568.

SIMPSON, G. G., 1940, Review of the mammal-bearing Tertiary of South America: American Philosophical Society, Proceedings, v. 83, p. 649–709.

STANISTREET, I. G. AND MCCARTHY, T. S., 1993, The Okavango Fan and the classification of subaerial fan systems: Sedimentary Geology, v. 85, p. 115–133.

ULIANA, M. A., 1979, Geología de la región comprendida entre los ríos Colorado y Negro, Provincias de Neuquén y Río Negro: Tésis Doctoral, Universidad Nacional de La Plata. 122 p.

ULIANA, M. A. AND BIDDLE, K. T., 1988, Mesozoic–Cenozoic Paleogeographic and Geodynamic evolution of southern South America: Revista Brasileira de Geociencias, v.18, p.172–190.

VAIL, P. R., 1987, Seismic stratigraphy interpretation procedure, in Bally, A. W., ed., Seismic Stratigraphy Atlas: Tulsa, American Association of Petroleum Geologists Studies in Geology 27, p. 1–10.

VAIL, P. R., MITCHUM, R. M., JR., AND THOMPSON, S., 1977, Seismic Stratigraphy and global sea level, Part III: Relative changes of sea level from coastal onlap, in Payton, Ch. E., ed., Seismic Stratigraphy— Applications to Hydrocarbon Exploration: Tulsa, American Association of Petroleum Geologists Memoir 26, p. 63–97.

VAN WAGONER, J. C., MITCHUM, R.M., JR., CAMPION, K. M., AND RAHMANIAN, V. D. 1990, Siliciclastic sequence stratigraphy in well logs, core, and outcrops: concepts for high-resolution correlation of time and facies: Tulsa, American Association of Petroleum Geologists Methods in Exploration Series 7, 55 p.

VAN WAGONER, J. C., POSAMENTIER, H. W., MITCHUM, R. M., VAIL, P. R., SARG, J. F., LOUTIT, T. S., AND HARDENBOL, J., 1988, An overview of sequence stratigraphy and key definitions, in Wilgus, C. W, Hastings, B. S., Kendall, C. G. St. C., Posamentier, H. W., Ross, C. A., and Van Wagoner, J. C., eds., Sea-level Change: An Integrated Approach: Tulsa, Society of Economic Paleontologists and Mineralogists Special Publication 42, p. 39–45.

WHEELER, H .E., 1964, Baselevel transit cycle, in Merriam, D. F., ed., Symposium on cyclic sedimentation: Kansas Geological Survey Bulletin 169, p. 623–630.

WRIGHT, V. P. AND MARRIOT, S. B., 1993, The sequence stratigraphy of fluvial depositional systems: the role of floodplain sediment storage: Sedimentary Geology, v. 86, p. 203–210.

ZAMBRANO, J. J., 1981, Distribución y evolución de las cuencas sedimentarias en el continente sudamericano durante el Jurásico y el Cretácico: Cuencas sedimentarias de América del Sur, v. I, p. 9–44.

EVOLUTION OF A BRAIDED RIVER SYSTEM: THE SALT WASH MEMBER OF THE MORRISON FORMATION (JURASSIC) IN SOUTHERN UTAH

JOHN W. ROBINSON*

Department of Geology and Geological Engineering, Colorado School of Mines, Golden, CO 80401

AND

PETER J. McCABE

U.S. Geological Survey, Federal Center MS 972, PO Box 25046, Denver, CO 80225

ABSTRACT: The Salt Wash Member of the Upper Jurassic Morrison Formation in the Henry Mountain region of southern Utah is up to 160 m thick and consists of sandstones, interpreted as fluvial-channel deposits, and mudrocks, interpreted as overbank or abandoned channel-fill deposits. The strata are interpreted to have been deposited by a braided river system based on the high sandstone:mudrock ratio, the paucity of ripple lamination in the upper part of fining-upward units, the coarse grain size of many of the sandstones, a lack of lateral accretion bedding, the sheet-like nature of the sandstone bodies and the low dispersion of paleocurrent vectors. The rivers appear to have had a highly flashy discharge with relatively little preservation of falling-stage and low-stage sedimentary features. There is an up-section change in stratal geometry from thin, highly amalgamated, sheet sandstone bodies, to thicker, more isolated, sheet sandstone bodies. This change suggests an increase in the rate of creation of accommodation over time— this may be related to expansion of lacustrine systems downstream. There may also have been a climatic change from arid to semi-arid conditions that resulted in larger streams and more stabilized banks. Perpendicular to the paleoflow direction there is a marked change in thickness of the strata with an increasing percentage of overbank/flood-plain mudstone in thinner sections— this change is interpreted to be the product of differential subsidence within the basin of deposition. An understanding of fluvial strata like the Salt Wash Member requires a holistic view of allocyclic controls on sedimentation including temporal and spatial variations in climate, tectonism and base level.

INTRODUCTION

Many studies of fluvial strata have investigated the lateral and vertical changes in lithofacies and related them to allocyclic and autocyclic controls (e.g., Marzo et al., 1988; Miall and Tyler, 1991; Shanley and McCabe, 1993, 1994; Puigdefàbregas, 1993). Understanding the controls on facies architecture is important because it allows prediction of the distribution and packaging of lithofacies in regions with limited data (Shanley and McCabe, 1993). Prediction of the architecture of fluvial strata is important in economic endeavors such as the search for hydrocarbons and ore minerals and the investigation of aquifers. Studies of fluvial architecture have been applied in a quantitative sense in the development of probabilistic models of the distribution of hydrocarbon reservoirs and sandstone-body interconnectedness (e.g., Budding et al., 1988; Henriquez et al., 1990). The development of predictive models of facies architecture requires analysis of high-quality data sets rather than just descriptive analogs (Alexander, 1993). Such data is best collected from extensive three-dimensional outcrops where lateral and vertical variations in stratal geometry can be documented. The Upper Jurassic Salt Wash Member of the Morrison Formation in south central Utah is an ideal unit for such a study because it is exceptionally well exposed in cliffs up to 6 km long and 300 m high, eroded parallel and perpendicular to paleodepositional strike. This stratigraphic unit has long been recognized as predominantly fluvial in nature (Craig et al., 1955; Mullens and Freeman, 1957; Tyler and Ethridge, 1983; Peterson, 1980, 1984).

The Salt Wash Member crops out over an area of about 150,000 km² on the Colorado Plateau in Utah, Colorado, New Mexico and Arizona. Comprehensive regional studies have determined the age, depositional extent and gross facies distribution of the Salt Wash Member (Craig et al., 1955; Mullens and Freeman, 1957; Peterson, 1980, 1984, 1986, 1994; Tyler and Ethridge, 1983). This study deals with outcrops of the Salt Wash Member at the southern end of the Henry Mountains (Fig. 1). The area is bounded to the west by the Circle Cliffs Uplift,

to the east by the Monument Upwarp and to the north by the San Rafael Swell. In this paper, we describe the facies and facies geometry of the Salt Wash Member in this region. We will then interpret the relative role of various controls on sedimentation of the strata and discuss how the system appears to have evolved over time.

STRATIGRAPHIC SETTING OF THE SALT WASH MEMBER

The present outcrop configuration in the Henry Mountains area was formed by a combination of intrusion of laccoliths during Oligocene time (Hunt et al., 1953; Nelson et al., 1992) and uplift of the entire Colorado Plateau Province during late Miocene–early Pliocene time (Luchitta, 1979). Late Mesozoic strata in the area dip at less than 2°, though locally they are disrupted by laccolithic intrusions of diorite porphyry in the Henry Mountains (Hunt et al., 1953). The base of the Morrison Formation (Lupton, 1914) is marked by the J–5 unconformity (Peterson, 1980, 1988a; O'Sullivan, 1981) and within the study area is underlain by the Middle Jurassic Summerville Formation. The Summerville Formation was deposited during the final withdrawal of the Middle Jurassic sea from Utah. By contrast, the Dakota Formation, which overlies the Morrison Formation, was deposited during the initial transgression of the Upper Cretaceous Western Interior Seaway into southern Utah (Kirschbaum and McCabe, 1992). Peterson (1980) defined three members of the Morrison Formation— Tidwell, Salt Wash, and Brushy Basin (Figs. 2, 3). The Tidwell Member, which underlies the Salt Wash Member, consists of a basal transgressive sandstone overlain by sandstone, limestone, gypsum and lacustrine mudstone that represent a mixture of lacustrine and fluvial environments (Peterson, 1980, 1994). The Brushy Basin Member, which overlies the Salt Wash Member, is composed of varicolored mudstone and conglomerate units of lacustrine and fluvial origin (Peterson, 1980, 1994; Turner and Fishman, 1991).

The Salt Wash Member is Late Jurassic (Kimmeridgian) in age (Craig et al., 1955; Kowallis and Heaton, 1987; Peterson,

*Present address: Snyder Oil Corporation, 1625 Broadway, Suite 2200, Denver, CO 80202
Relative Role of Eustasy, Climate, and Tectonism in Continental Rocks, SEPM Special Publication No. 59

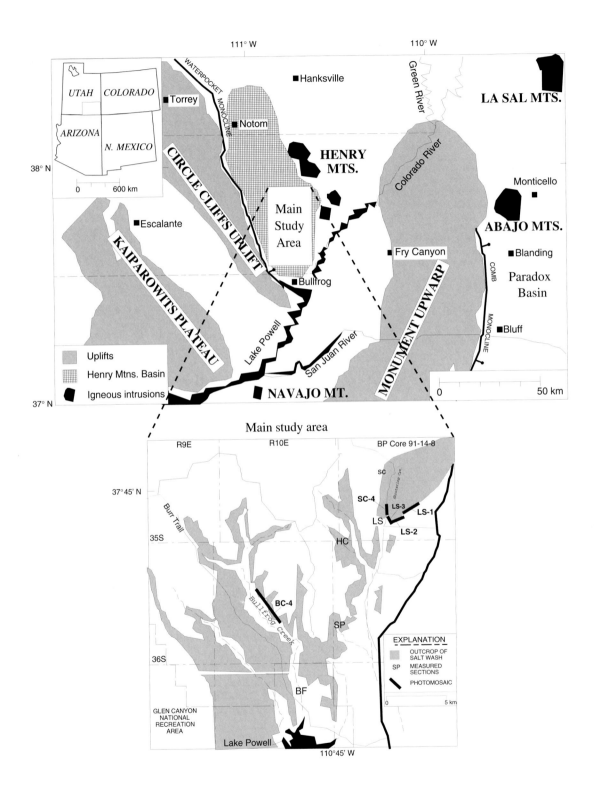

FIG. 1.—Regional location map of southern Utah showing the main study area near the Henry Mountains. SC = Shootering Canyon, LS = Lost Springs, HC = Hansen Creek, SP = Saleratus Point and BF = Bullfrog Creek. The Post measured section is 10–km northwest of the left margin of the map on the Burr Trail.

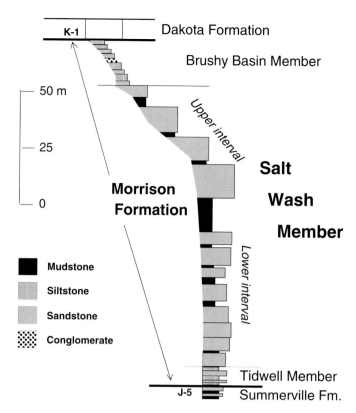

FIG. 2.—Lithostratigraphic column and schematic erosion profile for Middle and Late Jurassic strata in the study area in southern Utah. Upper and lower intervals of the Salt Wash Member are informal terms used in this study. Lower interval strata are composed of cliff-forming, amalgamated sandstone bodies (>80%) with subordinate amounts of mudstone. Upper interval strata are composed of recessive-weathering, isolated sandstone bodies (<80%) and mudstone. K–1 is the Jurassic/Cretaceous unconformity; J–5 is the Middle Jurassic/Late Jurassic unconformity (Peterson, 1988b).

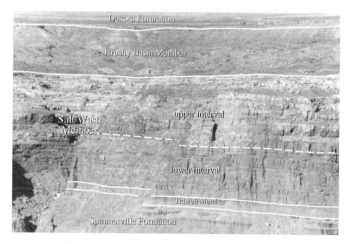

FIG. 3.—The stratigraphic section, as depicted on Figure 2, is shown on this photograph from Bullfrog Creek, Utah. The Middle Jurassic Summerville Formation forms the slope at the bottom of the photo. Contacts between the Tidwell, Salt Wash and Brushy Basin Members of the Morrison Formation are superimposed on the photo. The Salt Wash Member is approximately 150 m thick in this area. The Brushy Basin Member forms a hummocky slope in the upper part of the photo and is overlain by a resistant ledge of Cretaceous Dakota Formation.

1994). Within the study area, the member is composed predominantly of sandstones that are interbedded with heterolithic strata (Figs. 3, 4). The member ranges from 55 to 160 m in thickness. Although the Salt Wash Member extends over a considerable part of Utah, Colorado, New Mexico and Arizona, the thickest and sandiest portion is present within the study area and in adjoining parts of northern Arizona (Mullens and Freeman, 1957, Peterson, 1994). Paleocurrents indicate flow towards the northeast. The member thins to the northeast and also contains a decreasing proportion of sandstones in a distal direction. Uranium ore is present in sandstones near the base of the member and was mined from the 1950s to the 1980s. Because of the economic interest, the Salt Wash Member was extensively studied in the study area and adjoining areas by the U.S. Geological Survey, and Peterson (1978, 1980, 1984, 1988b) documented the regional extent of the member and clarified the stratigraphic terminology of Jurassic strata in the region.

The Salt Wash Member is terrestrial in aspect; at least within the study area, there is no evidence of marine influence, such as fossils or tidal indicators. Dinosaur tracks are present at several stratigraphic levels (Lockley et al., 1992). The vast majority of sedimentary structures within the sandstones are typical of those created by the migration of subaqueous bedforms. Furthermore, the pebbly nature of many of the sandstones suggests deposition by fluvial processes; an eolian element, if present at all, would constitute only a very small proportion of the facies. Many of the mudrocks contain roots. The base of the Salt Wash Member is generally marked by a sharp transition, with the eroded base of the lowest sandstone sitting on the finer grained sediments of the underlying Tidwell Member (Fig. 3). However, in Lost Springs Wash and Bullfrog Creek lithologies typical of the two members interfinger over a 2–m–thick interval (Robinson, 1994). The contact between the Salt Wash Member and the overlying Brushy Basin Member is placed at the top of the last laterally extensive sandstone in the Salt Wash Member (Peterson, 1980), but this is not a sharp facies change and the nature of the contact is somewhat enigmatic (Peterson, 1988a).

The average composition of Salt Wash Member sandstones is 69% grains (quartz, chert), 23% cement (calcite), 1% clay matrix and 7% porosity (Jensen, 1982). The petrology of sandstones within the Salt Wash Member is uniform throughout its depositional extent with a decrease in grain size in a distal direction (Mullens and Freeman, 1957). Jensen (1982) and Peterson (1987) interpreted the source area of Salt Wash Member sandstones to be Paleozoic and early Mesozoic strata exposed in western Utah, southern Nevada, northwest Arizona and southeast California. During Late Jurassic time, southern Utah was located east of the incipient Sevier Highlands and north of the Mogollon Highlands (Peterson, 1986, 1994). These highlands resulted from east-directed compression associated with a discontinuous subduction zone/magmatic arc complex to the west (Dickinson, 1976, 1981). Central and eastern Utah were located along the western margin of the North American craton within an incipient foreland basin, and low-relief erosional remnants of the ancestral Rocky Mountains lay to the east. Peterson (1984, 1986, 1988b) suggested that mild deformation of the craton, caused by compression from the west, generated periodic reactivation and adjustment of crustal blocks. At the time of deposition of the Salt Wash Member, rivers drained northeast

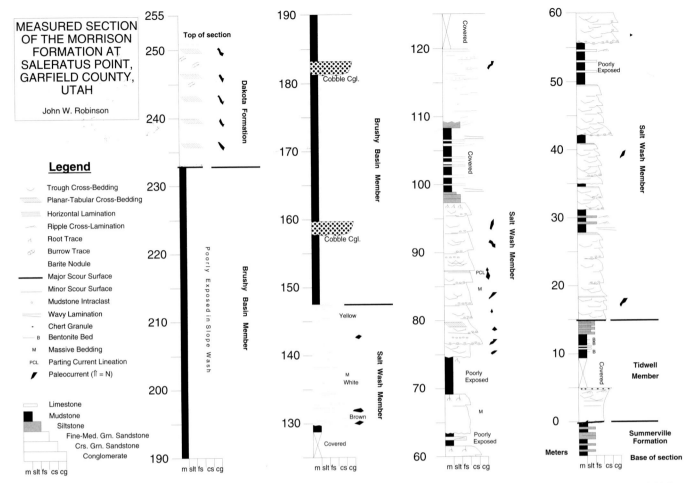

FIG. 4.—Representative measured section of the Tidwell, Salt Wash, and Brushy Basin Members of the Morrison Formation, at Saleratus Point, Garfield Co., Utah. Location of section shown on Figures 1 and 11.

towards northeast Utah and northwest Colorado, an area that had been at the southern end of a seaway that extended south from Canada in Callovian and Oxfordian times (Brenner, 1983; Peterson, 1994, Anderson and Lucas, 1995). By Kimmeridgian time, that seaway had retreated to the north with the shoreline in Alberta (Poulton, 1984). Based on paleomagnetic reconstructions (Van Fossen and Kent, 1992), Utah was located at a latitude of 30–35°N during Middle to Late Jurassic (Oxfordian–Kimmeridgian) deposition. The paleoclimate for western North America during this time was generally warm and dry which resulted in the deposition of widespread red beds, evaporites and shallow-water carbonates (Brenner, 1983). However, petrified logs (up to 1 m in diameter) and carbonaceous debris in the Salt Wash Member indicate that the highlands to the west supported conifer forests.

FACIES AND FACIES ARCHITECTURE

Measured sections of facies observed in outcrop and core drilled for uranium exploration are presented in Robinson (1994); one representative section is shown in Figure 5. Photomosaics were taken of cliff exposures oriented parallel and perpendicular to the predominant paleocurrent direction. Stratal boundaries that mark major facies changes were traced at dif-

ferent scales over these photomosaics while making field observations. These tracings were used to quantify the three-dimensional geometry of sandstone and shale bodies. However, measured widths are, in some cases, minimum values because units extend beyond the end of a photomosaic. This type of data can be used to evaluate changes in fluvial architecture and the ratio of net sandstone to gross composition and place them within the context of controls on the fluvial system. The approach is similar to that used by Blakey and Gubitosa (1984), Fielding and Crane (1987), Cowan (1991), Alexander and Gawthorpe (1993) and in the SEPM Concepts in Sedimentology volume edited by Miall and Tyler (1991). A complete set of photomosaics and tracings and tabulated width/thickness data are included in Robinson (1994). The definitions of Potter (1967) were used for 'sand body'— "a single, interconnected . . . body of sand"— and 'multistorey sandstone bodies'— "the superposition of a sand body of one cycle upon one or more earlier ones, the two or more forming an unusually thick sand section". We also use the term "storey" in a similar fashion to Friend et al. (1979) to describe a sandstone body that shows a general fining-upward trend and is bounded above and below, at least in places, by finer grained facies.

The Salt Wash Member can be divided into two lithosomes: thick sandstones, interpreted to have been deposited within ac-

IDEALIZED FACIES SUCCESSION IN THE SALT WASH MEMBER OF THE MORRISON FM.

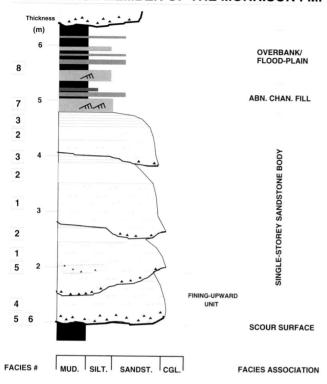

FIG. 5.—Idealized facies succession for the Salt Wash Member of the Morrison Formation in the Henry Mountains area, Utah. Fining-upward units stack to form an overall fining-upward facies succession, which was constructed by observations within measured sections and cores. Facies on section correspond to numbers in text: (1) trough cross-bedded sandstone, (2) planar-tabular cross-bedded sandstone, (3) horizontally laminated sandstone, (4) massive sandstone, (5) granule-pebble conglomerate, (6) mudstone intraclast conglomerate, (7) ripple cross-laminated sandstone and (8) siltstone and mudstone. Channel-fill (single storey sandstone body), abandoned channel-fill and overbank/flood-plain facies associations are shown in column to right.

tive fluvial channels, and finer grained strata, interpreted to have been deposited in abandoned channels and overbank areas. Each lithosome is further divided into facies based on lithology and stratification type (Table 1). The thick sandstone lithosome comprises up to 90% of the Salt Wash Member in the Henry Mountains area and consists of trough cross-bedded sandstone (Facies 1), planar-tabular cross-bedded sandstone (Facies 2), horizontally laminated sandstone (Facies 3), massive sandstone (Facies 4), granule-pebble conglomerate (Facies 5), mudstone intraclast conglomerate (Facies 6) and ripple cross-laminated sandstone (Facies 7). The fine-grained lithosome comprises up to 40% of the Salt Wash Member in the study area and consists of ripple cross-laminated sandstone (Facies 7) and siltstone and mudstone (Facies 8). Trough cross-bedding (79%), planar-tabular cross-bedding (16%) and horizontal lamination (5%) are the dominant stratification types within sandstones. Ripple cross-lamination and massive bedding comprise less than 5% of the stratification. Sedimentological descriptions of facies are summarized in Table 1.

An idealized succession through a single-storey sandstone-body is shown in Figure 5. This succession attempts to depict typical facies relationships and is based on observations of numerous outcrops and cores (Robinson, 1994). Although facies are present in a complex pattern, certain trends are clearly evident. Major erosion surfaces, with relief of several meters over distances of 1 km, are present at the base of thick sandstones that have an overall fining-upward nature. Most of these erosion surfaces are overlain by lags of granule-pebble conglomerate (Facies 5) or mudstone intraclast conglomerate (Facies 6). Each thick sandstone contains several minor fining-upward units (Fig. 6) within any vertical profile. The minor fining-upward units overlie a scour surface with considerably lower relief than the basal scour surface of the succession— generally <1 m over 100 m. Many of these minor scour surfaces are overlain by mudclast conglomerates (Facies 6). If traced laterally, many of the storeys of the Salt Wash Member can be seen, in some places, to overlie and be overlain by finer-grained heterolithic strata. However, many storeys are cut into one another, producing multistorey sandstone bodies that consist of up to eight storeys in a vertical profile. Storeys have an average thickness of 6.8 m (range 1.0–18.0 m) and a width of 391 m (range 31–1200 m). All of the storeys can be classified as sandstone sheets using the criterion (width vs. thickness >15:1) of Friend et al. (1979). Multistorey sandstone bodies are generally bounded above and below by finer-grained heterolithic strata, though locally two multistorey sandstone bodies may be separated by an erosion surface. Five laterally extensive multistorey sandstone bodies were identified and correlated on the photomosaics. Like storeys, multistorey sandstones also form sheets. Five multistorey bodies were measured; they average 24.6 m in thickness and are up to 1600 m wide (Fig. 7) with a range of width:thickness ratios from 24:1 to 80:1.

Massive sandstone (Facies 4), present at the base of some of the minor fining-upward units within storeys, probably represents bed load that was deposited rapidly during floods. Higher in a storey (Fig. 5), trough and planar-tabular cross-bedding predominate. Trough cross-beds (Facies 1) are 0.24–1.4 m thick and 3.25–16.0 m wide and are interpreted as the product of three-dimensional dunes (terminology of Ashley, 1990). Trough sets stack to form cosets several meters thick (Fig. 8) that extend laterally for hundreds of meters. Paleocurrents from trough cross-bedding are towards the northeast (Fig. 9). Planar-tabular cross-beds (Facies 2) are 0.31–0.68 m thick and 3.0–14.0 m wide and are interpreted as the deposits of straight- to sinuous-crested bedforms. Many planar-tabular cross-beds have topsets of horizontal lamination with primary current lineation, suggesting that flow was near the threshold to upper flow-regime conditions. Paleocurrent vectors from planar-tabular cross-bedding (Fig. 9) are similar in direction to that of trough cross-beds and have only a slightly greater dispersion. The upper part of most of the minor fining-upward units (Fig. 5) contain horizontally laminated sandstone (Facies 3) which is present as laterally extensive beds with low-angle (<5°) upper and lower contacts. This facies commonly has primary current lineation and fines upward into ripple-laminated sandstone (Facies 7) or siltstone and mudstone (Facies 8). The horizontal lamination is interpreted to have formed during upper flow-regime conditions, by migration of low-amplitude bedwaves (Best and Bridge, 1992).

TABLE 1.—SUMMARY OF FACIES DESCRIBED IN THE SALT WASH MEMBER OF THE MORRISON FORMATION IN THE HENRY MOUNTAINS AREA, UTAH. FACIES 1 THROUGH 6 ARE PRESENT IN FLUVIAL CHANNEL STRATA AND FACIES 7 AND 8 ARE PRESENT IN NON-CHANNEL STRATA.

Facies	Sedimentary Structures	Grain Size	Dimensions of Facies	Comments	Interpretation
(1) Trough cross-bedded sandstone	Trough cross beds.	Fine- to coarse-grained sandstone with granules and/or pebbles on toesets.	Set thickness 0.73 to 0.88 m. Set width 6.18 to 9.16 m. Increase in set size up section	Some troughs have upper transition to horizontal lam. Foreset vector mean N.53°E	Subaqueous 3-dimensional dunes.
(2) Planar-tabular cross-bedded sandstone	Planar-tabular cross beds.	Fine- to coarse-grained sandstone with granules and/or pebbles on toesets.	Average set thickness 0.54 m. Average set width 8.08 m.	Multiple reactivation surfaces. Some sets have upper transition to horizontal lamination. Foreset vector mean N.59°E.	Straight- to sinuous-crested dunes.
(3) Horizontally-laminated sandstone	Horizontal laminations. Low angle (<5°) terminations.	Fine- to medium-grained sandstone. Occasional granules and pebbles.	Avg. bed thickness 0.85 m. Avg. bed width 13.80 m.	Common at top of sedimentation units. Parting current lineation NE/SW. Sandfilled burrows.	Upper flow-regime plane beds.
(4) Massive sandstone	Massive bedding.	Medium- to coarse-grained sandstone.	Variable.	Present at base of channels. Not common.	Upper flow-regime and antidunes or rapid deposition.
(5) Granule-pebble conglomerate	Poorly stratified conglomerate with sandstone matrix.	Pebbles up to 6 cm in diameter, most <2 cm. Intra- and extra-formational clasts. Matrix of fine- to coarse-grained sandstone.	Bed thickness 1.0 to 2.0 m. Bed width 10 m to 300 m.	Multicolored chert and quartzite ganules and pebbles. Ang. mudstone intraclasts.	Basal lag in channel thalweg. Some beds may have formed in longitudinal bars.
(6) Mudstone intraclast conglomerate	Poorly to unstratified, matrix-supported, green mudstone intraclast conglomerate.	Pebbles-cobbles <10 cm in diameter. Matrix of fine- to coarse-grained sandstone.	Bed thickness up to 0.5 m. Bed width up to 300 m.		Basal lag in channel.
(7) Ripple cross-laminated sandstone	Symmetrical and asymmetrical ripple cross-lamination. Ripple indices from 5–12.	Very-fine to fine grained sandstone.	Set thickness 0.5 to 3.0 cm.	Deformed by compaction. Sand-filled burrows.	Lower flow regime.
(3) Siltstone and Mudstone	Ripple, horizontal and wavy lamination. Mottled texture common.	Equal proportion of mudstone and siltstone.	Bed thickness 0 to 1.0 m.	Pedogenic features common. Rooting and burrowing common.	Vertical accretion in overbank/floodplain.

FIG. 6.—Outcrop of a single-storey sandstone body in the Salt Wash Member of the Morrison Formation, Bullfrog Creek, Utah. This photograph depicts multiple, fining-upward units separated by small-scale scour surfaces. This sandstone storey is 8.5 m thick, measured from the bottom of the tape measure up to where geologist is seated. The base of the storey is eroded into heterolithic fine-grained strata, in the dark recessive near bottom of photograph, and the top of the storey is gradational into fine-grained strata exposed in slope above geologist.

Non-channel, fine-grained strata at the top of a storey are composed of two facies: ripple cross-laminated sandstone (Facies 7) and siltstone and mudstone (Facies 8). Ripple forms are not common, but both asymmetric and symmetric types are present. Current ripples were generated during falling stage conditions, while wave ripples most likely formed in residual standing bodies of water after flow had ceased. Siltstone and mudstone (Facies 8) occur as irregular, discontinuous beds that are interstratified with rippled sandstone. Fining-upward rippled sandstone and siltstone units above scour surfaces are interpreted as crevasse channels and flood-plain drainage channels. Overbank and levee mudstone and siltstone formed by successive flooding outside of active channels. Flood-plain deposits accumulated by episodic (seasonal?) progradation of crevasse splays and vertical accretion of mud. Heterolithic strata present as erosional remnants between single-storey and multistorey sandstone bodies are interpreted as overbank/flood-plain and abandoned channel-fill deposits. These two environments have different dimensions due to deposition in different geomorphic settings. Overbank/flood-plain deposits average 9 m thick (range 1–13 m) and 264 m wide (range 10–1500 m; Fig. 10) but become thicker upward in the member. Abandoned channel-fill deposits were interpreted on the photomosaics based on stratal position, lateral extent and overall shape. They average 3.1 m thick (range 1.6–4.5 m) and 86 m wide (range 24–186 m; Fig. 10).

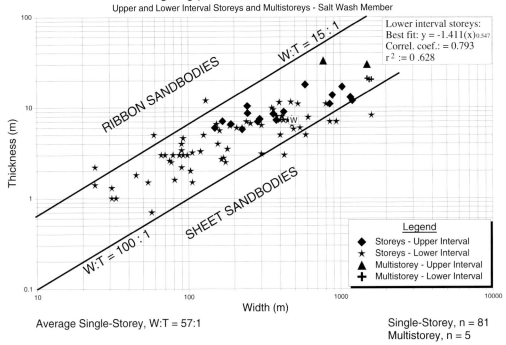

Log/Log plot of Width vs. Thickness
Upper and Lower Interval Storeys and Multistoreys - Salt Wash Member

Lower interval storeys:
Best fit: y = -1.411(x)$_{0.547}$
Correl. coef.: = 0.793
r^2 := 0 .628

W:T = 15 : 1

RIBBON SANDBODIES

SHEET SANDBODIES

W:T = 100 : 1

Legend
◆ Storeys - Upper Interval
★ Storeys - Lower Interval
▲ Multistorey - Upper Interval
✛ Multistorey - Lower Interval

Thickness (m)

Width (m)

Average Single-Storey, W:T = 57:1

Single-Storey, n = 81
Multistorey, n = 5

FIG. 7.—Log/log cross plot of width vs. thickness (W:T) for single-storey and multistorey sandstone bodies in the Salt Wash Member of the Morrison Formation. Data are from photomosaics of cliffs eroded perpendicular to paleoflow (northeastward) in Bullfrog Creek, Shootering Canyon and Lost Springs. Data are plotted with respect to stratigraphic position in the upper or lower interval of the Salt Wash Member. The ''W'' is average width and thickness value (W:T = 71:1) for the Westwater Canyon Member of the Morrison Formation in New Mexico (Cowan, 1991). Sandstone bodies with W:T values >15:1 are defined as sheets, while values < 15:1 are considered to be ribbons (Friend et al., 1979). A W:T = 100:1 is shown for reference purposes.

FIG. 8.—Exposure of pebbly, amalgamated trough cross-bedded sandstone, Bullfrog Creek, Utah. These troughs are 20 cm to 1.0 m thick and up to 5 m wide. Perspective of photograph is oblique to paleocurrent. Hammer handle at lower right is 28 cm long.

A BRAIDED RIVER

Like earlier workers (e.g., Mullens and Freeman, 1957; Peterson, 1984), we interpret the Salt Wash Member in the study area to have been deposited by a braided river complex. Evidence for this interpretation includes: (a) the high sandstone:mudrock ratio; (b) the paucity of ripple lamination in the upper part of fining-upward units; (c) the coarse-grain size of many of the sandstones; (d) a lack of lateral accretion bedding; (e) the sheet-like geometry of the sandstone bodies and (f) the low dispersion of paleocurrent vectors over a thick stratigraphic section. While none of these factors are unique to braided rivers, they are all typical of such systems and collectively argue against either meandering or anastomosed systems. The absence of bedding that is indicative of sediment-gravity flow and the preponderance of stratification types of the lower flow regime argues against a classic alluvial fan as defined by the restrictive terminology of Blair and McPherson (1994).

The facies sequence within the Salt Wash Member shows strong similarities with some of the classic ancient sandy braided systems such as the Devonian Battery Point Sandstone (Cant and Walker, 1976) and the Carboniferous Cannes de Roche Formation (Rust, 1978) of eastern Canada, the Permian Hawkesbury Sandstone of Australia (Conaghan and Jones, 1975), the Carboniferous Rough Rock of England (Bristow, 1993), and the Cretaceous Castlegate Member of the Blackhawk Formation of Utah (Van Wagoner et al., 1990; Miall, 1993). Like those units, the Salt Wash is dominated by trough cross-beds but also has well developed planar-tabular cross-beds. Interestingly, the percentage of the Battery Point succession that consists of trough cross-beds (75%) and planar-tabular cross-beds (16%) is very similar to that of the Salt Wash succession, 79% and 16% respectively. The low spread in vectors of the planar-tabular cross-beds (Fig. 9), however, is interesting in the light of several other studies that have suggested that fluvial planar-tabular beds may show a high variance in dip directions. Smith (1972) suggested that the transverse bars of

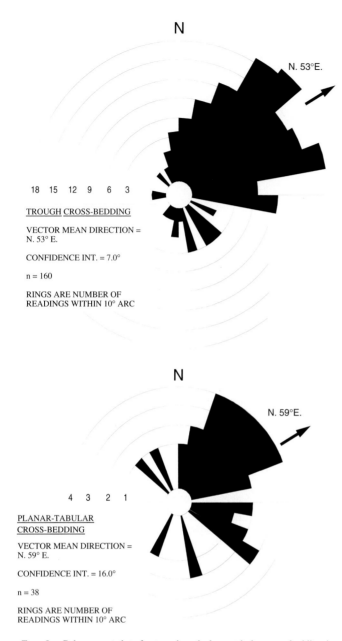

N

N. 53°E.

18 15 12 9   6   3

<u>TROUGH CROSS-BEDDING</u>

VECTOR MEAN DIRECTION =
N. 53° E.

CONFIDENCE INT. = 7.0°

n = 160

RINGS ARE NUMBER OF
READINGS WITHIN 10° ARC

N

N. 59°E.

4   3   2   1

<u>PLANAR-TABULAR
CROSS-BEDDING</u>

VECTOR MEAN DIRECTION =
N. 59° E.

CONFIDENCE INT. = 16.0°

n = 38

RINGS ARE NUMBER OF
READINGS WITHIN 10° ARC

FIG. 9.—Paleocurrent data for trough and planar-tabular cross-bedding in the Salt Wash Member of the Morrison Formation, Henry Mountains area, Utah. Paleocurrent vectors are dominantly northeast and are consistent with data from Jensen (1982) and Peterson (1984).

In braided rivers, the degree of dispersion of cross-beds produced may vary between high and low stages (Jones, 1977). At low stages, many bars are created as a result of the nonuniform and unsteady flow that are associated with flow around bends and between interconnected channels. By contrast the large bedforms formed at high stage, when the braided river becomes a single channel, may be more closely related to the classic uniform/steady flow conditions of the flow regime concept (Middleton and Southard, 1984). These high-stage bedforms may uniformly migrate in a downstream direction. Such major differences in types of bedforms and bedform migration patterns have been documented from the Platte River (Smith, 1971; Blodgett and Stanley, 1980; Crowley, 1983). Banks and Collinson (1974) also suggested that the migration of linguoid bars at high stage in rivers can produce planar-tabular cross-beds with a low variance in dip direction as the bedforms migrate in a downstream direction with the majority of the volume of the bar resulting from build-out of the foresets at the downstream end of the bar. Jones (1977) suggested that the degree of dispersion of preserved cross-beds may be a function of the shape of the falling stage hydrograph, where the amount of lowstand reworking depends on the rate of stage decline.

The low dispersion of Salt Wash cross-beds indicates that there was no extensive low stage modification of bedforms. It seems unlikely that there was relatively even flow in the Salt Wash channels because the oxidized nature of the strata, the incipient caliche and small root traces within the overbank strata, and the paucity of preserved plant material suggest at least a seasonally dry climate. More likely, the discharge of the Salt Wash rivers was flashy with relatively little reworking during low stage and with extensive reworking of any lowstand bedforms during subsequent high stages. The planar-tabular cross-beds probably formed from two-dimensional dunes or linguoid bars that existed coevally with the three-dimensional dunes that created the trough cross-beds.

Single-storey sandbodies in the Salt Wash Member probably represent the deposits of one braided stream whereas the internal small-scale scour surfaces and fining-upward units are interpreted to be the product of repeated episodes of channel cutting, infilling and abandonment within the braided stream. The disorganized nature of the internal scours within the storeys suggests a random pattern of channel development typical of a braided river. The width of storeys is somewhat smaller than that of two well documented modern sandy braided rivers: the 400–1000 m wide South Saskatchewan River, Canada (Cant and Walker, 1978) and the 600–2000 m wide Tana River, Norway (Collinson, 1970).

EVOLUTION OF A FLUVIAL SYSTEM

This section will discuss the regional variations in stratigraphic architecture and discuss how the fluvial architecture of the Salt Wash Member evolved through time. We will suggest how autocyclic and allocyclic controls may have influenced the development of the fluvial system. We provide a regional perspective of the fluvial system and discuss how the facies and facies architecture varied through time within the study area. Finally, we discuss the application of sequence stratigraphic concepts to a fluvial succession, such as the Salt Wash Member, that shows no clear connection to coeval marine strata.

the Platte River produce cross-beds with a high dispersion of azimuths whereas Cant (1978) and Cant and Walker (1978) demonstrated that the cross-channel bars and expansion bars of the South Saskatchewan River often produce planar tabular cross-beds with dip directions that are highly divergent from the overall stream flow direction. Jackson (1976) noted that scroll bars in the meandering Lower Wabash River migrate up onto the point bar, producing planar tabular cross-beds that dip in directions that trend up to 90° from the flow in the main channel. In the braided river strata of the Battery Point Formation, planar-tabular cross-beds commonly dip normal to the flow direction as indicated by other paleocurrent indicators (Cant and Walker, 1976).

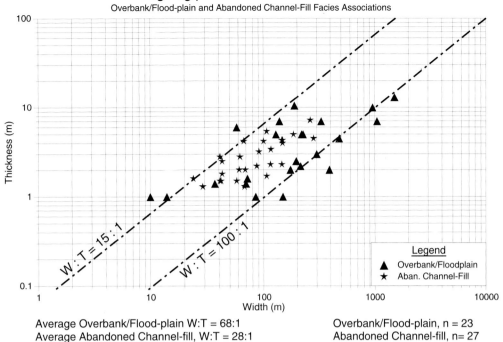

Log/Log plot of Width vs. Thickness
Overbank/Flood-plain and Abandoned Channel-Fill Facies Associations

Average Overbank/Flood-plain W:T = 68:1   Overbank/Flood-plain, n = 23
Average Abandoned Channel-fill, W:T = 28:1   Abandoned Channel-fill, n= 27

FIG. 10.—Log/log cross plot of width vs. thickness (W:T) for overbank/flood-plain and abandoned channel-fill heterolithic units in the Salt Wash Member of the Morrison Formation. Overbank/flood-plain deposits average W:T = 68:1, and abandoned channel-fill deposits average W:T = 28:1. Identification of abandoned channel-fill facies is based on stratigraphic position, lateral extent, and overall shape. Data is from photomosaics of cliffs eroded perpendicular to paleoflow direction (northeastward) in Bullfrog Creek (BF), Shootering Canyon (SC) and Lost Springs (LS). A W:T = 15:1 is shown to illustrate that these deposits have sheet geometries, as defined by Friend et al. (1979). The W:T = 100:1 is shown for reference purposes.

On the basis of isopachs of the member and on the distribution of sandstone:mudrock ratio, early studies (Craig et al., 1955; Mullens and Freeman, 1957; Tyler and Ethridge, 1983) suggested that the Salt Wash Member was deposited as a fan-shaped deposit that was sourced from a single point. The sandy, proximal portion of the Salt Wash Member fluvial system encompasses about 30,000 km^2 and is comparable in areal extent to the so-called alluvial fan of the modern day braided Kosi River of India (Wells and Dorr, 1987; Gohain and Parkash, 1990), and it is tempting to compare the two systems. However, recent studies suggest that the Salt Wash Member was not sourced from a single point (Peterson, 1980, 1988b, 1994). Paleocurrents do not show a radiating pattern, rather, measurements from locations on the Waterpocket Monocline and around the periphery of the Henry Mountains (Peterson, 1980; Peterson and Roylance, 1982; this study) show a uniform paleocurrent towards the northeast. Peterson's (1994) reconstruction of the middle Kimmeridgian paleogeography suggests that Salt Wash-type facies were fed by a multiple sources and extended along the eastern edge of highlands from southern Idaho to central Arizona. Regional paleogeographic reconstructions show the Salt Wash braided fluvial system draining down to an area of widespread small meandering rivers (Tyler and Ethridge, 1983), "distal alluvial plain" and "mud flats" (Peterson, 1986, 1994), and eolian dunes (Kocurek and Dott, 1983). Clearly, the braided rivers were not tributaries of major rivers (as is the Kosi, which feeds into the Ganges) but instead appear to have dispersed out into a broad, arid lowland or ephemeral

lake complex forming a braid delta (definition of McPherson et al., 1987). The situation may have been comparable with that of some of the wide, relatively flat, interior basins of central Asia (such as the Tarim Basin of western China) that have extensive fluvial systems that drain surrounding mountains but where the flow of the rivers does not persist to the lowest part of the basin. Our interpretation of ephemeral flow in the braided rivers is compatible with such a setting. A similar interpretation has been made for the Upper Cretaceous and Paleocene of the Provence Basin in France (Cojan, 1993).

Vertical variations in facies and facies architecture within the Henry Mountain area provide further clues as to the nature of the Salt Wash braided system. Detailed studies (Robinson, 1994) in the Bullfrog–Saleratus Point-Hansen Creek area (Fig. 1) show an upward change in fluvial architecture from thinner, highly amalgamated sandstone bodies near the base of the Salt Wash Member to thicker, more isolated sandstone bodies in the upper part (Fig. 7). Strata of the lower part of the Salt Wash Member are dominantly cliff-forming sandstones (>80%) with only thin mudstone intervals. The upper part of the member is composed of recessive-weathering sandstones (<80%) and has thicker mudstones (Fig. 2). Sixty-four storeys measured in the lower part of the Salt Wash Member have an average thickness of 5.8 m (range 1.0–12.0 m) and width of 343 m (range 31–1600 m; Fig. 7). Seventeen storeys measured in the upper interval average 10.7 m in thickness (range 5.8–18.0 m) and 572 m in width (range 149–1200 m; Fig. 7). Multistorey sandstone bodies are also thicker in the upper interval (average 31.5 m)

of Economic Paleontologists and Mineralogists, Rocky Mountain Section, Rocky Mountain Paleogeography Symposium 2, p. 119–132.

BRIDGE, J. S. AND LEEDER, M. R., 1979, A simulation model of alluvial stratigraphy: Sedimentology, v. 26, p. 617–644.

BRIDGE, J. S. AND MACKEY, S. D., 1993a, A theoretical study of fluvial sandstone body dimensions, *in* Bryant, I. D. and Flint, S. S., eds., The Geological Modelling of Hydrocarbon Reservoirs: Oxford, International Association of Sedimentologists Special Publication 15, Blackwell Scientific Publications, p. 213–236.

BRIDGE, J. S. AND MACKEY, S. D., 1993b, A revised alluvial stratigraphy model, *in* Marzo, M. and Puigdefàbregas, C., eds., Alluvial Sedimentation: Oxford, International Association of Sedimentologists Special Publication 17, Blackwell Scientific Publications, p. 319–336.

BRISTOW, C. S., 1993, Sedimentology of the Rough Rock: a Carboniferous braided river sheet sandstone in northern England, *in* Best, J. L. and Bristow, C. S., eds., Braided Rivers: London, Geological Society of London Special Publication 75, p. 291–304.

BUDDING, M. C., EASTWOOD, K. M., HERWEIJER, J. C., LIVERA, S. E., PAARDEKAM, A. H. M., AND REGTIEN, J. M. M., 1988, Probabalistic modelling of discontinuous reservoirs: Jakarta, Proceedings Indonesian Petroleum Association, 7th Annual Conference, p. 15–24.

CANT, D. J., 1978, Development of a facies model for sandy braided river sedimentation: Comparison of the South Saskatchewan River and the Battery Point Formation, *in* Miall, A.D., ed., Fluvial Sedimentology: Calgary, Canadian Society of Petroleum Geologists Memoir 5, p. 627–640.

CANT, D. J. AND WALKER, R. G., 1976, Development of a braided fluvial facies model for the Devonian Battery Point Sandstone, Quebec: Canadian Journal of Earth Science, v. 13, p. 102–119.

CANT, D. J. AND WALKER, R. G., 1978, Fluvial processes and facies sequences in the sandy braided South Saskatchewan River, Canada: Sedimentology, v. 25, p. 625–648.

COJAN, I., 1993, Alternating fluvial and lacustrine sedimentation: tectonic and climatic controls (Provence basin, S. France, Upper Cretaceous/Paleocene), *in* Marzo, M. and Puigdefábregas, C., eds., Alluvial Sedimentation: Oxford, International Association of Sedimentologists Special Publication 17, Blackwell Scientific Publications, p. 425–438.

COLLINSON, J. D., 1970, Bed forms of Tana River, Norway: Geographica Annaler, v. 52A, p. 31–55.

CONAGHAN, P. J. AND JONES, J. G., 1975, The Hawkesbury Sandstone and the Brahmaputra: a depositional model for continental sheet sandstones: Journal of the Geological Society of Australia, v. 22, p. 275–283.

COWAN, E. J., 1991, The large-scale architecture of the fluvial Westwater Canyon Member, Morrison Formation (Upper Jurassic), San Juan basin, New Mexico, *in* Miall, A. D. and Tyler, N., eds., The Three-dimensional Facies Architecture of Terrigenous Clastic Sediments and its Implication for Hydrocarbon Discovery and Recovery: Tulsa, Society of Economic Paleontologists and Mineralogists Concepts in Sedimentology and Paleontology 3, p. 80–93.

CRAIG, L. C., HOLMES, C. N., CADIGAN, R. A., FREEMAN, V. L., MULLENS, T. E., AND WEIR, G. W., 1955, Stratigraphy of the Morrison and related formations, Colorado Plateau region, a preliminary report: United States Geological Survey Bulletin 1009–E, p. 125–168.

CRAIG, L. C., HOLMES, C. N., FREEMAN, V. L., MULLENS, T. E., AND OTHERS, 1959, Measured sections of the Morrison and adjacent formations: United States Geological Survey Open File Report 485, 231 p.

CROWLEY, K. D., 1983, Large scale bed configurations (macroforms) Platte River Basin, Colorado and Nebraska: primary structures and formative processes: Geological Society of America Bulletin, v. 94, p. 117–133.

DICKINSON, W. R., 1976, Sedimentary basins developed during evolution of Mesozoic–Cenozoic arc-trench system in western North America: Canadian Journal of Earth Science, v. 13, p. 1268–1287.

DICKINSON, W. R., 1981, Plate tectonics and the continental margin of California, *in* Ernst, W. G., ed., Geotectonic Development of California: Englewood Cliffs, Prentice Hall, p. 1–28.

DUBIEL, R. F., 1994, Triassic deposystems, paleogeography, and paleoclimate of the Western Interior, *in* Caputo, M., Peterson, J. A., and Franczyk, K.J., eds., Mesozoic Systems of the Rocky Mountain Region, USA: Denver, Society of Economic Paleontologists and Mineralogists, Rocky Mountain Section, p. 133–168.

FIELDING, C. R. AND CRANE, R. C., 1987, An application of statistical modelling to the prediction of hydrocarbon recovery factors in fluvial sequences, *in* Ethridge, F. G., Flores, R. M., and Harvey, M. D., eds., Recent Developments in Fluvial Sedimentology: Tulsa, Society of Economic Paleontologists and Mineralogists Special Publication 39, p. 321–327.

FRIEND, P. F., SLATER, M. J., AND WILLIAMS, R. C., 1979, Vertical and lateral building of river sandstone bodies, Ebro basin, Spain: The Geological Society of London, v. 136, p. 39–46.

GEEHAN, G. W., LAWTON, T. F., SAKURAI, S., KLOB, H., CLIFTON, T. R., INMAN, K. F., AND NITZBERG, K. E., 1986, Geologic prediction of shale continuity, Prudhoe Bay Field, *in* Lake, L. W., and Carroll, H. B., eds., Reservoir Characterization: Orlando, Academic Press, p. 63–82.

GOHAIN, K. AND PARKASH, B., 1990, Morphology of the Kosi megafan, *in* Rachocki, A. H. and Church, M., eds., Alluvial Fans: A Field Approach: New York, John Wiley and Sons, Ltd., p. 151–178.

HENRIQUEZ, A., TYLER, K. J., AND HURST, A., 1990, Characterization of fluvial sedimentology for reservoir simulation modeling: Society of Petroleum Engineers Formation Evaluation, v. 5, p. 211–216.

HUNT, C. B., AVERITT, P., AND MILLER, R. L., 1953, Geology and geography of the Henry Mountains region: Washington D.C., United States Geological Survey Professional Paper 228, 234 p.

JACKSON, R. G., II, 1976, Depositional model of point bars in the Lower Wabash River of Illinois and Indiana: Journal of Sedimentary Petrology, v. 46, p. 579–594.

JAMISON, H. C., BROCKETT, L. D., AND MCINTOSH, R. A., 1980, Prudhoe Bay: A 10–year perspective, *in* Halbouty, M. T., ed., Giant Oil and Gas Fields of the Decade 1968–1978: Tulsa, American Association of Petroleum Geologists Memoir 30, p. 289–314.

JENSEN, T., 1982, Petrology of the Salt Wash Member of the Morrison Formation, Henry basin, Utah: Unpublished Master's Thesis, Northern Arizona University, Flagstaff, 82 p.

JONES, C. M., 1977, Effects of varying discharge regimes on bedform sedimentary structures in modern rivers: Geology, v. 5, p. 567–570.

KIRSCHBAUM, M. A. AND MCCABE, P. J., 1992, Controls on the accumulation of coal and the development of anastomosed fluvial systems in the Cretaceous Dakota Formation of southern Utah: Sedimentology, v. 39, p. 581–598.

KOCUREK, G. AND DOTT, R. H., JR., 1983, Jurassic paleogeography and paleoclimate of the central and southern Rocky Mountains region, *in* Reynolds, M. W. and Dolly, E. D., eds., Mesozoic Paleogeography of the West-central United States: Denver, Society of Economic Paleontologists and Mineralogists, Rocky Mountain Section, Rocky Mountain Paleogeography Symposium 2, p. 101–116.

KOWALLIS, B. J. AND HEATON, J. S., 1987, Fission-track dating of bentonites and bentonitic mudstones from the Morrison Formation in central Utah: Geology, v. 15, p. 1138–1142.

LAWRENCE, D. A. AND WILLIAMS, B. P. J., 1987, Evolution of drainage systems in response to Acadian deformation: the Devonian Battery Point Formation, eastern Canada, *in* Ethridge, F. G., Flores, R. M., and Harvey, M. D., eds., Recent Developments in Fluvial Sedimentology: Tulsa, Society of Economic Paleontologists and Mineralogists Special Publication 39, p. 287–300.

LAWTON, T. F., GEEHAN, G., AND VOORHEES, B. J., 1987, Lithofacies and depositional environments of the Ivishak Formation, Prudhoe Bay Field, *in* Tailleur, I. and Weimer, P., eds., Alaska North Slope Geology: Bakersfield, Society of Economic Paleontologist and Mineralogists, Pacific Section, and Alaskan Geological Society, v. 1, p. 61–76.

LOCKLEY, M., HUNT, A., CONRAD, K., AND ROBINSON, J., 1992, Tracking dinosaurs and other extinct animals at Lake Powell: Park Science, v. 12, no. 3, p. 16–17.

LUCCHITTA, I., 1979, Late Cenozoic uplift of the southwestern Colorado Plateau and adjacent lower Colorado River region: Tectonophysics, v. 61, p. 63–95.

LUPTON, C. T., 1914, Oil and gas near Green River, Grand County, Utah: United States Geological Survey Bulletin 541, p. 115–133.

MACDONALD, A.C., AND HALLAND, E. K., 1993, Sedimentology and shale modeling of a sandstone-rich fluvial reservoir: Upper Stratfjord Formation, Statfjord Field, northern North Sea: American Association of Petroleum Geologists Bulletin, v. 77, p. 1016–1040.

MARZO, M., NIJMAN, W., AND PUIGDEFÁBREGAS, C., 1988, Architecture of the Castissent fluvial sheet sandstones, Eocene, south Pyrenees, Spain: Sedimentology, v. 35, p. 719–738.

MCCABE, P. J., 1991, Tectonic controls on coal accummulation: Bulletin de la Société Géologique de France, 8e Série, v. 162, p. 277–282.

MCPHERSON, J. G., SHANMUGAM, G., AND MOIOLA, R. J., 1987, Fan-deltas and braid deltas: Geological Society of America Bulletin, v. 99, p. 331–340.

MIALL, A. D., 1993, The architecture of fluvial-deltaic sequences in the Upper Mesaverde Group (Upper Cretaceous), Book Cliffs, Utah, *in* Best, J. L. and Bristow, C. S., eds., Braided Rivers: London, Geological Society of London Special Publication 75, p. 305–332.

MIALL, A. D. AND TYLER, N., eds., 1991, The Three-dimensional Facies Architecture of Terrigenous Clastic Sediments and its Implication for Hydrocarbon Discovery and Recovery: Tulsa, Society of Economic Paleontologists and Mineralogists, Concepts in Sedimentology and Paleontology 3, 309 p.

MIDDLETON, G. V. AND SOUTHARD, J. B., 1984, Mechanics of Sediment Movement: Binghamton, Society of Economic Paleontologists and Mineralogists, Eastern Section, Short Course 3, 401 p.

MULLENS, T. E. AND FREEMAN, V. L., 1957, Lithofacies of the Salt Wash Member of the Morrison Formation, Colorado Plateau: Geological Society of America Bulletin, v. 68, p. 505–528.

NELSON, S. T., DAVIDSON, J. P., AND SULLIVAN, K. R., 1992, New age determinations of central Colorado Plateau laccoliths, Utah: Recognizing disturbed K–Ar systematics and re-evaluating tectonomagmatic relationships: Geological Society of America Bulletin, v. 104, p. 1547–1560.

O'SULLIVAN, R. B., 1981, The Middle Jurassic San Rafael Group and related rocks in east-central Utah, *in* Epis, R. C. and Callender, J., eds., Western Slope, Colorado: Albuquerque, New Mexico Geological Society, 32nd Field Conference and Guidebook, p. 89–95.

PETERSON, F., 1978, Measured sections of the lower member and Salt Wash Member of the Morrison Formation (Upper Jurassic) in the Henry Mountains mineral belt of southern Utah: Washington D.C., United States Geological Survey Open–File Report 78–1094, 95 p.

PETERSON, F., 1980, Sedimentology of the uranium-bearing Salt Wash Member and Tidwell unit of the Morrison Formation in the Henry and Kaiparowits basins, Utah: Utah Geological Association Publication 8, Henry Mountains Symposium, p. 305–322.

PETERSON, F., 1984, Fluvial sedimentation on a quivering craton: Influence of slight crustal movements on fluvial processes, Upper Jurassic Morrison Formation, western Colorado Plateau: Sedimentary Geology, v. 38, p. 21–49.

PETERSON, F., 1986, Jurassic paleotectonics in the western part of the Colorado Plateau, Utah and Arizona, *in* Peterson, J. A., ed., Paleotectonics and Sedimentation in the Rocky Mountain Region, United States: Tulsa, American Association of Petroleum Geologists Memoir 41, p. 563–596.

PETERSON, F., 1987, The search for source areas of Morrison (Upper Jurassic) clastics on the Colorado Plateau: Geological Society of America Abstracts with Programs, v. 19, no. 7, p. 804.

PETERSON, F., 1988a, Revisions to stratigraphic nomenclature of Jurassic and Cretaceous rocks of the Colorado Plateau: United States Geological Survey Bulletin 1633–B, p. 17–56.

PETERSON, F., 1988b, A synthesis of the Jurassic system in the southern Rocky Mountain region, *in* Sloss, L. L., ed., Sedimentary Cover—North American Craton; U.S.: Boulder, Geological Society of America, The Geology of North America, v. D–2, p. 65–76.

PETERSON, F., 1994, Sand dunes, sabkhas, streams and shallow seas: Jurassic paleogeography in the southern part of the Western Interior basin, *in* Caputo, M. V., Peterson, J. A., and Franczyk, K. J., eds., Mesozoic Systems of the Rocky Mountain Region, U.S.A.: Denver, Rocky Mountain Section, SEPM (Society for Sedimentary Geology), p. 233–272.

PETERSON, L. M. AND ROYLANCE, M. M., 1982, Stratigraphy and depositional environments of the Upper Jurassic Morrison Formation near Capitol Reef National Park, Utah: Brigham Young University Studies in Geology, v. 29, p. 1–12.

POTTER, P. E., 1967, Sand bodies and sedimentary environments— a review: American Association of Petroleum Geologists Bulletin, v. 51, p. 337–365.

POULTON, T. P., 1984, The Jurassic of the Canadian Western Interior, from 49°N latitude to Beaufort Sea, *in* Stott, D. F. and Glass, D. J., eds., The Mesozoic of Middle North America: Calgary, Canadian Society of Petroleum Geologists Memoir 9, p. 15–41.

PUIGDEFÁBREGAS, C., 1993, Controls on fluvial sequence architecture, *in* Yu, B. and Fielding, C. R., eds., Modern and Ancient Rivers and their Importance

to Mankind: Brisbane, 5th International Conference on Fluvial Sedimentology, Conference Proceedings, p. K42–K48.

ROBINSON, J. W., 1994, Facies architecture of the Salt Wash Member of the Morrison Formation in southern Utah: a model for reservoir heterogeneity in fluvial strata: Unpublished Ph.D. Thesis, Colorado School of Mines, Golden, 321 p.

RUST, B. R., 1978, Depositional models for braided alluvium, *in* Miall, A. D., ed., Fluvial Sedimentology: Calgary, Canadian Society of Petroleum Geologists Memoir 5, p. 605–625.

SADLER, S. P. AND KELLY, S. B., 1993, Fluvial processes and cyclicity in terminal fan deposits: an example from the Late Devonian of southwest Ireland: Sedimentary Geology, v. 85, p. 375–386.

SHANLEY, K. W. AND MCCABE, P. J., 1991, Predicting facies architecture through sequence stratigraphy— an example from the Kaiparowits Plateau, Utah: Geology, v. 19, p. 742–745.

SHANLEY, K. W. AND MCCABE, P. J., 1993, Alluvial architecture in a sequence stratigraphic framework: a case history from the Upper Cretaceous of southern Utah, USA, *in* Bryant, I. D. and Flint, S. S., eds., The Geological Modelling of Hydrocarbon Reservoirs: Oxford, International Association of Sedimentologists Special Publication 15, Blackwell Scientific Publications, p. 21–56.

SHANLEY, K. W. AND MCCABE, P. J., 1994, Perspectives on the sequence stratigraphy of continental strata: American Association of Petroleum Geologists Bulletin, v. 78, p. 544–568.

SHUSTER, M. W. AND STEIDTMANN, J. R., 1987, Fluvial-sandstone architecture and thrust-induced subsidence, northern Green River basin, Wyoming, *in* Ethridge, F. G., Flores, R. M., and Harvey, M. D., eds., Recent Developments in Fluvial Sedimentology: Tulsa, Society of Economic Paleontologists and Mineralogists Special Publication 39, p. 279–285.

SMITH, N. D., 1971, Transverse bars and braiding in the lower Platte River, Nebraska: Geological Society of America Bulletin, v. 82, p. 3407–3420.

SMITH, N. D., 1972, Some sedimentological aspects of planar cross-stratification in a sandy braided river: Journal of Sedimentary Petrology, v. 42, p. 624–634.

STEEL, R. J. AND AASHEIM, S. M., 1978, Alluvial sand deposition in a rapidly subsiding basin (Devonian, Norway), *in* Miall, A. D., ed., Fluvial Sedimentology: Calgary, Canadian Society of Petroleum Geologists Memoir 5, p. 385–412.

STEEL, R. J., MÆHLE, S., NILSØN, J., RØE, S. L., SPINNANGR, A., 1977, Coarsening-upward cycles in the alluvium of Hornelen basin (Devonian) Norway: Sedimentary response to tectonic events: Geological Society of America Bulletin, v. 88, p. 1124–1134.

TURNER, C. E. AND FISHMAN, N. S., 1991, Jurassic Lake T'oo'dichi': A large alkaline, saline lake, Morrison Formation, eastern Colorado Plateau: Geological Society of America Bulletin, v. 103, p. 538–558.

TYLER, N. AND ETHRIDGE, F. G., 1983, Depositional setting of the Salt Wash Member of the Morrison Formation, southwest Colorado: Journal of Sedimentary Petrology, v. 53, p. 67–82.

ULMISHEK, G. AND HARRISON, W., 1981, Uzen development gives new insight into projecting future Soviet oil output: Oil and Gas Journal, August 24, p. 148–154.

VAN FOSSEN, M. C. AND KENT, D. V., 1992, Paleomagnetism of the Front Range (Colorado) Morrison Formation and an alternative model of Late Jurassic North American apparent polar wander: Geology, v. 20, p. 223–226.

VAN WAGONER, J. C., MITCHUM, R. M., CAMPION, K. M., AND RAHMANIAN, V. D., 1990, Siliciclastic sequence stratigraphy in well logs, cores, and outcrops: American Association of Petroleum Geologists Method in Exploration Series 7, 55 p.

WELLS, N. A. AND DORR, J. A., JR., 1987, A reconnaissance of sedimentation of the Kosi alluvial fan of India, *in* Ethridge, F. G., Flores, R. M., and Harvey, M. D., eds., Recent Developments in Fluvial Sedimentology: Society of Economic Paleontologists and Mineralogists Special Publication 39, p. 51–61.

SEDIMENTOLOGICAL EVOLUTION OF A FAULT–CONTROLLED EARLY PALEOCENE INCISED–VALLEY SYSTEM, NUUSSUAQ BASIN, WEST GREENLAND

GREGERS DAM AND MARTIN SØNDERHOLM

The Geological Survey of Denmark and Greenland, Thoravej 8, DK-2400, Copenhagen NV, Denmark

ABSTRACT: The Nuussuaq Basin of West Greenland is a Late Cretaceous–early Tertiary extensional basin, related to the opening of the Labrador Sea and Baffin Bay. An extensive Early Paleocene valley system was incised following a period of active faulting and uplift of the basin. The paleovalley system is exposed along the south coast of the Nuussuaq peninsula and includes five broadly synchronous valleys that embrace sandstones assigned to the Quikavsak Member. The valleys are 1–2 km wide and up to 190 m deep. The valleys were incised during a major relative sea-level fall and were cut into Cretaceous and earliest Paleocene sediments. The valley-fill deposits represent a transgressive systems tract comprising a uniform succession of fluvial sandstones attributed to a very rapid rise in relative sea level. Where the relative sea-level rise exceeded sediment supply, the increased accommodation space caused the valleys to turn into estuaries. A mudstone separating the fluvial and estuarine sandstones was deposited in the mid-estuary funnel. A sharp boundary between the estuary-mouth sandstone and the mid-estuary funnel mudstone probably formed as the ebb tidal delta retreated into the estuary, reflecting a major transgressive erosional surface. The fluvial and estuarine valley-fill sandstones are succeeded abruptly by shelf mudstones.

INTRODUCTION

Incised valleys are commonly associated with sequence boundaries (Van Wagoner et al., 1990), although there is a Recent example of incision caused by tectonic uplifting of river headwaters (Leckie, 1994). Most studies describe Quaternary examples of valley incision along passive margins, attributed to fall in relative sea level (e.g., Suter and Berryhill, 1985; Suter et al., 1987; Coleman and Mixon, 1988; Allen and Posamentier, 1993; Ashley and Sheridan, 1994; Belknap et al., 1994; Clifton, 1994; Kindinger et al., 1994; Thomas and Anderson, 1994). Most ancient examples of incised valleys are based on seismic and well-log data (e.g., Keith et al., 1988; Ranger and Pemberton, 1988; Smith, 1988; Weimer and Sonnenberg, 1989; Krystinik and Blakeney-DeJarnett, 1990; Van Wagoner et al., 1990; Allen and Posamentier, 1991; Boreen and Walker, 1991; Mattison, 1992; Wood and Hopkins, 1992; MacEachern and Pemberton, 1994). Outcrop studies documenting the relationship to alluvial and coastal systems are well-known (e.g., Bluck, 1974; Evanoff, 1990; Shanley et al., 1992; Shanley and McCabe, 1993; Gibling and Wightman, 1994), but there are relatively few outcrop studies of ancient incised valleys cutting down into shelf deposits (e.g., Land and Weimer, 1978; Baum and Vail, 1988; Archer et al., 1994; Dyson and Borch, 1994; Kvale and Barnhill, 1994; Martinsen, 1994). In experimental studies, Wood et al. (1993, 1994) showed the importance of incised valleys as morphologic features on shelves during base-level falls. This scarcity of ancient outcrops seems therefore mainly to be a result of limited and poor exposures preventing the recognition of critical criteria.

Continuously exposed outcrops of three Late Maastrichtian–Early Paleocene incised-valley systems occur along the south coast of Nuussuaq. The Early Paleocene valleys of the Quikavsak Member resemble pearls on a string and are a well-exposed and spectacular example of an obviously tectonically controlled incised-valley system. The aim of this paper is to describe the origin, morphology and filling processes of one of the Early Paleocene valley systems.

REGIONAL SETTING AND STRATIGRAPHY

The margin of West Greenland was formed by extensional opening of the Labrador Sea in late Mesozoic–early Cenozoic time. A complex of linked rift basins stretch from the Labrador Sea to northern Baffin Bay recording a scissor-like opening (Rolle, 1985; Chalmers, 1991; Chalmers and Pulvertaft, 1993; Chalmers et al., 1993). A conspicuous element of this tectonic framework is the Ungava transform fault system. Disposal of the movement along this left-lateral wrench system was accomodated by numerous splays. Locally restraining bends along these splays have formed compressional structures (Chalmers et al., 1995).

Extensional subsidence of the Nuussuaq Basin is believed to have been intermittent, as evidenced by unconformable sequence boundaries. A period of basin uplift occurred in the mid-Paleocene resulting in major valley incision and reorganisation of dispersal patterns. This period of basin uplift and valley incision was followed by rapid subsidence and voluminous extrusion of hayaloclastites and lavas (Pedersen and Dueholm, 1992; Pedersen, 1993; Dam et al., 1998).

In the Labrador Sea–Baffin Bay region, Mesozoic–Early Tertiary sediments are only exposed at Cape Dyer on Baffin Island (Burden and Langille, 1990) and in West Greenland (Rosenkrantz, 1970; Fig. 1). The sedimentary succession in West Greenland is in places 6–8 km (Christiansen et al., 1995) and ranges in age from Early Cretaceous (Albian) to Early Tertiary (Paleocene; Pulvertaft, 1987). The outcrops are bounded to the east by Precambrian basement rocks against which Cretaceous sediments have a faulted contact (Rosenkrantz and Pulvertaft, 1969; Pedersen and Pulvertaft, 1992; Fig. 1).

The Lower Paleocene Quikavsak Member is the basal member of the Upper Atanikerdluk Formation (Koch, 1959; Fig. 2). The member rests unconformably upon the Albian–Campanian Atane Formation and the Upper Maastrichtian–Lower Paleocene Kangilia Formation and is succeeded by marine mudstones. Macroplant evidence suggests an Early Paleocene age for the Quikavsak Member (Kock, 1959). The dinoflagellate cyst assemblage indicates that the uppermost part of the member and the mudstones above have a latest Early to Late Paleocene age (marine nannoplankton zones NP 4–7; H. Nøhr-Hansen, pers. commun., 1993). The marine mudstone above the Quikavsak Member appears to be regionally widespread and marks the drowning of the valley system. Filling of the valley system and deposition of the marine mudstone were followed by eruption of picritic hyaloclastite breccias building up giant Gilbert-type delta structures with foresets up to 700 m high indicating continued subsidence

Relative Role of Eustasy, Climate, and Tectonism in Continental Rocks, SEPM Special Publication No. 59

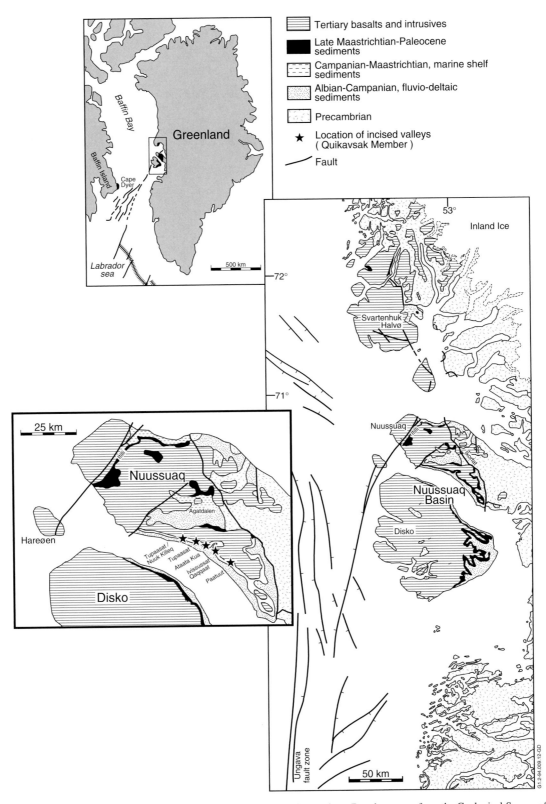

FIG. 1.—Geological map of central West Greenland showing location of studied sections. Based on maps from the Geological Survey of Greenland.

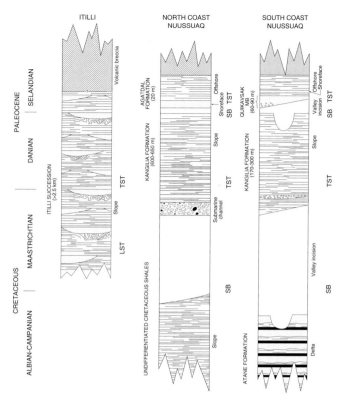

FIG. 2.—Stratigraphy of the Cretaceous–Lower Paleocene sedimentary succession of western, northern and southern Nuussuaq, central West Greenland. SB–Sequence boundary; LST–Lowstand systems tract; TST–Transgressive systems tract.

after valley filling (Pedersen and Dueholm, 1992; Pedersen, 1993). On the basis of macroplant fossils, the Quikavsak Member has been directly correlated with the marine beds of the Agatdalen Formation in central Nuussuaq and fluvial deposits corresponding to the Quikavsak Member are also known from the Disko island (Koch, 1959, 1964; Fig. 1). A new nannoplanctonic collection indicates that the first stage of magmatic activity started close to the Lower and Upper Paleocene boundary (NP4–5; T. Jürgensen and L. R. Wilken, pers. commun., 1995), suggesting that a correlation of the Quikavsak Member with the upper part of the Itilli turbidite succession of Dam and Sønderholm (1994) is possible (Fig. 2). This is confirmed by new ^{40}Ar/^{39}Ar datings suggesting that the bulk of these onshore volcanics were erupted during a short time interval in the Late Paleocene (61–59.5 Ma; Storey et al., 1998).

The sediments below the Quikavsak Member record an episode of faulting before valley excavation took place. Fault displacement of several hundred meters has been recorded (Rosenkrantz and Pulvertaft, 1969). Evidence of this tectonic event and of a substantial Late Mesozoic–Early Paleocene unconformity has also been recorded from offshore West Greenland (Bate et al., 1995). An episode of Early Paleocene extensional tectonism and valley excavation prior to eruption of basalts has also been recorded from the Cape Searle Formation on southeastern Baffin Island (Burden and Langille, 1990), the equivalent deposits to the Quikavsak Member on the western side of Baffin Bay.

SEDIMENTOLOGY OF INCISED VALLEY FILL

Valley Morphology

The Quikavsak Member embraces fills of five incised paleovalleys exposed along the south coast of Nuussuaq from Paatuut to Nuuk Killeq (Fig. 1). These rocks were first interpreted as palaeovalley fills by Koch (1959). The incised valleys are either broad U–shaped (Ivissussat Qaqqaat, Ataata Kuua, Tupaasat and Tupaasat/Nuuk Killeq) or less commonly have a "Texas-longhorn" geometry (Paatuut; Fig. 3). The paleovalleys are 1–2 km wide and 60–190 m deep (Fig. 4). The valley walls show large variations in slope and may be nearly vertical (80°) in the U–shaped valleys. The paleovalleys are exposed only in transverse section, but paleocurrent directions show considerable variability from valley to valley, suggesting slightly sinuous geometries.

The valleys were cut into deltaic and shelf sandstones, mudstones and paraconglomerates of the Atane and Kangilia Formations or are erosively stacked on older valley deposits (at Tupaasat/Nuuk Killeq). Locally, the valley floors have a step-like form reflecting variable footwall lithologies of cohesive mudstones and less cohesive sandstones and paraconglomerates.

Basal Conglomerate

Conglomerates overlie the erosive valley floors at Ivissussat Qaqqaat and Ataata Kuua. They consist mainly of (>90%) sub-

Fig. 3.—Incised valley from Paatuut at the south coast of Nuussuaq (see Fig. 1 for location). The valley is 190 m deep and cuts down into Lower Cretaceous deltaic deposits of the Atane Formation. Arrow is placed at mid-estuarine mudstone.

rounded pebble- to boulder-size gneiss clasts up to 70 cm in diameter, with subordinate mudstone clasts (Fig. 5). The basal conglomerates range in thickness from a single clast at Ataata Kuua to more than 5 m in the central part of the Ivissussat Qaqqaat paleovalley. The conglomerates are clast supported and the matrix is composed of poorly sorted pebbly coarse sand. The clast-supported conglomerates are ungraded, but commonly display a-axis imbrication (Fig. 5B).

The gneiss conglomerates probably originated by erosion of paraconglomerates of the underlying Kangilia Formation. The one clast thick conglomerates along the valley floor in Ataata Kua are interpreted as a lag deposit, while the thicker beds in Ivissussat Qaqqaat are interpreted as longitudinal bars (cf. Miall, 1978).

Fluviatile Sandstone

Nearly all the paleovalley-fill deposits consist of fluvial sandstones, except for the uppermost part of the Paatuut valley (Fig. 6). The fluviatile deposits range from 65 m to more than 120 m in thickness and consist of monotonous successions of well-sorted dominantly planar and subordinate trough cross-bedded medium- to very-coarse-grained pebbly sandstone (Figs. 4, 6). Set thickness varies from 0.1–3 m (average 60 cm). Foresets are generally angular or tangential, less commonly sigmoidal and convex-upward. Foreset dip is 18–34°, and coal spar com-

monly drapes the avalanche foresets. Preserved topsets show cross-lamination typical of ripple migration (Fig. 7). Extraformational gneiss and quartz pebbles are generally less than 2 cm in diameter. Mudstone clasts as large as 1 m, and logs are common.

Penecontemporaneous deformation structures are widespread in some horizons or locally occur in single cross-bedded sets (Fig. 8). An intriguing characteristic of the fluvial valley fill successions is the absence of repetitive genetic units such as stacked fining-upward successions. The only vertical trend is an overall upward decrease in grain size, clast size and set thickness (Fig. 4).

The foresets dip unimodally within each valley fill toward the west-southwest, west and northwest (Fig. 9).

This unimodal cross-bed distribution, together with the absence of marine fossils or any evidence of tidal current processes (e.g., clay drapes, flasers, bundles-wise upbuilding of foresets, bimodal foreset orientation of cross-beds), suggest that these sandstones were deposited in a fluvial environment or in the fluvial-dominated innermost part of an estuary. The cross-bedded sandstones were deposited from fields of subaqueous linear and sinuous dunes migrating along the valley floor. The low-angle sigmoidal and convex-upward foresets suggest deposition from bedforms formed in the dune to upper-stage plane-bed transition (e.g., Saunderson and Lockett, 1983; Røe, 1987). The convolute bedding is attributed to rapid accumulation and

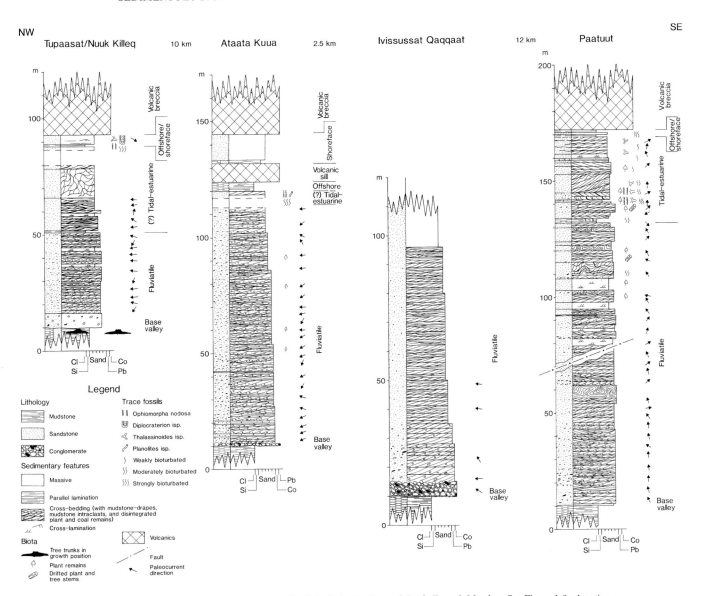

FIG. 4.—Sedimentological logs through each of the incised valleys of the Quikavsak Member. See Figure 1 for location.

FIG. 5.—(A) Gneiss conglomerate from the base of the Ivissussat Qaqqaat valley (see Fig. 1 for location; person for scale). (B) Gneiss conglomerate from the base of the Ivissussat Qaqqaat valley, showing a dominantly a-axis imbrication (see Fig. 1 for location; hammer in lower right corner for scale (arrow)).

FIG. 6.—Uniform cross-bedded fluvial sandstone without any repeated facies successions. Person for scale. From Paatuut (see Fig. 1 for location).

dewatering of the sands (cf. Lowe, 1975). The overall uniform composition of the well-sorted, cross-bedded fluvial sandstones and the absence of stacked channel units suggest that a single flow occupied the full width of the valley without segregation into multiple channels. It is, however, difficult to explain these uniform sandstones by normal fluvial processes alone. The monotoneous paleovalley fills showing no indication of down-cutting in previous deposited sediments are therefore suggested to record very rapid or possibly catastrophical vertical aggradation, caused by an over supply of arenaceous sediments.

Tidal-estuarine Sandstone

The fluvial sandstone at Paatuut is overlain by estuarine deposits which make up the uppermost 26 m of the Quikavsak Member at this locality. The estuarine fill comprises four lithofacies: bioturbated sandstone, cross-bedded sandstone, megascale cross-bedded sandstone and a single mudstone bed.

FIG. 8.—Soft-sediment folds in cross-bedded sandstone. Hammer for scale. From Paatuut (see Fig. 1 for location).

FIG. 7.—Fluvial cross-bedded sandstone with preserved cross-laminated topset. Hammer for scale. From Paatuut (see Fig. 1 for location).

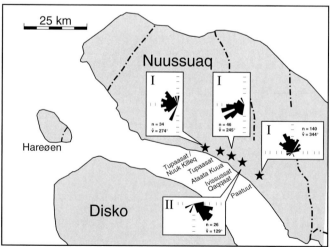

FIG. 9.—Rose diagrams showing: I) paleocurrent directions based on foreset dip directions of cross-bedded sandstones; II) a-axis imbrication of gneiss clasts indicate paleocurrents towards northwest. See Figure 1 for locations.

The mudstone bed occurs at the base of the estuarine unit (Figs. 3, 4, 10). It is about one meter thick and is laterally continuous (Fig. 3). It contains marine dinoflagellates (H. Nøhr-Hansen, pers. commun., 1993) and some plant fossils (Koch, 1959). The mudstone is erosionally overlain by cross-bedded sandstone, showing planar to trough cross-bedding similar in style to that of the fluvial sandstone below. Average set thickness is 20 cm (range 0.1–1.6 m). Foresets are tangential, and foreset dip is 20°–34°. Structures typical of tidal activity (double mudstone-drapes, reactivation surfaces draped by mudstone clasts and local bidirectional cross-beds), convolute bedding, and tree trunks are common. The sets may be separated by mudstone drapes in the upper part of the unit. *Ophiomorpha nodosa* and *Thalassinoides* isp. are common (Fig. 11), and marine mollusks, including oysters, have been found in places (Koch, 1959).

The mega-scale cross-bedded medium- to coarse-grained sandstone includes a solitary 7-m-thick, tabular low-angle cross-bedded set, which is continuous for at least 300 m perpendicular to the valley axis (Fig. 12). Foresets are tangential and dip towards north, parallel to the valley axis. The foresets contain smaller scale intrasets with a unimodal northward foreset dip direction, similar to that of the master bedding. Convolute bedding is present locally. Toesets are bioturbated with *Ophiomorpha nodosa* and *Thalassinoides* isp. The upper boundary is sharp and erosional and is succeeded by cross-bedded sandstone.

The indications of bipolar paleocurrent orientations, the presence of cross-sets with double mudstone-drapes, abundant *Ophiomorpha nodosa*, together with marine dinoflagellates in the mudstones indicate a high-energy, shallow subtidal to intertidal environment. The cross-bedded sandstones were deposited from small- and medium-scale dunes. The general northward paleocurrent directions are similar to those of the underlying fluvial deposits, suggesting an ebb-dominated tidal current regime. The mega-foresets probably formed by a northward progradation of a terminal lobe at the mouth of an ebb-dominated estuary.

FIG. 11.—(A) *Ophiomorpha nodosa* from strongly bioturbated estuarine sandstone. Pen for scale. (B) Sandstone beds strongly bioturbated with *Ophiomorpha nodosa* at the base of megascale cross-bedded sandstone (arrows). Encircled hammer for scale. From Paatuut (see Fig. 1 for location).

FIG. 10.—Mid-estuary funnel mudstone. Bed is 1 m thick. From Paatuut (see Figs. 1 and 3 for location).

Upper Boundary/Marine Mudstone

The incised valley fill sandstones are blanketed by a widespread marine mudstone (Fig. 13) marking the final inundation of the valley fill. The dinoflagellate cyst assemblage of this transgressive mudstone dates the drowning of the valleys as early Late Paleocene (NP5–6; Piasecki et al., 1992).

DEPOSITIONAL ENVIRONMENT

The five incised valleys now exposed along the south coast of Nuussuaq were formed synchronously just after an episode of Early Paleocene faulting. All the paleovalleys cross-cut faults within the Cretaceous and Lower Paleocene deltaic and marine sediments of the Atane and Kangilia Formations suggesting that faulting associated with uplift of the basin may have controlled not only the establishment but also the trend of the valleys. The exact downstream extension of the valleys is unknown. Similar correlation between faults (structural lows) and valleys have been described from the Lower Cretaceous Muddy Sandstone in the Rocky Mountain region, USA (Wei-

FIG. 12.—Megascale cross-bedded sandstone deposited from a terminal lobe at the mouth of an ebb-dominated estuary. Megascale cross-bedded sandstone is 7 m thick. Arrows mark beds strongly bioturbated with *Thalassinoides* isp. and *Ophiomorpha nodosa* at the base of the megascale cross-bedded sandstone (see Fig. 13). From Paatuut (see Fig. 1 for location).

FIG. 13.—Estuarine valley-fill sandstone sharply overlain by shelf mudstone. Hammer for scale (arrow). From Paatuut (see Fig. 1 for location).

mer, 1992) and for the Lower Cretaceous Fall River Sandstone in the Powder River Basin, Wyoming, USA (Ethridge et al., 1994).

During uplift of the basin and the resulting relative sea-level fall incised valleys were formed by erosion of the underlying fine-grained shelf and deltaic sediments of the Kangilia and Atane Formations. The only record of sedimentation in the valleys during the lowstand period are the paraconglomerates of the Kangilia Formation, which were concentrated as a basal lag in the Ataata Kuua and Ivissussat Qaqqaat paleovalleys (Fig. 14A) as finer grained sediments were flushed through the valley system. During the lowstand period, the valleys probably supplied sediments to Paleocene turbidite slope channels occuring in the uppermost part of a thick Maastrichtian–Paleocene turbidite succession in the Itilli valley area (Dam and Sønderholm, 1994). A lowstand shoreline developed in the central and north-

ern part of Nuussuaq and is represented by fossiliferous, coarse-grained pebbly sandstones of the Lower Paleocene Agatdal Formation (Fig. 15).

The succeeding phase of fluvial sandstone aggradation was caused by a rise in relative sea level. A balance between a rapidly rising relative sea level, creating new accommodation space and high river discharge and sedimentation rates resulted in an extremely uniform facies composition of the fluvial succession in which evidence of organised channel flow or repetitive stacking of genetic units are lacking. The extremely uniform stacking pattern of cross-bedded sandstones, the lack of internal fining-upward successions, erosive surfaces and interchannel areas is suggestive of a single river that covered the full width of the valleys during valley filling (Fig. 14B) and that deposition was very rapid, possibly catastrophic.

The presence of tidal-estuarine deposits overlying the fluvial valley fill at Paatuut indicates that relative sea-level rise exceeded the rate of fluvial sediment supply at the late stage of infilling of the valleys. Consequently, the sediments transported down the valley system were reworked by tidal-estuarine processes and deposited as aggrading transgressive facies onlapping the fluvial succession. A mudstone bed separating the fluvial and estuarine deposits was probably deposited in the mid-estuary funnel (Fig. 14C). The sharp boundary between the estuary mouth-sandstone and the mid-estuary funnel mudstone probably formed as the ebb-tidal-delta migrated up the estuary thus forming a tidal ravinement surface (cf. Allen and Posamentier, 1993). The presence of estuarine sediments in the upper part of the Paatuut valley suggest deposition took place in the middle segment of the valley (Segment 2 of Zaitlin et al., 1994), whereas deposition in the other valleys took place further upstream in the inner segment of the vallleys (Segment 3 of Zaitlin et al., 1994). The drowning of the valleys coincided with their final infilling. Fluvial and estuary valley-fill sandstones are, throughout the area, sharply overlain by a marine mudstone (Figs. 13, 14D). This surface reflects a major landward shift of facies, probably formed during continued relative sea-level rise and is a major transgressive surface (cf. Dam et al., 1998).

The presence of picritic hyaloclastites with foresets up to 700 m just above the marine mudstone is suggesting that the relative sea level continued to rise after deposition of the mudstone.

DISCUSSION AND CONCLUSIONS

This study documents the relationship of Early Paleocene valley incision to faulting, uplift and magmatism in the Nuussuaq Basin of West Greenland (see also Dam et al., 1998). The paleovalley system includes five synchronously developed valleys exposed along the south coast of Nuussuaq. Each of the valleys are 1–2 km wide and up to 190 m deep. The valley fill comprises a succession of fluvial and estuarine deposits.

The episode of Early Paleocene tectonism and uplift that preceded valley incision indicates vertical displacement of several hundreds of meters and probably resulted in a pronounced structural relief. All the Quikavsak paleovalleys cut across Early Paleocene faults in the basin, suggesting that the incision of the paleovalleys was related to this tectonic episode, rather than global eustatic changes. It has been suggested that the generation of the West Greenland volcanic continental margin was

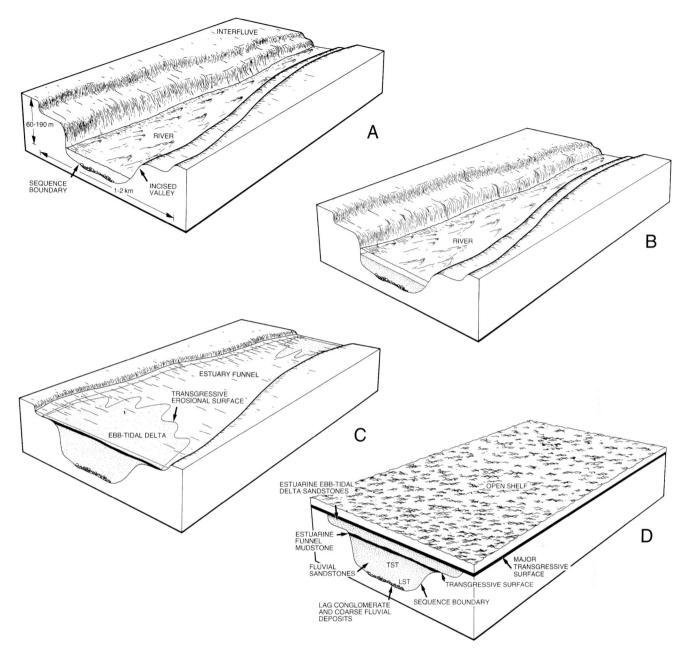

FIG. 14.—Schematic illustration of the geological evolution of the Quikavsak incised valleys. (A) Lowstand: Valley incision; sediments derived from the shelf and deltaic deposits of the Kangilia and Atane Formations bypassed the incised valleys to be deposited along a lowstand shoreline located in central and northern Nuussuaq and in lowstand slope channels in the western part of Nuussuaq. When present, the lowstand systems tract is expressed as either one clast lag conglomerate or a thicker fluvial conglomerate deposited from longitudinal bars. (B) Transgression: During transgression the incised valleys became a zone of sediment trapping and a thick uniform succession of fluvial sandstone was deposited, onlapping the valley floor and sides. During valley filling a single channel covered the full width of the valleys. A delicate balance between relative sea-level rise, river discharge and sedimentation resulted in an extremely uniform valley fill deposited by fluvial aggradation without any repeated genetic facies successions, probably indicating a very rapid rise in relative sea level. Deposition was very rapid, probably catastrophic (C) In one of the valleys the fluvial sandstone is succeeded by estuarine sandstones, indicating that the rate of relative sea level rise was greater than the rate of fluvial sedimentation. The fluvial and estuarine sandstones are separated by a thin mid-estuary funnel mudstone. The sharp boundary between the mudstone and the above estuary mouth sandstone formed as the the ebb-tidal-delta migrated up the estuary and is a tidal ravinement surface. (D) A major transgressive surface is coinciding with the final infilling of the valleys.

caused by extension related to regional uplift associated with a thermal anomaly created by a hot mantle plume beneath East Greenland (White and Mckenzie, 1989; Campbell and Griffith, 1990). Using a model of plate motions relative to major hot-spots underneath the African, Indian, North American, South American, and Australian plates to compute the track of the Iceland hotspot, Lawver and Müller (1994), suggested that the Nuussuaq region was situated above the Iceland hotspot 60 Ma ago and that the volcanic activity in the region was related to hotspot activity, rather than a hot mantle plume situated beneath

Qaarsutjaegerdal

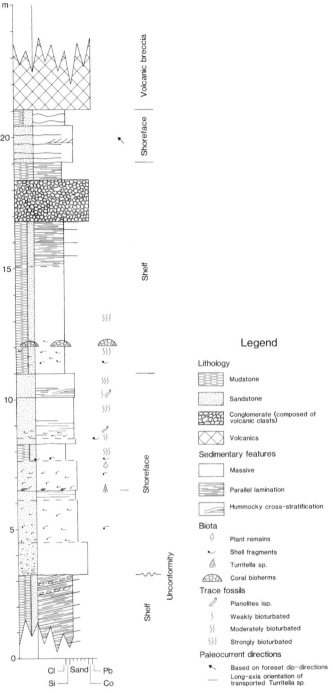

FIG. 15.—Sedimentological log through the lowstand shoreline present in the central and northern Nuussuaq. From Agatdalen in the central part of Nuussuaq (see Fig. 1 for location).

from shallow magma chambers. Alternatively, the uplift can be attributed to structural reactivation and uplift (inversion) related to a reorganisation of stress fields in the early Tertiary (A. J. Tankard, pers. commun., 1995). Extrusive centers are then related to dilation of releasing bends of strike-slip faults as well as intersecting fault systems. This model is supported by observations of A. K. Pedersen (pers. commun., 1995), who reported that many of the volcanic extrusive centers in the Nuussuaq Basin are related to major faults.

In a sequence stratigraphic interpretation of the Quikavsak succession, the sequence boundary is placed at the base of the valleys (Figs. 14A, 16). The overlying fill comprises a rapidly deposited aggradational succession of fluvial and tidal-estuarine sandstones.

In the Ivissussat Qaqqaat and Ataata Kuua sections, the valley surfaces are overlain by thin sheets of fluvial gneiss boulder conglomerates derived from erosion of the Kangilia Formation and deposited as lags and longitudinal bars, while sediments otherwise bypassed the valley during lowstand times. Most of

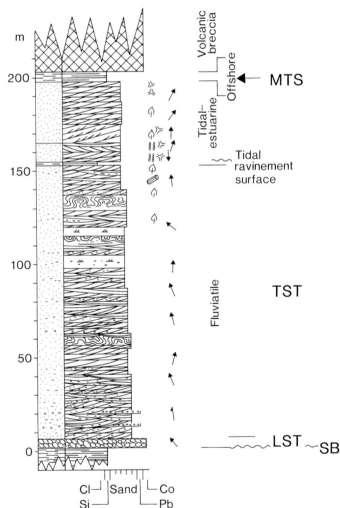

FIG. 16.—Schematic section through an incised valley of the Quikavsak Member, showing high-resolution sequence stratigraphic interpretation. SB–Sequence boundary; LST–Lowstand systems tract; TST–Transgressive systems tract; MTS–Major transgressive surface (see Fig. 4 for legend).

East Greenland as suggested by White and McKenzie (1989) and Campbell and Griffith (1990). Both these models provide an explanation for the relationship between widespread magmatism and uplift of the area at the time of valley incision followed by major subsidence caused by removal of magma these lowstand deposits were deposited seaward of the valleys as lowstand shoreface sandstones (Agatdal Formation) in the central and northern part of Nuussuaq and as lowstand slope channels in the western part of Nuussuaq (upper part of the Itilli succession of Dam and Sønderholm, 1994). The boulder conglomerates at the base of the valleys are interpreted as the only deposits of the lowstand systems tract within the valleys (Fig. 16).

Rapid subsidence in Early Paleocene time was generally balanced by discharge and sedimentation and resulted in a uniform fluvial valley-fill succession. During valley filling, a single channel occupied the entire width of the valleys. In the Paatuut valley, the relative sea-level rise exceeded the fluvial sediment flux, so that transgressive tidal-estuarine deposits onlapped the fluvial sandstone. The fluvial and estuarine sandstones are separated by a thin mid-estuary funnel mudstone. The sharp boundary between the mudstone and the overlying estuary mouth sandstone is a tidal ravinement surface and formed as the ebb tidal delta migrated up the estuary. Both the fluvial and the tidal-estuarine deposits are placed within the transgressive systems tract (Fig. 16), because the uniform fluvial sandstone deposits suggest a rapid rise in relative sea level and because the shift from fluvial to estuarine sedimentation indicates that a marine transgression occurred within the valleys. The drowning of the valleys coincided with the final infilling of the valleys, and valley-fills throughout the area are sharply overlain by marine mudstones. The base of the marine mudstone reflects a major landward shift in facies and is a major transgressive surface (Fig. 16).

Volcanic hyaloclastites, present just a few meters above the marine mudstone, indicates deposition at water depths in places exceeding 700 m (Dueholm and Pedersen, 1988; Pedersen, 1993), suggesting that the uplift was followed by very rapid subsidence possibly due to removal of magma from shallow magma chambers (cf. Campbell and Griffith, 1990).

ACKNOWLEDGMENTS

This study was generously supported by the Carlsberg Foundation and the Danish Ministry of Energy Research Programme (EFP–92). We thank F. G. Ethridge, G. K. Pedersen, T. C. R. Pulvertaft and A. J. Tankard for suggestions and constructive criticism of early manuscript versions as well as placing their own ideas of the basin development at our disposal. J. Halskov and C. Thuesen are thanked for drafting, and J. Lautrup and A. Winther for darkroom work. The paper is published with permission of the Geological Survey of Denmark and Greenland.

REFERENCES

ALLEN, G. P. AND POSAMENTIER, H. W., 1991, Facies and stratal patterns in incised valley complexes: examples from the Recent (Canada) (abs.): American Association of Petroleum Geologists Annual Convention 1991, p. 70.

ALLEN, G. P. AND POSAMENTIER, H. W., 1993, Sequence stratigraphy and facies model of an incised valley fill: The Gironde Estuary, France: Journal of Sedimentary Petrology, v. 63, p. 378–391.

ARCHER, A. W., LANIER, W. P., AND FELDMAN, H. R., 1994, Stratigraphy and depositional history within incised-paleovalley fills and related facies, Douglas Group (Missourian/Virgilian; Upper Carboniferous) of Kansas, U.S.A., in Dalrymple, R. W., Boyd, R., and Zaitlin, B. A., eds., Incised-valley Systems: Origin and Sedimentary Sequences: Tulsa, SEPM Special Publication 51, p. 175–190.

ASHLEY, G. M. AND SHERIDAN, R. E., 1994, Depositional model for valley fills

on a passive continental margin, in Dalrymple, R. W., Boyd, R., and Zaitlin, B. A., eds., Incised-valley Systems: Origin and Sedimentary Sequences: Tulsa, SEPM Special Publication 51, p. 285–301.

BATE, K. J., WHITTAKER, R. C., CHALMERS, J. A., AND DAHL-JENSEN, T., 1995, The Fylla structural complex: possible very large gas reserves offshore southern West Greenland: Rapport Grønlands geologiske Undersøgelse, v. 165, p. 22–27.

BAUM, G. R. AND VAIL, P. R., 1988, Sequence stratigraphic concepts applied to Paleogene outcrops, Gulf and Atlantic basins, in Wilgus, C. K., Hastings, B. S., Kendall, C. G. S. T. C., Posamentier, H. W., Ross, C. A., and Van Wagoner, J. C., eds., Sea-level Changes: An Integrated Approach: Tulsa, Society of Economic Paleontologists and Mineralogists Special Publication 42, p. 309–327.

BELKNAP, D. F, KRAFT, J. C., AND DUNN R. K., 1994, Transgressive valley-fill lithosomes: Delaware and Maine, in Dalrymple, R. W., Boyd, R., and Zaitlin, B. A., eds., Incised-valley Systems: Origin and Sedimentary Sequences: Tulsa, SEPM Special Publication 51, p. 303–320.

BLUCK, B. J., 1974, Structure and directional properties of some valley sandur deposits in southern Iceland: Sedimentology, v. 21, p. 533–554.

BOREEN, T. D. AND WALKER, R. G., 1991, Definition of allomembers and their facies assemblages in the Viking Formation, Willesden Green area, Alberta: Bulletin of Canadian Petroleum Geology, v. 39, p. 123–144.

BURDEN, E. T. AND LANGILLE, A. B., 1990, Stratigraphy and sedimentology of Cretaceous and Paleocene strata in half-grabens on the southeast coast of Baffin Island, Northwest Territories: Bulletin of Canadian Petroleum Geology, v. 38, p. 185–195.

CAMPBELL, I. H. AND GRIFFITH, R. W., 1990, Implications of mantle plume structure for the evolution of flood basalts: Earth and Planetary Science Letters, v. 99, p. 79–93.

CHALMERS, J. A., 1991, New evidence on the structure of the Labrador Sea, Greenland continental margin: Journal of the Geological Society of London, v. 148, p. 899–908.

CHALMERS, J. A., DAHL-JENSEN, T., BATE, K. J., AND WHITTAKER, R. C., 1995, Geology and petroleum prospectivity of the region offshore southern West Greenland—a summary: Rapport Grønlands geologiske Undersøgelse, v. 165, 13–21.

CHALMERS, J. A. AND PULVERTAFT, T. C. R., 1993, The southern West Greenland continental shelf—was petroleum exploration abandoned prematurely?, in Vorren, T. O., ed., Arctic Geology and Petroleum Potential: Amsterdam, Elseveir for Norwegian Petroleum Society, p. 55–66.

CHALMERS, J. A., PULVERTAFT, T. C. R., CHRISTIANSEN, F. G., LARSEN, H. C., LAURSEN, K. H., AND OTTESEN, T. G., 1993, The southern West Greenland continental margin: rifting history, basin development, and petroleum potential, in Spencer, A. M., ed., Proceedings of the 4th conference on the Petroleum Geology of North West Europe: London, The Geological Society, p. 915–931.

CHRISTIANSEN, F. G., MARCUSSEN, C., AND CHALMERS, J. A., 1995, Geophysical and petroleum geological activities in the Nuussuaq–Svartenhuk Halvø area 1994—promising results for an onshore exploration potential: Rapport Grønlands geologiske Undersøgelse, v. 165, p. 32–41.

CLIFTON, H. E., 1994, Preservation of transgressive and highstand Late Pleistocene valley-fill/estuary deposits, Willapa Bay, Washington, in Dalrymple, R. W., Boyd, R., and Zaitlin, B. A., eds., Incised-valley Systems: Origin and Sedimentary Sequences: Tulsa, SEPM Special Publication 51, p. 321–333.

COLEMAN, S. M. AND MIXON, R. B., 1988, The record of major Quaternary sea-level changes in a large coastal plain estuary, Chesapeake Bay, eastern United States: Palaeogeography, Palaeoclimatology, Palaeoecology, v. 8, p. 99–116.

DAM, G., LARSON, M., AND SONDERHOLM, M., 1998, Sedimentary response to mantle plumes: Implications from Paleocene onshore successions, West and East Greenland: Geology, in press.

DAM, G. AND SØNDERHOLM, M., 1994, Lowstand slope channels of the Itilli succession (Maastrichtian–Lower Paleocene), Nuussuaq, West Greenland: Sedimentary Geology, v. 94, p. 49–71.

DUEHOLM, K. S. AND PEDERSEN, A. K., 1988, Geological photogrammetry using oblique aerial photographs: Rapport Grønlands geologiske Undersøgelse, v. 140, p. 33–38.

DYSON, I. A. AND BORCH, C. VON DER, 1994, Sequence stratigraphy of an incised-valley fill: the Neoproterozoic Seacliff Sandstone, Adelaide Geosyncline, south Australia, in Dalrymple, R. W., Boyd, R., and Zaitlin, B. A., eds., Incised-valley Systems: Origin and Sedimentary Sequences: Tulsa, SEPM Special Publication 51, p. 209–222.

ETHRIDGE, F. G., KELLISON, L. B., AND JUMP-WARE, C.J., 1994, Unconformity related hydrocarbon production: Lower Cretaceous, Fall River Sandstone, Power and Buck Draw Fields, Southern Powder River Basin, Wyoming, in Dolson, J. C., Hendricks, W. L., and Wescott, W. A., eds, Unconformity Related Hydrocarbons in Sedimentary Sequences: Denver, Rocky Mountain Association of Geologists, p. 149–156.

EVANOFF, E., 1990, Early Oligocene paleovalleys in southern and central Wyoming: Evidence of high local relief on the late Eocene unconformity: Geology, v. 18, p. 443–446.

GIBLING, M. AND WIGHTMAN, W. G., 1994, Palaeovalleys and protozoan assemblages in a Late Carboniferous cyclothem, Sydney Basin, Nova Scotia: Sedimentology, v. 41, p. 699–719.

KEITH, D. A. W., WIGHTMAN, D. M., PEMBERTON, S. G., MACGILLIVRAY, J. R., BEREZNIUK, T., AND BERHANE, H., 1988, Sedimentology of the Mc-Murray Formation and Wabiskaw Member (Clearwater Formation), Lower Cretaceous, in the central region of the Athabasca oil sands area, notheastern Alberta, in James, D. P. and Leckie, D. A., eds., Sequences, Stratigraphy, Sedimentology: Surface and Subsurface: Calgary, Canadian Society of Petroleum Geologists Memoir 15, p. 309–324.

KINDINGER, J. L., BALSON, P. S., AND FLOCKS, J. G., 1994, Stratigraphy of the Mississippi–Alabama shelf and the Mobile River incised-valley system, in Dalrymple, R. W., Boyd, R., and Zaitlin, B. A., eds., Incised-valley Systems: Origin and Sedimentary Sequences: Tulsa, SEPM Special Publication 51, p. 83–95.

KOCH, B. E., 1959, Contribution to the stratigraphy of the non-marine Tertiary deposits on the south coast of the Nûgssuaq peninsula northwest Greenland with remarks on the fossil flora: Bulletin Grønlands geologiske Undersøgelse, v. 22, 100 p.

KOCH, B. E., 1964, Review of fossil floras and nonmarine deposits of West Greenland. Geological Society of America Bulletin, v. 75, p. 535–548.

KRYSTINIK, L. F. AND BLAKENEY-DEJARNETT, B. A., 1990, Sedimentology of the upper Morrow Formation in eastern Colorado and western Kansas, in Sonnenberg, S. A., Shannon, L. T., Rader, K., Drehle, W. F. von, and Martin, G. W., eds., Morrow Sandstones of Southeast Colorado and Adjacent Areas: Denver, Rocky Mountain Association of Geologists, p. 37–50.

KVALE, E. P. AND BARNHILL, M. L., 1994, Evolution of Lower Pennsylvanian estuarine facies within two adjacent paleovalleys, Illinois, Indiana, in Dalrymple, R. W., Boyd, R., and Zaitlin, B. A., eds., Incised-valley Systems: Origin and Sedimentary Sequences: Tulsa, SEPM Special Publication 51, p. 191–207.

LAND, C. B. AND WEIMER, R. J., 1978, Peoria Field, Denver Basin, Colorado—J Sandstone distributary channel reservoir: Denver, Rocky Mountain Association of Geologists Symposium, p. 81–104.

LAWVER, L. A. AND MÜLLER, R. D., 1994, Iceland hotspot track: Geology, v. 22, p. 311–314.

LECKIE, D. A., 1994, Canterbury Plains, New Zealand—Implications for sequence stratigraphic models: American Association of Petroleum Geologists Bulletin, v. 78, p. 1240–1256.

LOWE, D. R., 1975, Water escape structures in coarse-grained sediments: Sedimentology, v. 2, p. 157–204.

MACEACHERN, J. A. AND PEMBERTON, S. G., 1994, Ichnological aspects of incised-valley fill systems from the Viking Formation of the Western Canada Sedimentary Basin, in Dalrymple, R. W., Boyd, R., and Zaitlin, B. A., eds., Incised-valley Systems: Origin and Sedimentary Sequences: Tulsa, SEPM Special Publication 51, p. 129–157.

MARTINSEN, O. J., 1994, Evolution of an incised-valley fill, the Pine Ridge Sandstone of southeastern Wyoming, U.S.A., in Dalrymple, R. W., Boyd, R., and Zaitlin, B. A., eds., Incised-valley Systems: Origin and Sedimentary Sequences: Tulsa, SEPM Special Publication 51, p. 109–128.

MATTISON, B. W., 1992, Recognition of paleovalley systems and associated type 1 unconformities within the Upper and Middle Mannville Subgroups of east central Alberta (abs.): American Association of Petroleum Geologists Annual Convention 1992, p. 83.

MIALL, A. D., 1978, Lithofacies types and vertical profile models in braided river deposits: a summary, in Miall, A. D., ed., Fluvial Sedimentology: Calgary, Canadian Society of Petroleum Geology Memoir 5, p. 597–604.

PEDERSEN, A. K., 1993, Geological section along the south coast of Nuussuaq, central West Greenland: Copenhagen, The Geological Survey of Greenland.

PEDERSEN, A. K. AND DUEHOLM, K. S., 1992, New methods for the geological analysis of Tertiary volcanic formations on Nuussuaq and Disko, central West Greenland, using multi-model photogrammetry: Rapport Grønlands Geologiske Undersøgelse, v. 156, p. 19–34.

PEDERSEN, G. K. AND PULVERTAFT, T. C. R., 1992: The nonmarine Cretaceous

of the West Greenland Basin, onshore West Greenland: Cretaceous Research, v. 13, p. 263–272.

PIASECKI, S., LARSEN, L. M., PEDERSEN, A. K., AND PEDERSEN, G. K., 1992, Palynostratigraphy of the Lower Tertiary volcanic and marine clastic sediments in the southern part of the West Greenland Basin: implications for the timing and duration of the volcanism: Rapport Grønlands geologiske Undersøgelse, v. 54, p. 13–31.

PULVERTAFT, T. C. R., 1987, Status review of the results of stratigraphical and sedimentological investigations in the Cretaceous–Tertiary of West Greenland, and recommendations for new GGU activity in these fields: Open File Grønlands Geologiske Undersøgelse, 18 p.

RANGER, M. J. AND PEMBERTON, S. G. 1988, Marine influence on the McMurray Formation in the Primrose area, Alberta, in James, D. P. and Leckie, D. A., eds., Sequences, Stratigraphy, Sedimentology: Surface and Subsurface: Calgary, Canadian Society of Petroleum Geologists Memoir 15, p. 439–449.

ROLLE, F., 1985, Late Cretaceous–Tertiary sediments offshore central West Greenland: lithostratigraphy, sedimentary evolution, and petroleum potential: Canadian Journal of Earth Sciences, v. 22, p. 1001–1019.

ROSENKRANTZ, A., 1970, Marine Upper Cretaceous and lowermost Tertiary deposits in West Greenland: Meddelelser fra Dansk Geologisk Forening, v. 19, p. 406–453.

ROSENKRANTZ, A. AND PULVERTAFT, T. C. R. 1969, Cretaceous–Tertiary stratigraphy and tectonics in northern West Greenland, in Kay, M, ed., North Atlantic—Geology and Continental Drift: Tulsa, American Association of Petroleum Geologists Memoir 12, p. 883–898.

RØE, S-L., 1987, Cross-strata and bedforms of probable transitional dune to upper-stage plane-bed origin from a Late Precambrian fluvial sandstone, northern Norway: Sedimentology, v. 34, p. 89–101.

SAUNDERSON, H. C. AND LOCKETT, F. P. J., 1983, Flume experiments on bedforms and structures at the dune plane bed transition, in Collinson, J. D. and Lewin, I., eds., Modern and Ancient Fluvial Systems: Oxford, International Association of Sedimentologists Special Publication 6, p. 49–58.

SHANLEY, K. W. AND MCCABE, P. J., 1993, Alluvial architecture in a sequence stratigraphic framework; a case history from the Upper Cretaceous of southern Utah, USA: Oxford, International Association of Sedimentologists Specical Publication 15, 21–56.

SHANLEY, K. W., MCCABE, P. J., AND HETTINGER, R. D., 1992, Tidal influence in Cretaceous fluvial strata from Utah, USA: a key to sequence stratigraphic interpretation: Sedimentology, v. 39, p. 905–930.

SMITH, D. G., 1988, Modern point bar deposits analogous to the Athabasca oil sands, Alberta, Canada, in De Boer, P. L., Van Gelder, A., and Nio, D., eds., Tide-influenced Sedimentary Environments and Facies: Dordrecht, D. Reidel Publishing, p. 417–432.

STOREY, M., DUNCAN, R. A., PEDERSEN, A. K., LARSEN, L. M., AND LARSEN, H. C., 1998, ^{40}Ar/^{39}Ar geochronology of the West Greenland Tertiary volcanic province: Earth and Planetary Science Letters, in press.

SUTER, J. R. AND BERRYHILL, H. L., 1985, Late Quaternary shelf-margin deltas, northwest Gulf of Mexico: American Association of Petroleum Geologists Bulletin, v. 69, p. 77–91.

SUTER, J. R., BERRYHILL, H. L., AND PENLAND, S., 1987, Late Quaternary sea-level fluctuations and depositional sequences, southwest Lousiana continental shelf, in Nummedal, D., Pilkey, O. H., and Howard, J. D., eds., Sea-level Fluctuation and Coastal Evolution: Tulsa, Society of Economic Paleontologists and Mineralogists Special Publication 41, p. 199–219.

THOMAS, M. A. AND ANDERSON, J. B., 1994, Eustatic controls on the facies architecture of the Trinity—Sabine incised valley system, Texas continental shelf, in Dalrymple, R. W., Boyd, and R., Zaitlin, B. A., eds., Incised-valley Systems: Origin and Sedimentary Sequences: Tulsa, SEPM Special Publication 51, p. 63–82.

VAN WAGONER, J. C., MITCHUM, R. M., CAMPION, K. M., AND RAHMANIAN, V. D., 1990, Siliciclastic sequence stratigraphy in well logs, cores, and outcrops: Tulsa, American Association of Petroleum Petrologists Methods in Exploration Series 7, p. 1–55.

WEIMER, R. J., 1992, Developments in sequence stratigraphy: foreland and cratonic basins: American Association of Petroleum Geologists Bulletin, v. 76, p. 965–982.

WEIMER, R. J. AND SONNENBERG, S. A., 1989, Sequence stratigraphic analysis, Muddy (J) sandstone reservoir, Wattenberg Field, Denver Basin, Colorado, in Coalson, E., ed., Sandstone Reservoirs: Denver, Colorado, Rocky Mountain Association of Geologists, p. 197–220.

WHITE, R. AND MCKENZIE, D., 1989, Magmatism at rift zones: the generation of volcanic continental margins and flood basalts: Journal of Geophysical Research, v. 94, p. 7685–7729.

WOOD, J. M. AND HOPKINS, J. C., 1992, Traps associated with paleovalleys and interfluves in an unconformity bounded sequence: Lower Cretaceous Glauconitic Member, Southern Alberta, Canada: American Association of Petroleum Geologists Bulletin, v. 76, p. 904–926.

WOOD, L. J., ETHRIDGE, F. G., AND SCHUMM, S. A., 1993, The effects of rate of base-level fluctuation on coastal-plain, shelf and slope depositional systems: An experimental approach, *in* Posamentier, H. W., Summerhayes, C. P., Haq, B. U., and Allen, G. P., eds, Sequence Stratigraphy and Facies Associations: Oxford, International Association Sedimentologists Special Publication 18, p. 43–53.

WOOD, L. J., ETHRIDGE, F. G., AND SCHUMM, S. A., 1994, An experimental study of the influence of subaqueous shelf angles on coastal-plain and shelf deposits, *in* Weimer, P. and Posamentier, H. W., eds., Siliciclastic Sequence Stratigraphy, Recent Developments and Applications: Tulsa, American Association of Petroleum Geologists Memoir 58, p. 381–391.

ZAITLIN, B. A., DALRYMPLE, R. W., AND BOYD, R., 1994, The stratigraphic organization of incised-valley systems associated with relative sea-level change, *in* Dalrymple, R. W., Boyd, R. and Zaitlin, B. A., eds, Incised-valley Systems: Origin and Sedimentary Sequences: Tulsa, SEPM Special Publication 51, p. 45–60.

EVIDENCE FOR SUBTLE UPLIFT FROM LITHOFACIES DISTRIBUTION AND SEQUENCE ARCHITECTURE: EXAMPLES FROM LOWER CRETACEOUS STRATA OF NORTHEASTERN NEW MEXICO

JOHN M. HOLBROOK AND DAVID C. WHITE

Southeast Missouri State University, Cape Girardeau, Missouri 63701

ABSTRACT: Reactivation of pre-existing zones of crustal weakness commonly results in structural generation of low-relief ($\leq 10^1$ m) topographic bulges, here called welts, which affect contemporary depositional processes. Lower Cretaceous (Albian) Glencairn, Mesa Rica and Pajarito strata above Sierra Grande basement structure in northeastern New Mexico reveal a spectrum of depositional effects that are best attributed to pene-contemporaneous elevation of such a low-relief welt and argue for Albian reactivation of underlying Sierra Grande basement structure.

Transgressive and regressive offshore marine strata of the Glencairn section pinchout in association with the Sierra Grande axis reflecting non-deposition and/or erosion of marine strata above higher Sierra Grande topography. Fluvial Mesa Rica Sandstone overlies a sequence-bounding unconformity above marine Glencairn strata, and represents deposition from predominantly straight, single-channel streams. Increased abundance of lateral-accretion and abandoned-channel-fill architectural elements in Mesa Rica Sandstone on the southeastern (paleo-downstream) flank of the Sierra Grande uplift infers an increase in channel sinuosity that is restricted to this location. Increased sinuosity of Mesa Rica channels on the southeastern flank of the Sierra Grande uplift occurred in response to increased stream gradients on the downstream flank of a topographically positive Sierra Grande welt.

INTRODUCTION

A growing list of stratigraphic and geomorphic studies reveal that individual low-relief ($\leq 10^1$ m) topographic bulges of structural origin, or welts (cf. Bucher, 1933), are a common and pervasive feature of both modern and ancient, otherwise tectonically quiescent, plate interiors (e.g., Burnett and Schumm, 1983; De Miranda and Da Boa Hora, 1986; Letouzey et al, 1990; Holbrook, 1996a). These welts apparently form because of reactivation of pre-existing crustal weakness trends by sufficient levels of tectonic stress (Holbrook, 1992, 1996a; Meyers et al., 1992; Heller et al., 1993).

Although such welts tend to have only subtle topographic and structural expression, they commonly exert a profound and, collectively, regional tectonic control on concurrent surface processes and their resultant sedimentary deposits. Typical effects include diversion of river drainages around higher welt topography (Goodrich, 1898), preserved as thicker paleovalley deposits in tectonic low areas (Weimer, 1984) and shallowing of the water column above tectonically elevated welts (Letouzey et al., 1990), resulting in local shoaling of marine deposits (Hattin, 1986).

Sedimentary data support the presence in northeastern New Mexico of an Early Cretaceous low-relief topographic welt which was most likely generated by uplift on Sierra Grande structures (Fig. 1). Depositional patterns in Lower Cretaceous Glencairn and Mesa Rica strata (Fig. 2) that are discussed in this paper both argue for subtle tectonic effects on sedimentation by such welt generation and offer examples which can be used to detect other ancient welts from sedimentary data.

THE SIERRA GRANDE UPLIFT

The Sierra Grande uplift is identified by a northeast/southwest trending high in Precambrian basement of northeastern New Mexico and has a southeast extension which is referred to as the Bravo Dome (Fig. 1). Most structure of the Sierra Grande uplift predates proposed Early Cretaceous deformation, as the Sierra Grande was a major uplift within the late Paleozoic Ancestral Rockies (King, 1959; Baltz, 1965). Likewise, Baldwin and Muelberger (1959) have mapped structures revealing Tertiary reactivation and uplift of Sierra Grande basement struc-

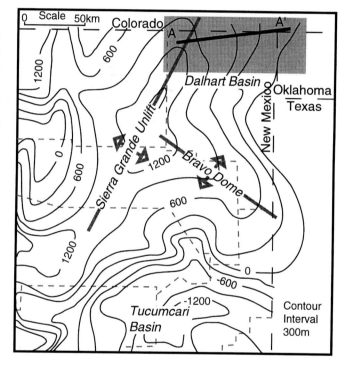

FIG. 1.—Structural contour map of Precambrian basement in northeastern New Mexico, and location of cross-section A–A' (see FIG. 3) which runs the length of the Dry Cimarron valley. The shaded area represents the region depicted in Figure 4 (after Suleiman and Keller, 1985).

tures in positions similar to Paleozoic activity and present basement topography. Any Early Cretaceous, tectonically generated topography above the Sierra Grande uplift likely represents a similar episode of reactivation on Sierra Grande structures.

REGIONAL DEPOSITIONAL/SEQUENCE–STRATIGRAPHIC MODEL

Albian (latest Early Cretaceous) Glencairn, Mesa Rica and Pajarito formations (Fig. 2) are targeted in this study and record deposition in northeastern New Mexico during the Albian Kiowa–Skull Creek transgressive/regressive cycle and the Albian part of the latest Albian through mid-Turonian (early Late Cretaceous) Greenhorn transgressive/regressive cycle (Holbrook and Wright Dunbar, 1992). The following discussion of depositional history and sequence stratigraphy for these strata is extracted from Holbrook (1992, 1996b) and Holbrook and Wright Dunbar (1992) in order to establish a regional framework for depositional and sequence-stratigraphic relationships discussed throughout this paper.

The basal Long Canyon sandstone bed of the Glencairn Formation is separated from underlying Lytle Sandstone by a regional lowstand surface of erosion (surface SB2) and reflects aggradation of coastal and terrestrial strata during Kiowa–Skull Creek transgression (Fig. 2). Long Canyon strata are separated from overlying marine transgressive shale of the upper Glencairn Formation by a widespread transgressive surface of erosion (surface TSE), which was formed by coastal erosional/ravinement processes during transgression (Fig. 2).

The uppermost part of the Glencairn Formation records progradation of nearshore marine sediments during Kiowa–Skull Creek regression. Exposure and subaerial erosion of these regressive marine deposits during maximum Kiowa–Skull Creek regression are recorded as a regional lowstand surface of erosion/sequence boundary cut into the upper Glencairn Formation (surface SB3; Fig. 2). The dominantly fine-grained marine deposits of the Glencairn Formation that are bound between surfaces TSE and SB3 have not been formally named in previous studies and will be referred to informally as the shale member of the Glencairn Formation in this paper.

Streams flowing southward and southeastward across surface SB3 debouched into the lowstand sea and are preserved as fluvial deposits of the overlying Mesa Rica Sandstone (Fig. 2). Transgression of the Greenhorn cycle followed lowstand deposition and resulted in burial of Mesa Rica fluvial deposits beneath aggrading Pajarito coastal-plain strata and thin and localized Romeroville shoreface deposits.

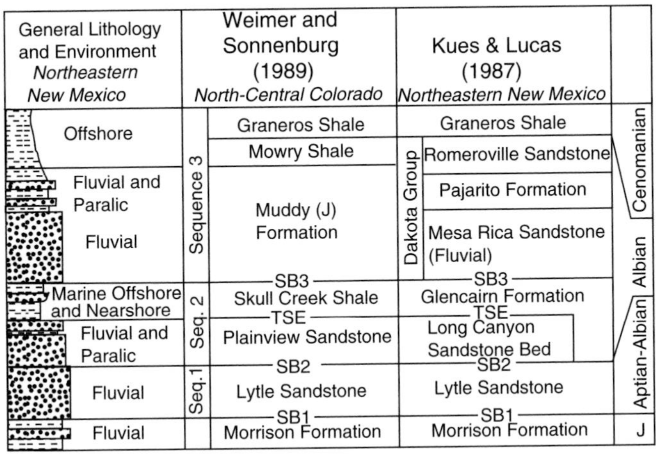

TECTONIC EFFECTS ON THE GLENCAIRN "SHALE MEMBER"

Lithofacies

The shale member of the Glencairn Formation is dominantly composed of intervals of black to light-gray fissile shale that contain up to 20% thin (1–3 cm) silty sandstone interbeds. Typical macrofossils in these shaley intervals include *Nuculana, Nucula, Breviarca, Turritella* and *Drepanochilus* (Scott, 1970). Shaley intervals are separated by sandstone-rich intervals, which are gradational with shaley intervals at their base, and are in abrupt contact with shaley intervals at their top. Sandstone-rich intervals comprise thin to medium beds of fine- to very-fine-grained quartz arenite, which increase in thickness and grain size upward in the individual sandstone-rich intervals and are interbedded with thin to very-thin beds of dark-gray fissile shale. Sandstone beds are typically massive, owing to bioturbation, but may locally reveal current or oscillation ripples and hummocky cross-stratification. Such sandstone beds contain an abundant ichnofauna that includes *Skolithos, Arenicolites, Ophiomorpha, Thalassinoides, Rhizocorallium* and *Planolites*. Typical macrofossils from these sandstone beds include *Corbula, Breviarca, Turritella, Trachycardium, Lopha quadriplicata, Brachidontes,* and *Protocardia* (Scott, 1970). Individual, sandstone-rich intervals are typically less than 7–10

General Lithology and Environment *Northeastern New Mexico*		Weimer and Sonnenburg (1989) *North-Central Colorado*	Kues & Lucas (1987) *Northeastern New Mexico*	
Offshore	Sequence 3	Graneros Shale	Graneros Shale	Cenomanian
		Mowry Shale	Romeroville Sandstone	
Fluvial and Paralic		Muddy (J) Formation	Pajarito Formation	Albian
Fluvial			Mesa Rica Sandstone (Fluvial)	
		SB3	SB3	
Marine Offshore and Nearshore	Seq. 2	Skull Creek Shale	Glencairn Formation	Aptian-Albian
Fluvial and Paralic		TSE	TSE	
		Plainview Sandstone	Long Canyon Sandstone Bed	
		SB2	SB2	
Fluvial	Seq. 1	Lytle Sandstone	Lytle Sandstone	
		SB1	SB1	
Fluvial		Morrison Formation	Morrison Formation	J

FIG. 2.—Correlation of strata and sequence-stratigraphically significant surfaces between the study area and the Colorado Front Range foothills, with primary lithofacies relationships. That part of the Glencairn Formation above the Long Canyon sandstone bed (Glencairn strata between surfaces TSE and SB3) is informally referred to as the Glencairn shale member in this paper (after Holbrook and Wright Dunbar, 1992).

m thick and are rarely repeated more that twice in a single Glencairn section.

Scott (1970) concludes that shaley intervals of the Glencairn Formation were deposited in a fully open-marine environment, based on lithologic similarities to bay-center sediments in Port Philip Bay, Victoria, Australia and occurrence of an offshore macrofaunal association (*Nucula–Nuculana*). Ripple to hummocky cross-stratified sandstone beds within sandstone-rich intervals reflect moderately high-energy conditions and contain a nearshore marine fossil assemblage. These strata have been attributed accordingly to deposition in an upper to lower shoreface setting (Long, 1966; Scott, 1970; Gustason and Kauffman, 1985; Holbrook, 1992). Gustason and Kauffman (1985) further asserted that the hummocky cross-bedded sandstones common to these sand-rich intervals reflect occasional reworking of Glencairn sands during storms.

Tectonic Influence

Deflection of the shale member zero isopach contour, and other thickness contours, approximates the shape of Precambrian relief on the combined Sierra Grande and Bravo Dome basement structure (Figs. 1, 3). Pinchout of these strata coincident with the eastern and northwestern flanks of this basement structure implies non-deposition of marine strata over an elevated Sierra Grande welt during Kiowa–Skull Creek marine flooding and/or preferential erosion of previously deposited Kiowa–Skull Creek marine strata above a rising Sierra Grande welt during lowstand erosion of surface SB3. In either case, such close coincidence of basement structure and marine stratal thickness infers effects on marine Glencairn strata by topography generated above a Sierra Grande uplift which was rising either prior to, during or shortly after Kiowa–Skull Creek marine flooding.

An alternative, non-tectonic, explanation for absence of the Glencairn shale member above Sierra Grande structure is that

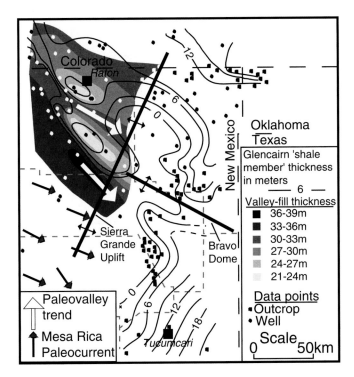

FIG. 3.—Isopach map of Glencairn strata between surfaces TSE and SB3, the Glencairn shale member, in northeastern New Mexico. In the northwestern part of the map area where valley-fill strata are complete, isopach contours of valley-fill strata are superimposed on the TSE—SB3 isopach map. Paleocurrents are from Gilbert and Asquith (1976), and each represent the average of 25 paleocurrent measurements from trough and planar cross beds in Mesa Rica Sandstone.

lack of these strata in this area represents arbitrary incision of a valley into the shale member and underlying strata along the Sierra Grande axis during Kiowa–Skull Creek lowstand. As such, pinchout of the shale member above Sierra Grande structure might reflect only localized deep erosion and offer no clear indication of an Albian Sierra Grande welt.

Aggradation of valley-fill and coastal plain strata following lowstand incision resulted in burial of topography on surface SB3 beneath Mesa Rica, Pajarito and Romeroville strata, as well as their collective equivalent, the Muddy Sandstone (Weimer, 1984; Dolson et al., 1991; Holbrook, 1992; Fig. 2). Thickening trends in combined Mesa Rica, Pajarito and Romeroville strata should, thus, coincide in position with those thinning trends in Glencairn strata that are reflective of major valley incision. Unfortunately, modern erosion into the upper Pajarito Formation above and eastward of the Sierra Grande axis makes assessment of thickening trends in combined Mesa Rica, Pajarito and Romeroville strata impossible for these regions. Complete Lower Cretaceous sections are present in the Raton Basin, however, and reveal two valley trends where minor local thinning of the Glencairn shale member is coincident with minor thickening of valley-fill strata (Fig. 3). These valleys are of minimal incised depth and are oriented northwest/southeast perpendicular, and in close proximity, to the Sierra Grande axis. Paleocurrent sets from Mesa Rica Sandstone confirm continuation of these paleoslope trends across the Sierra Grande structure (Figs. 3, 4).

In the absence of Sierra Grande topography, thinner sections of the Glencairn shale member in paleovalleys and thicker sections of the Glencairn shale member in paleointerfluves should have continued their southeast trend across the Sierra Grande axis with no significant deflection of thickness contours coincident with Sierra Grande basement structure. Instead, paleovalley thins and paleointerfluve thicks of the Raton Basin both terminate abruptly along a northward deflection of the shale member pinchout, coincident with the Sierra Grande axis (Fig. 3). This argues that elevated topography present above Sierra Grande basement structure most likely presented an obstacle across the path of regional paleoflow, resulting in preferential erosion of the shale member and/or underlying strata as rivers passing roughly perpendicular over the Sierra Grande axis leveled their course.

Magnitude and extent of proposed Sierra Grande topography is difficult to constrain independent of sedimentary data, as clearly defined structures associated with Early Cretaceous uplift on this feature are not apparent. Mesa Rica and Pajarito strata are continuous across the Sierra Grande axis to a maximum thickness of at least 33 m (Gilbert and Asquith, 1976), indicating that any Early Cretaceous topographic relief above this structure was sufficiently minor to be buried by Lower Cretaceous deposition ($\leq 10^1$ m). Modern examples of similarly minor tectonic uplifts are often expressed at the surface as simple warping of strata, with only minimal, if any, distinctive faulting and folding (e.g., Russ, 1982). Such was apparently the case for the Sierra Grande uplift. Locus of major shale member thinning, and thus maximum Early Cretaceous uplift, appears to be consistent with elevated position of modern (Fig. 1) and Paleozoic structure (Fig. 4 of Mallory, 1972).

TECTONIC EFFECTS ON MESA RICA AND PAJARITO FORMATIONS

Lithofacies

Mesa Rica Sandstone.—

Mesa Rica Sandstone within the study area is divided into a dominantly sandstone lithofacies (Stp; Fig. 5) and a mixed sandstone and mudstone lithofacies (Sfil; Fig. 6).

Lithofacies Stp consists almost entirely of trough and planar cross-bedded (sets 0.2–0.6 m thick), yellow to orange, fine- to medium-grained quartz arenite, with only minor conglomerate and mudstone present. Paleocurrents from trough and planar cross beds in lithofacies Stp are generally unimodal (Gilbert and Asquith, 1976). Lithofacies Stp is internally partitioned by abundant channel-shaped scours and scattered epsilon cross bedding (cf. Allen, 1963). Some scour surfaces are associated with thin (<8 cm) pebble lags (composed mostly of chert, quartzite, and mud rip-up clasts) or thinly laminated siltstone and silty mudstone linings (2–12 cm thick); however, most channel scours lack these features. A few random epsilon cross lamina in local lamina sets are draped by thin (<8 cm) silty mudstone beds similar to those draping some channel scours.

Lithofacies Sfil is represented by thin- to thick-bedded silty mudstone intercalated with medium-bedded, yellow to light-gray, fine- to medium-grained quartz arenite. Sandstone beds are dominated by either medium to thick parallel laminations or current-ripple and, rarely, trough and planar cross-stratification. Silty mudstone is typified by medium to thick parallel

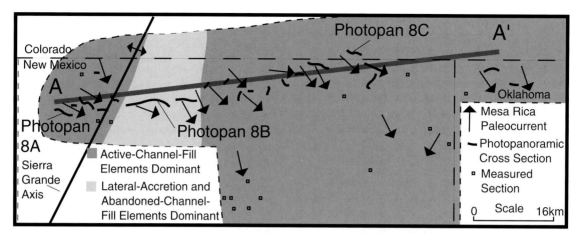

FIG. 4.—Distribution of architectural elements and paleocurrent trends in Mesa Rica Sandstone of the Dry Cimarron valley and environs. Locations of photopanoramic cross sections and other data points used to establish architectural elements and construct cross section A–A' are also provided. Paleocurrents each represent the average of 25 paleocurrent measurements from trough and planar cross beds in Mesa Rica Sandstone.

FIG. 5.—Cross-bedded sandstone of lithofacies Stp in Mesa Rica Sandstone.

FIG. 6.—Rocks of lithofacies Sfil in channel-fill element of Mesa Rica Sandstone.

lamentations; however, current ripple cross stratification is a common feature of these rocks. Lithofacies Sfil crops out as lenticular, channel-shaped bodies with a scoured basal contact and typically comprises between 30% and 60% mudstone.

Wood fragment impressions are abundant in both Stp and Sfil lithofacies. Dinosaur footprints (dominantly *Amblydactylus* and *Caririchnium*; Gillette and Thomas, 1985; Lucas et al., 1987; Lockley et al., 1992) also occur locally along the upper Mesa Rica contact. Mesa Rica Sandstone is otherwise devoid of macrofossils, and rare trace fossils are generally associated only with the Pajarito contact.

Several authors have previously interpreted Mesa Rica Stp sandstone in northeastern New Mexico to represent a widespread complex of amalgamated fluvial channels, based on presence of such features as abundant channel scours, channel lag and drape deposits, dominance by trough and planar cross-bedded sandstone, unimodal paleocurrent indicators, abundant plant fragments, dinosaur tracks, and lack of marine fauna in most all locations (Long, 1966; Jacka and Brand, 1972; Taylor, 1974; Gilbert and Asquith, 1976; Atalik, 1984; Holbrook and Wright Dunbar, 1992; Holbrook, 1996b). Scoured basal contact, channel-shaped geometry and internal characteristics of lithofacies Sfil are here interpreted to reflect filling of river channels following abandonment of the channel by the primary river flow (see Allen, 1965). Silty mudstone interbeds probably reflect deposition of fine-grained sediment during waning stages of channel reoccupation and/or overbank events related to flooding of a neighboring channel. Sandstone beds in most Sfil exposures are interpreted to represent movement of fine sand in ripples and dunes during flooding and reoccupation of abandoned channels by the primary river flow (cf. Hopkins, 1985).

Pajarito Formation.—

Rocks in the lower half of the Pajarito Formation (basal 10–17 m) are dominated by bluish-gray, thickly laminated claystone and mudstone, interbedded with dispersed, thin (typically <1 cm) beds of rippled or parallel-laminated, fine-grained quartz arenite (Fig. 7). Locally, these more common mud-rich strata interfinger with wavy- to lenticular-bedded sandstone, which possesses clay-draped and non-clay-draped current to oscillation-ripple lamina. Individual or amalgamated channel-shaped bodies of lithofacies Stp and/or Sfil rocks (1–17 m thick) are also commonly encased in these lower Pajarito mud-rich strata (Fig. 7) and are locally incised into the top of underlying Mesa Rica Sandstone (Fig. 8C). The upper zero to 17 m of the Pajarito Formation comprise rocks of lithofacies Stp. The basal 2 m of the Pajarito Formation are typified by medium-bedded, current-rippled (commonly climbing ripples) and parallel-laminated sandstone, interbedded with thin to medium beds of dark-gray siltstone and mudstone (Fig. 9). These sandstone beds commonly bear root impressions, contain thin (<.3 m) channel-shaped sandstone lenses and are laterally continuous with both Pajarito mud-rich lithofacies and underlying Mesa Rica Sandstone.

Skolithos, Plainolites, Thalassinoides, and *Arenicolites* burrows, as well as *Teredo*-bored wood impressions, occur locally within all Pajarito lithofacies and are typical of localized, wavy-bedded lithofacies. Plant fragment fossils are abundant throughout Pajarito strata.

FIG. 7.—''Mud-rich'' rocks in lower half of Pajarito Formation. Rocks of lithofacies Stp, which are also in the lower Pajarito Formation, are incised into these mud-rich strata.

The mud-rich lithofacies, locally with a coastal marine ichnofauna assemblage (Seilacher, 1964), which dominates the lower half of the Pajarito Formation, represents undifferentiated flood-plain and lacustrine deposits (Long, 1966; Atalik, 1984), as well as marine-influenced marsh (Taylor, 1974) and/or lagoonal (Atalik, 1984) strata. Sections of wavy-bedded sandstone with abundant ichnofauna represent local tidal flat development (Long, 1966). Rocks of lithofacies Stp and Sfil in Pajarito strata resemble rocks of these same lithofacies in underlying Mesa Rica Sandstone and are interpreted similarly to represent deposition from fluvial channels and channel belts (Waage, 1953; Long, 1966; Jacka and Brand, 1972, 1973; Bejnar and Lessard, 1976; Atalik, 1984). Interbedded sandstone and siltstone/mudstone commonly exposed in the basal 2 m of the Pajarito Formation preserve the remnants of a levee/crevasse splay complex associated with Mesa Rica channel deposition (Taylor, 1974; Holbrook, 1992, 1996b).

Lower Pajarito strata are best considered to represent deposition from a collection of environments operating in a marine-influenced coastal-plain setting.

Architectural Elements

Additional insights into Mesa Rica and, to a more limited extent, Pajarito deposition are gained through architectural element analysis. Architectural element analysis divides stratal units into geometric building blocks, called architectural elements, with consistent internal lithofacies and distinct bounding-surface geometries (Miall, 1985). Elements in Mesa Rica and Pajarito strata reflect distinct depositional origins.

Fluvial Mesa Rica strata are composed almost entirely of three architectural elements: active-channel-fill, abandoned-channel-fill and lateral-accretion elements. Active-channel-fill elements account for most Mesa Rica exposure in the Dry Cimarron valley and tend to crop out as pronounced cliffs. Individual active-channel-fill elements are bounded by a channel-shaped basal scour (width-depth ratios from 7.9–11.6) and are composed of lithofacies Stp rocks (Figs. 8A, C). Lithofacies Stp strata within this element are commonly partitioned by a series of nested channel scours (Fig. 8C).

Dominance of active-channel-fill elements by cross-bedded sandstone of lithofacies Stp infers that migration of sands in dunes persisted throughout fill of original channels. Maintenance of relatively high-energy flow in the channel as discharge waned and the channel bottom aggraded, best explains complete filling of channels with such a coarse-grained lithofacies. Furthermore, partitioning of fills primarily by nested channel scours, as opposed to inclined accretion surfaces, is highly suggestive of filling by gradual aggradation of channel-bottom deposits, instead of filling by emplacement of large bar forms. Active-channel-fill elements most likely reflect filling of channels while primary flow gradually shifted into a new channel and discharge was proportionally reduced in the old channel. Similar gradual abandonment processes have been cited previously as a mechanism for generating analogous coarse-grained channel fills (Fisk, 1955; Hopkins, 1985).

Abandoned-channel-fill elements are less abundant in Mesa Rica Sandstone than active-channel-fill elements and typically weather to partly covered slopes. These elements are bounded by a channel-shaped scour (width-depth ratios from 10.9–15.6)

supports lateral Mesa Rica deposition mostly through numerous cycles of avulsion, incision and abandonment of individual low-sinuosity rivers, as opposed to point-bar growth during wide lateral migration of meander loops (Holbrook, 1996b). Such preponderance of amalgamated symmetrical to near-symmetrical channel fills is typical of deposits from single-channel straight to low-sinuosity rivers (Moody-Stuart, 1966; Tyler and Ethridge, 1983; Eschard et al., 1991).

Alternatively, Mesa Rica architecture could be interpreted to represent deposition from a braided system. Low channel width-depth ratios (Schumm et al., 1987), conspicuous lack of architectural elements typically associated with braided deposits (e.g., foreset macroforms) and indications of channel confinement by a basal Pajarito levee/splay complex, however, argue against such a braided interpretation. Likewise, Mesa Rica channel-fill elements aggrade upwardly into overlying Pajarito fines as independent bodies (Fig. 8C), rather than the wider complexes of amalgamated bars and channel-fills one would expect if individual Mesa Rica channel-fill elements originated by filling of minor channels within wider braided rivers. Mesa Rica strata in these areas are, thus, interpreted to reflect deposition from single-channel straight to low-sinuosity rivers.

Mesa Rica exposure from the Sierra Grande axis eastward for 20 km is dominated by abandoned-channel-fill and lateral-accretion elements, and active channel fills are common but comparatively subordinate in this area (Figs. 4, 8C). The observed increase in lateral-accretion element size and abundance reflects more extensive point bar growth locally, and abundance of abandoned-channel-fill elements infers a local increase in chute and/or neck cutoff of stream meanders. This clear change in Mesa Rica architecture signifies a relative increase in sinuosity for Mesa Rica streams in reaches extending over the southeastern/downstream flank of the Sierra Grande uplift.

Fluctuations in Mesa Rica channel sinuosity on the southeastern Sierra Grande flank could be due to local increase in fine-grained sediment load, increases in regional slope or changes in stream discharge (see Schumm et al., 1987). Both climate change and stream piracy could have altered discharge and/or sediment load in Mesa Rica rivers and, thus, affected their sinuosity. A substantial fluctuation in either of these variables by such events, however, should have influenced long segments of individual Mesa Rica rivers, and thus broad segments of the Mesa Rica fluvial plain. In contrast, the proposed sinuosity increases are prolific for the short channel reaches on the downstream flank of the Sierra Grande structure only and are insignificant in other areas. Study of Mesa Rica Sandstone revealed no mechanism which could have changed load and/or discharge in most all Mesa Rica rivers for this short reach, and then consistently changed these variables back to the original conditions just 20 km farther downstream. Eustasy could have affected slope of Mesa Rica streams, but a mechanism for such a localized effect is similarly not forthcoming. Such strong coincidence between increased Mesa Rica channel sinuosity and the Sierra Grande east flank is best attributed to elevated gradients encountered by Mesa Rica rivers on the southeastern/downstream flank of a topographic welt rising penecontemporaneously above reactivated Sierra Grande basement structures.

Sinuosity increase as an adjustment of stream channels to such newly increased slope is well documented in experimental flume studies (Friedkin, 1945; Schumm and Khan 1972;

Schumm et al., 1987). Flume studies also reveal that subtle uplifts induced perpendicular to the median of stable meandering channels will prompt sinuosity increase on the downstream uplift flank in a matter analogous to that proposed for the Mesa Rica system (Ouchi, 1985). Such sinuosity effects are also seen on the modern Mississippi River of southeastern Missouri, where the river passes over the crest of the active Lake County uplift (10–m topographic relief) and increases in sinuosity on the steepened downstream flank of this structurally generated topographic welt (Russ, 1982).

The present data do not reveal or refute substantial effects on Pajarito deposition by the Sierra Grande uplift. The photo-panoramic techniques used in this study are not highly effective at revealing details of highly covered strata, such as that of the Pajarito Formation. More focused study on the Pajarito Formation, possibly using different techniques, might reveal tectonic effects missed in this study.

CONCLUSIONS

Early Cretaceous reactivation of the Sierra Grande uplift in northeastern New Mexico resulted in generation of a subtle topographic welt above Sierra Grande basement structure. Two of the many possible effects on coeval depositional processes which could be imposed by such structurally generated topography are exemplified in lithofacies distribution and sequence architecture of overlying Lower Cretaceous strata. These effects are: (1) thinning of marine Glencairn strata coincident with Sierra Grande basement structure, owing to non-deposition and/or preferential erosion of marine strata above Sierra Grande topography and (2) local increase in proportion of abandoned-channel-fill and lateral-accretion architectural elements coincident with the Sierra Grande east flank, owing to locally increased gradient, and thus sinuosity, in Mesa Rica channels on the southeastern/paleodownstream flank of the Sierra Grande uplift.

ACKNOWLEDGMENTS

This research was funded by grants from the Petroleum Research Fund of the American Chemical Society and Southeast Missouri State University. Thanks are extended to Donald Hattin, Lee Suttner, Michael Hamburger and John Droste, who each provided valuable comments and review throughout the course of this research. Thanks also go to Bob O'Dean who offered valuable comments and insight during field research and Betty Black, who helped in preparation of photography. Robert Weimer and Mark Kirschbaum both offered valuable comments in thorough reviews of this manuscript.

REFERENCES

ALLEN, J. R. L., 1963, The classification of cross-stratified units with notes on their origin: Sedimentology, v. 2, p. 93–114.

ALLEN, J. R. L., 1965, A review of the origin and characteristics of recent alluvial sediments: Sedimentology, v. 5, p. 89–191.

ATALIK, E., 1984, Depositional facies and environment of the Dakota Sandstone in northwestern Cimarron County, Oklahoma: Unpublished M.S. Thesis, University of Tulsa, Tulsa, 124 p.

BALDWIN, B. AND MUEHLBERGER, W. R., 1959, Geologic studies of Union County, New Mexico: New Mexico Bureau of Mines and Mineral Resources Bulletin, v. 63, 171 p.

BALTZ, E. H., 1965, Stratigraphy and history of Raton Basin, and notes on San Luis Basin, Colorado–New Mexico: American Association of Petroleum Geologists Bulletin, v. 49, p. 2041–2075.

BEJNAR, C. R. AND LESSARD, R. H., 1976, Paleocurrents and depositional environments of the Dakota Group, San Miguel and Mora counties, New Mexico: Albuquerque, New Mexico Geological Society, 27th Annual Field Conference Guidebook, p. 157–163.

BUCHER, W. H., 1933, The Deformation of the Earth's Crust: Princeton, Princeton University Press, 518 p.

BURNETT, A. W. AND SCHUMM, S. A., 1983, Active tectonics and river response in Louisiana and Mississippi: Science, v. 222, p. 49–50.

DE MIRANDA, F. P. AND DA BOA HORA, M. P. P., 1986, Morphostructural analysis as and aid to hydrocarbon exploration in the Amazonas Basin, Brazil: Journal of Petroleum Geology, v. 9, p. 163–178.

DOLSON, J. C., MULLER, D., EVETTS, M. J., AND STEIN, J. A., 1991, Regional paleotopograhic trends and production, Muddy Sandstone (Lower Cretaceous), central and northern Rocky Mountains: American Association of Petroleum Geologists Bulletin, v. 75, p. 409–435.

ESCHARD, R., RAVENNE, C., HOUEL, P., AND KNOX, R., 1991, Three dimensional reservoir architecture of a valley-fill sequence and a deltaic aggradational sequence: influences of minor sea-level variations (Scalby Formation, England), *in* Miall, A. D. and Tyler, N., eds., The Three Dimensional Facies Architecture of Terrigenous Clastic Sediments and its Implications for Hydrocarbon Discovery and Recovery: Tulsa, SEPM Concepts in Sedimentology and Paleontology 3, p. 134–147.

FISK, H. N., 1955, Sand facies of recent Mississippi delta deposits: Proceedings of the Forth World Petroleum Congress, Section 1/C, p. 337–398.

FRIEDKIN, J. F., 1945, A laboratory study of the meandering of alluvial rivers: Vicksburg, United States Waterways Experiment Station Mississippi River Commission, 18 p.

GILBERT, J. L. AND ASQUITH, G.B., 1976, Sedimentology of braided alluvial interval of Dakota Sandstone northeastern New Mexico: Socorro, New Mexico Bureau of Mines and Mineral Resources Circular 150, 16 p.

GILLETTE, D. D. AND THOMAS, D. S., 1985, Dinosaur tracks in the Dakota Formation (Aptian–Albian) at Clayton Lake State Park, Union County, New Mexico: Albuerque, New Mexico Geological Society, 36th Annual Field Conference Guidebook, p. 283–288.

GOODRICH, H. B., 1898, Geology of the Yukon Gold District, Alaska: Washington, D. C., United States Geological Society Annual Report, no. 18, Part III, 180 p.

GUSTASON, E. R. AND KAUFFMAN, E. G., 1985, The Dakota Group and the Kiowa–Skull Creek Cyclothem in the Canon City–Pueblo area, Colorado, *in* Pratt, L. M., Kauffman, E. G., and Zelt, F. B., eds., Socieity of Economic Paleontologists and Mineralologists Field Trip Guidebook No. 4, Mid-year Meeting: Golden, Socieity of Economic Paleontologists and Mineralologists, p. 72–89.

HATTIN, D. E., 1986, Interregional model for deposition of Upper Cretaceous pelagic rhythmites, United States Western Interior: Paleoceanography, v. 1, p. 483–494.

HELLER, P. L., BEEKMAN, F., ANGEVINE, C. L., AND CLOETINGH, S. A. P. L., 1993, Cause of tectonic reactivation and subtle uplifts in the Rocky Mountain region and its effect on the stratigraphic record: Geology, v. 21, p. 1003–1006.

HOLBROOK, J. M., 1996a, Structural noise in seemingly undeformed intraplate regions: Implications from welts raised in a shattered Aptian/Albian U.S. Western Interior: Contributions to Advanced Studies in Geology, v. 1, p 87–93.

HOLBROOK, J. M., 1996b, Complex fluvial response to low gradients at maximum regression: A genetic link between smooth sequence-boundary morphology and architecture of overlying sheet sandstone: Journal of Sedimentary Research, v. 66, p. 713–722,

HOLBROOK, J. M., 1992, Developmental and sequence-stratigraphic analysis of Lower Cretaceous sedimentary systems in the southern part of the United States Western Interior: Interrelationships between eustasy, local tectonics, and depositional environments: Unpublished Ph.D. Dissertation, Indiana University, Bloomington, 271 p.

HOLBROOK, J. M. AND WRIGHT DUNBAR, R., 1992, Depositional history of Lower Cretaceous strata in northeastern New Mexico: Implications for regional tectonics and depositional sequences: Geological Society of America Bulletin, v. 104, p. 802–813.

HOPKINS, J. C., 1985, Channel-fill deposits formed by aggradation in deeply scoured, superimposed distributaries of the lower Kootenai Formation (Cretaceous): Journal of Sedimentary Petrology, v. 55, p. 42–52.

JACKA, A. D. AND BRAND, J. P., 1972, An analysis of the Dakota Sandstone in the vicinity of Las Vegas, New Mexico and eastward to the Canadian River Valley: Albuquerque, New Mexico Geological Society, 23rd Annual Field Conference Guidebook, p. 105–107.

JACKA, A. D. AND BRAND, J. P., 1973, The San Jon section of the Tucumcari Shale, Mesa Rica Sandstone, and Pajarito Shale, *in* Phillips, D. A., ed., Guidebook of interpretations of depositional environments from selected exposures of Paleozoic and Mesozoic rocks in north-central New Mexico: Amarillo, Panhandle Geological Society, p. 44–48.

KING, P. B., 1959, The evolution of North America: Princeton, Princeton University Press, 99 p.

KUES, B. S. AND LUCAS, S. G., 1987, Cretaceous stratigraphy and paleontology in the Dry Cimarron valley, New Mexico, Colorado, and Oklahoma: Albuquerque, New Mexico Geological Society, 38th Annual Field Conference, Guidebook, p. 167–198.

LOCKLEY, M., HOLBROOK, J. M., HUNT, A., MATSUKAWA, M., AND MEYER, C., 1992, The Dinosaur Freeway: A preliminary report on the Cretaceous megatracksite, Dakota Group, Rocky Mountain Front Range, and High Plains, Colorado, Oklahoma, and New Mexico, *in* Flores, R. M., ed., Socieity of Economic Paleontologists and Mineralologists Field Trip Guidebook, Mid-year Meeting: Golden, Socieity of Economic Paleontologists and Mineralologists, p. 39–54.

LETOUZEY, J., WERNER, P., AND MARTY, A., 1990, Fault reactivations and structural inversion, backarc and intraplate compressive deformations, example of the eastern Sunda shelf (Indonesia): Tectonophysics, v. 183, p. 341–362.

LONG, C. S., JR., 1966, Basal Cretaceous strata, southeastern Colorado: Unpublished Ph.D. Dissertation, University of Colorado, Boulder, 479 p.

LUCAS, S. G., HOLBROOK, J. M., SULLIVAN, R. M., AND HAYDEN, S. N., 1987, Dinosaur footprints from the Cretaceous Pajarito Formation, Harding County, New Mexico: Albuquerque, New Mexico Geological Society, 38th Annual Field Conference Guidebook, p. 31–32.

MALLORY, W. M., 1972, Regional synthesis of the Pennsylvanian System, *in* Mallory, W. M., ed., Geologic Atlas of the Rocky Mountain Region: Denver, Rocky Mountain Association of Geologists, p. 111–127.

MEYERS, J. H., SUTTNER, L. J., FURER, L. C., MAY, M. T., AND SOREGHAN, M. J., 1992, Intrabasinal tectonic control on fluvial sandstone bodies in the Cloverly Formation (Early Cretaceous), west-central Wyoming, USA: Basin Research, v. 4, p. 315–333.

MIALL, A. D., 1985, Architectural-element analysis: A new method of facies analysis applied to fluvial deposits: Earth-Science Reviews, v. 22, p. 261–308.

MOODY-STUART, M., 1966, High- and low-sinuosity stream deposits, with examples from the Devonian of Spitsbergen: Journal of Sedimentary Petrology, v. 29, p. 1102–1117.

OUCHI, S., 1985, Response of alluvial rivers to slow active tectonic movement: Geological Society of America Bulletin, v. 6, p. 504–515.

RUSS, D. P., 1982, Style and significance of surface deformation in the vicinity of New Madrid, Missouri: Albuquerque, United States Geological Survey Professional Paper 1236, p. 45–114.

SCHUMM, S. A. AND KHAN, H. R., 1972, Experimental study of channel patterns: Geological Society of America Bulletin, v. 83, p. 1755–1800.

SCHUMM, S. A., MOSELY, M. P., AND WEAVER, W. E., 1987, Experimental Fluvial Geomorphology: New York, Wiley and Sons, 413 p.

SCOTT, R. W., 1970, Stratigraphy and sedimentary environments of Lower Cretaceous rocks, southern Western Interior: American Association of Petroleum Geologists Bulletin, v. 54, p. 1225–1244.

SEILACHER, A., 1964, Biogenic sedimentary structures, *in* Imbre, J. and Newell, N. D., eds., Approaches to Paleoecology: New York, Wiley and Sons, p. 296–315.

SULEIMAN, A. S. AND KELLER, G. R., 1985, A geophysical study of basement structure in northeastern New Mexico: Albuquerque, New Mexico Geological Society, 36th Annual Field Conference, Guidebook, p. 153–160.

TAYLOR, A. M., 1974, Stratigraphy and depositional environments of Upper Jurassic and Cretaceous rocks in Bent, Las Animas, and Otero Counties, Colorado: Unpublished Ph.D. Dissertation, Colorado School of Mines, Golden, 302 p.

THOMAS, R. G., SMITH, D. G., WOOD, J. M., VISSER, J., CALVERLEY-RANGE, E. A., AND KOSTER, E. H., 1987, Inclined heterolithic stratification—terminology, description, interpretation, and significance: Sedimentary Geology, v. 53, p. 123–179.

TYLER, N. AND ETHRIDGE, F. G., 1983, Depositional setting of the Salt Wash Member of the Morrison Formation, southwest Colorado: Journal of Sedimentary Petrology, v. 53, p. 67–82.

WAAGE, K. M., 1953, Refractory clay deposits of south-central Colorado: United States Geological Survey Bulletin, v. 993, 101 p.

WEIMER, R.J., 1984, Relation of unconformities, tectonics and sea level changes, Cretaceous of the Western Interior, United States of America, *in* Schlee, J. S., ed., Interregional Unconformities and Hydrocarbon Accumulation: Tulsa, American Association of Petroleum Geologists Memoir 36, p. 7–36.

WEIMER, R. J. AND SONNENBERG, S. A., 1989, Sequence stratigraphy, Lower Cretaceous, Denver Basin, Colorado, U.S.A, *in* Ginsburg, R. D. and Beaudoin, B., eds., NATO ASI Series, Volume on Cretaceous Resources, Events, and Rhythms: Dordrecht, The Netherlands, Kluwer Academic Publishers.

TECTONIC CONTROLS ON ALLUVIAL SYSTEMS IN A DISTAL FORELAND BASIN: THE LAKOTA AND CLOVERLY FORMATIONS (EARLY CRETACEOUS) IN WYOMING, MONTANA AND SOUTH DAKOTA

J. NATHAN WAY, PATRICK J. O'MALLEY, LEE J. SUTTNER, AND LLOYD C. FURER

Department of Geological Sciences, Indiana University, Bloomington, Indiana 47405

ABSTRACT: Integrated surface and subsurface stratigraphic and sedimentologic analysis of the nonmarine Lower Cretaceous rocks along the eastern margin of the Rocky Mountain foreland demonstrates the critical role of intrabasinal and cratonic deformation in controlling alluvial architecture in a distal foreland-basin setting.

Three coarse-grained intervals within the Lakota Formation can be recognized and correlated throughout the Black Hills and into the subsurface of the Powder River Basin. We informally designate these intervals from oldest to youngest as L1, L2 and L3. Detritus in the L1 interval was deposited in northward-flowing high-sinuosity rivers. The L2 interval lies atop of a regional intraformational angular unconformity which truncates rocks as old as the Middle Jurassic Sundance Formation. L2 detritus was deposited by northeastward-flowing braided rivers. The coincidence of location of these rivers and lineaments reflecting recurrent movement of basement-rooted structures is evidence of intrabasinal tectonic control. A variety of evidence suggests that this movement was associated with transpression along steeply dipping basement-rooted faults. As much as 65 m of stratigraphic throw may have occurred on one of these faults in the study area. Provenance analysis indicates that sediment deposited by L3 estuarine systems was derived from uplift of the Transcontinental Arch located within the craton, approximately 600–km southeast of the study area.

Preliminary chronostratigraphic correlation of Early Cretaceous alluvial deposits throughout Wyoming indicates that intrabasinal deformation may have started as early as 134 Ma in the western Wind River Basin and progressively advanced eastward through the foreland over approximately the next 25 my, culminating slightly less than 110 Ma with the uplift of the Transcontinental Arch. This uplift resulted in a 120° change in direction of paleoslope in the distal Rocky Mountain foreland.

INTRODUCTION

Foreland basins result from flexure of the lithosphere caused by tectonic loading associated with stacking of thrust slabs along the basin margin. These basins preserve a large proportion of nonmarine strata globally (Miall, 1981) and thus are an important setting for the study of alluvial stratigraphy. Many recent stratigraphic and sedimentologic studies of the Early Cretaceous Rocky Mountain foreland of the western U.S. have focused on the response of nonmarine depositional systems to tectonism in the adjacent Sevier fold-thrust belt (e.g., Heller and Paola, 1989; DeCelles and Burden, 1992). Moreover, most basin-fill modeling studies have emphasized either the temporal relationship between basin-margin uplift and sediment delivery to the basin, or the flexural response of the basin to thrust loading (Beaumont, 1981; Wiltschko and Dorr, 1983; Flemings and Jordan, 1990; Yingling and Heller, 1992). These models typically rely on a simplified basin stratigraphy which reflects an assumption of near total control on sediment derivation and dispersal throughout the basin by events in the adjacent fold-thrust belt. Recent investigations of the *medial* portion of the unusually wide Rocky Mountain foreland basin have elucidated the role of intrabasinal deformation on alluvial stratigraphy in this part of the basin (Schwartz, 1983; DeCelles, 1984, 1986; Meyers et al., 1992; Holbrook, 1992, 1993; May, et al., 1995). Our study expands on this recent work by examining alluvial stratigraphy in the *distal* part of the Rocky Mountain foreland.

The nonmarine to transitional marine Lakota and Cloverly Formations in northeastern Wyoming, southeastern Montana and western South Dakota preserve evidence of both intrabasinal and cratonic controls on alluvial systems and illustrate several problems in sequence-stratigraphic analysis of a far-inboard foreland basin setting. For example, the use of an unconformity-bounded interval as a third- or fourth-order stratigraphic sequence is complicated for these strata because erosion and non-deposition have resulted in a stratigraphic record punctuated by multiple local intraformational unconformities. Because of syndepositional intrabasinal tectonism, these unconformities split and merge, hampering recognition of basin-wide unconformities. Furthermore, diachronous deformation, both within the basin and the source areas, complicates interpretation of the depositional history of strata, and therefore limits the predictive value of nonmarine sequence stratigraphy. Although many of these problems can be resolved through detailed local studies, poor chronostratigraphic resolution and lateral discontinuity of strata limit our ability to determine how locally observed phenomena might affect regional stratigraphic interpretation. Consequently, stratigraphic complications recognized in local studies are commonly overlooked in regional, sequence-stratigraphic-scale analyses.

In this paper, we re-evaluate the Lakota stratigraphy of the Black Hills and eastern Powder River Basin in light of new chronostratigraphy and the recent advances in understanding the role of intraplate stresses in reactivating basement-rooted structures within the foreland. New constraints on interpretation of the timing of intra- and extrabasinal syndepositional tectonism allow us to provide a more accurate regional interpretation of the stratigraphy than past studies.

This paper complements the study by Holbrook and White (this volume), which is an account of intrabasinal structural control on distal foreland-basin alluvial systems in strata of the same age in Colorado and New Mexico. Holbrook and White (this volume) are able to document the influence of subtle tectonic movements on alluvial systems through detailed facies and paleocurrent analysis. In contrast, in the northern Black Hills syndepositional structural relief is great enough that the effects of intrabasinal tectonism are evident in thickness trends and erosional relationships between the Lakta Formation and older strata. Although our documentation of intrabasinal tectonism focuses on these thickness relationships, we also give examples of how differential interbasinal subsidence affected Lakota alluvial facies and regional depositional gradients. We use integrated outcrop, drill-hole geophysical and seismic reflection data to interpret the style of syndepositional intrabasinal tectonism. We also demonstrate that uplift of the Transcontinental Arch influenced Lakota and Cloverly stratigraphy in the distal part of the foreland basin. We expect that documentation and interpretation of the origin of complex stratigraphic relationships within a distal foreland-basin setting will guide future quantitative and sequence-stratigraphic modeling of the nonmarine fill of foreland basins.

REGIONAL SETTING AND METHODS OF STUDY

During Late Jurassic and Early Cretaceous time, a broad foreland basin began to develop in the western interior U.S. in response to the earliest phases of deformation within the Sevier fold-thrust belt (Fig. 1). Flexural subsidence caused by initial thrust stacking and thrust-derived sediment loading was concentrated in the proximal part of the foreland basin, extending about 200 km toward the craton from the thrust belt (Burbank et al., 1989). The area of study is located approximately 400 km inboard from this site of maximum subsidence and includes deposits in the distal portion of the "overfilled" Rocky Mountain foreland basin (DeCelles and Burden, 1992) (Fig. 1). The area of detailed subsurface study (Fig. 1C) is centered on the northern Black Hills and lies approximately 500 km northwest of the Transcontinental Arch in southwestern Nebraska. This

arch is an intracratonic uplift where onlap of over 330 m of Cretaceous strata onto early Paleozoic rocks has been documented by Shurr (1981) and Bunker (1981). Data for our study were collected from more than 120 wells in the eastern Powder River Basin and supplemented by detailed study of more than 75 surface sections in the northern Black Hills (Figs. 1B, C, D). Measured outcrop sections provided thickness, paleoflow direction and lithologic data which enabled three-dimensional mapping of fluvial channels.

REGIONAL EARLY CRETACEOUS STRATIGRAPHY

Throughout essentially all of Wyoming, Lower Cretaceous nonmarine strata (i.e., the coeval Lakota and Cloverly Formations) unconformably overlie nonmarine Jurassic strata. Farther east in South Dakota Lower Cretaceous rocks progressively onlap older sediments which in turn offlap the Transcontinental Arch (Fig. 2). Near its eastern limit, the Lakota Formation overlies Proterozoic metasedimentary rocks (Schoon, 1971; McGookey et al., 1972). In the study area near the Black Hills, the Lakota Formation overlies the Upper Jurassic Morrison For-

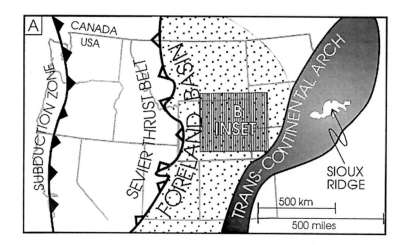

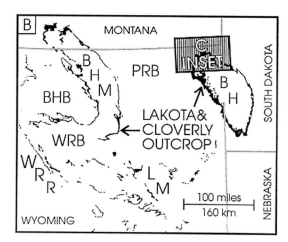

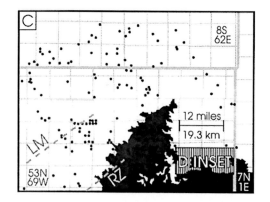

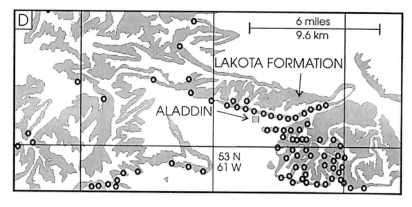

FIG. 1.—Maps showing location and geologic setting of study area. (A) Early Cretaceous tectonic setting of regional study area. Stippled pattern shows extent of nonmarine early Cretaceous sediment. Sioux Ridge is portion of Transcontinental Arch with over 330 m of onlap of Cretaceous strata. Vertical hachures shows location of Figure 1B. Geology from Bunker (1981), Heller and Paola (1989), Shurr (1981) and DeCelles and Burden (1992). (B) Areas of outcrop of Lakota and Cloverly Formations in and adjacent to study area. Vertical hachures shows area of Figure 1C. BH, Black Hills; BHB, Bighorn Basin; BHM, Bighorn Mtns.; LM, Laramie Mtns.; PRB, Powder River Basin; WRB, Wind River Basin; WRR, Wind River Range. (C) Location of drill holes forming subsurface data base. Black is area where Cretaceous rocks are absent over the Black Hills uplift. Gray dashed lines show the location of Little Missouri (LM) and Rozet (RZ) lineaments of Slack (1981). Vertical hachures shows area of Figure 1D. (D) Outcrop belt of the Lakota Formation in the Aladdin area. Circles mark locations of measured sections where paleocurrent data were collected. Outcrop base map modified from Darton and O'Harra (1905).

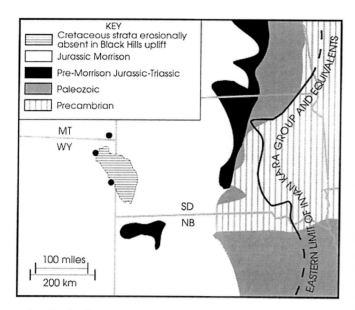

FIG. 2.—Pre-Cretaceous subcrop map modified from McGookey (1972), Anna (1986) and Mapel and Gott (1959). Over most of Wyoming, Cloverly and Lakota Formations rest on the Jurassic Morrison Formation. The three solid circles are sites where the Lakota Formation rests on the Jurassic Sundance Formation.

mation, except at a few localities where Lakota rocks are in angular unconformable contact with the Late Jurassic Sundance Formation (Fig. 2). The duration of time represented by the unconformity separating the Lakota and Cloverly Formations from the Morrison Formation is variable. However, this variability has a relatively minor effect on our correlations of Lower Cretaceous strata because we do not use the contact as an isochronous surface. Instead we rely on detailed integrated outcrop and subsurface data to discern the different stratigraphic levels of coarse clastic units of the Lakota and Cloverly interval and rely on dates from other studies for our chronostratigraphic control.

Our correlations are complicated, however, by lack of agreement among earlier workers on placing the Morrison/Lakota contact in the surface and subsurface. Most regional lithologic correlations place the contact at the base of the stratigraphically lowest dark chert-bearing, coarse-grained sandstone or conglomerate and equate this contact with a regional Jurassic–Cretaceous unconformity (Wilson, 1958; Wulf, 1959; Young, 1960; Haun and Barlow, 1962; Curry, 1962; Anna, 1986; Heller and Paola, 1989; Dolson and Muller, 1994). This unconformity corresponds to the K–1 sequence boundary of Pipiringos and O'Sullivan (1978). This correlation implies that the Lower Cretaceous coarse-grained sandstone or conglomerate body is a widespread sheet-like deposit covering most of Wyoming. But May et al. (1995) have demonstrated that in the eastern Wind River Basin, basal Cloverly conglomerates are absent and that the oldest chert-bearing, coarse sandstone or conglomerate overlies Cloverly mudstone. Where the lower conglomerate is absent, purple and gray smectitic mudstone of the Cloverly Formation directly overlies green and red illitic mudstone of the Morrison Formation. Similarly, in the Black Hills, coarse-grained, chert-bearing sandstone commonly overlies a mudstone-dominated interval within the Lakota Formation. Waage

(1959) noted this relationship and proposed that the contact in this area be placed at the base of the first carbonaceous mudstone, which is easy to distinguish in the field from calcareous mudstone of the Morrison Formation. We employ Wagge's (1959) criterion for placing the contact in outcrop, and we are able to extend this contact into the subsurface through correlation of closely spaced surface and subsurface sections.

Several workers recognize multiple stratigraphic levels of chert-bearing, coarse sandstone or conglomerate within the Cloverly and Lakota Formations (e.g., Mapel and Gott, 1959; Ostrom, 1970; May, 1992). May et al. (1995) observed three distinct stratigraphic intervals of the Cloverly Formation containing conglomerate or conglomeratic sandstone in the Wind River Basin. They informally designated these from oldest to youngest as the A–, B– and C–intervals. In the Bighorn Basin, Ostrom (1970) defined four Cloverly subdivisions which are similarly designated IV, V, VI and VII. Divisions IV and VI contain conglomerate or conglomeratic sandstone. In the northern Black Hills, we recognize three Lakota subdivisions which we designate informally from oldest to youngest as L1, L2 and L3 (Fig. 3). The lithologic and sedimentologic criteria which are the basis for these subdivisions are discussed in the next section. The L1, L2 and L3 intervals equate with the S_1, S_2, and S_3 intervals defined for the western Black Hills by Mapel and Gott (1959). The L2 and L3 intervals contain coarse-grained chert-bearing sandstone and conglomerate.

Our preliminary correlations equate Cloverly C–interval conglomerate of the eastern Wind River Basin defined by May et al. (1995) with Lakota L2 conglomeratic sandstone in the northern Black Hills (Way et al., 1994; Fig. 4). The basis for this correlation is a volcano-genic mudstone interval which separates the C-chert-pebble conglomerate from the underlying A– and B–intervals of the Cloverly Formation in the Wind River basin (May et al., 1995). We use this volcano-genic interval as a time-equivalent horizon that we can correlate in the subsurface from the Wind River and Bighorn Basins to the northern Black Hills where metabentonite occurs in well cuttings (Union, Govt. Lowe #1; sec.1, T8S, R56E; Carter Co., Montana) and to the southern Black Hills where tuff, ash and felsite beds have been reported in the oldest Lakota fluvial rocks (Gott et al., 1974). Magnetostratigraphy, radiometric dates and a few biostratigraphic dates in the Wind River and Bighorn Basins (May et al., 1995) and biostratigraphy (Pish, 1988; Rich et al., 1988) in the Black Hills support our chronostratigraphic correlations of the Lakota and Cloverly subdivisions shown in Figure 4.

The Cloverly Formation is overlain by the marginal marine "rusty beds" of the Thermopolis Formation in the Wind River Basin. These Albian-age beds (Dyman et al., 1994) are equivalent to the Sykes Mountain Formation in the Bighorn Basin (Moberly, 1960). In the Black Hills the Lakota Formation is separated from the overlying Albian Fall River Formation by the "transgressive disconformity" of Waage (1959), which we interpret as a flooding surface.

LAKOTA STRATIGRAPHY IN THE NORTHERN BLACK HILLS

The Lakota–Cloverly interval has been subdivided in the Wind River Basin (May, 1992), Bighorn Basin (Ostrom, 1970) and in the southern (Gott et al., 1974) and western Black Hills (Mapel and Gott, 1959). However, prior to this study, no such

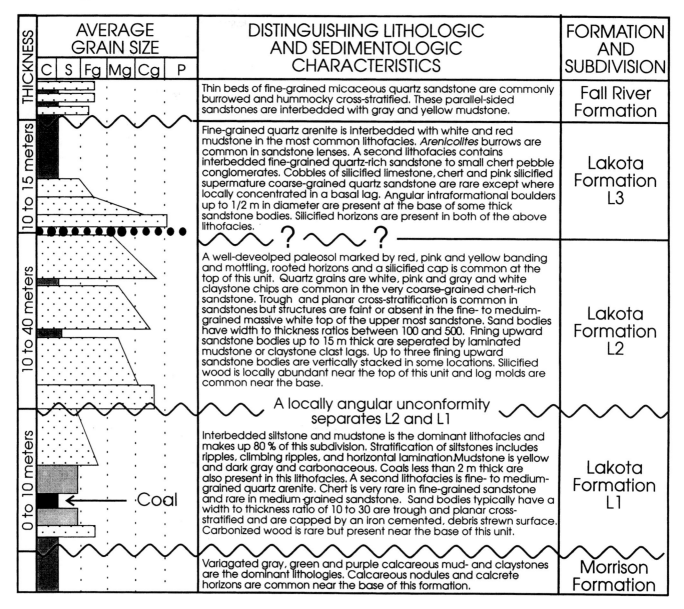

THICKNESS	AVERAGE GRAIN SIZE						DISTINGUISHING LITHOLOGIC AND SEDIMENTOLOGIC CHARACTERISTICS	FORMATION AND SUBDIVISION
	C	S	Fg	Mg	Cg	P		
							Thin beds of fine-grained micaceous quartz sandstone are commonly burrowed and hummocky cross-stratified. These parallel-sided sandstones are interbedded with gray and yellow mudstone.	Fall River Formation
10 to 15 meters							Fine-grained quartz arenite is interbedded with white and red mudstone in the most common lithofacies. *Arenicolites* burrows are common in sandstone lenses. A second lithofacies contains interbedded fine-grained quartz-rich sandstone to small chert pebble conglomerates. Cobbles of silicified limestone, chert and pink silicified supermature coarse-grained quartz sandstone are rare except where locally concentrated in a basal lag. Angular intraformational boulders up to 1/2 m in diameter are present at the base of some thick sandstone bodies. Silicified horizons are present in both of the above lithofacies.	Lakota Formation L3
10 to 40 meters							A well-deveolped paleosol marked by red, pink and yellow banding and mottling, rooted horizons and a silicified cap is common at the top of this unit. Quartz grains are white, pink and gray and white claystone chips are common in the very coarse-grained chert-rich sandstone. Trough and planar cross-stratification is common in sandstones but structures are faint or absent in the fine- to meduim-grained massive white top of the upper most sandstone. Sand bodies have width to thickness ratios between 100 and 500. Fining upward sandstone bodies up to 15 m thick are seperated by laminated mudstone or claystone clast lags. Up to three fining upward sandstone bodies are vertically stacked in some locations. Silicified wood is locally abundant near the top of this unit and log molds are common near the base.	Lakota Formation L2
							A locally angular unconformity separates L2 and L1	
0 to 10 meters						·Coal	Interbedded siltstone and mudstone is the dominant lithofacies and makes up 80 % of this subdivision. Stratification of siltstones includes ripples, climbing ripples, and horizontal lamination. Mudstone is yellow and dark gray and carbonaceous. Coals less than 2 m thick are also present in this lithofacies. A second lithofacies is fine- to medium-grained quartz arenite. Chert is very rare in fine-grained sandstone and rare in medium-grained sandstone. Sand bodies typically have a width to thickness ratio of 10 to 30 are trough and planar cross-stratified and are capped by an iron cemented, debris strewn surface. Carbonized wood is rare but present near the base of this unit.	Lakota Formation L1
							Variagated gray, green and purple calcareous mud- and claystones are the dominant lithologies. Calcareous nodules and calcrete horizons are common near the base of this formation.	Morrison Formation

FIG. 3.—Detailed stratigraphy of the Lakota Formation in the northern Black Hills. C–clay, S–silt, Fg–fine-grained sandstone, Mg–medium-grained sandstone, Cg–coarse-grained sandstone, P–pebbles.

subdivision has been reported for the northern Black Hills. Subdivision in this area is difficult because fluvial deposits of different ages are concentrated in a relatively thin stratigraphic interval. Previous paleoflow studies in the northern Black Hills have treated all Lakota fluvial deposits as part of a single fluvial system, thereby hindering interpretation of paleocurrent data and fluvial architecture. For example, Mapel and Pillmore (1962) grouped all paleocurrent data from the Lakota Formation in the northern Black Hills into a single population and concluded that the large azimuthal dispersion of data suggested " . . . a complicated local pattern of stream flow such as would result from meandering streams that flowed generally northward." As a result of our intensive outcrop study, we have recognized three Lakota subdivisions in the northern Black Hills based on sedimentologic criteria and clast composition

(Fig. 3). These subdivisions facilitate our interpretation of regional stratigraphy as discussed in the previous section and add to our understanding of the complex depositional and erosional history of these strata. These subdivisions also provide detailed stratigraphic control for our paleoflow analysis.

Lower Lakota – L1

The oldest Lakota subdivision (L1; Fig. 3) corresponds to the Chilson Member of the southern Black Hills (Mapel and Gott, 1959). It consists primarily of carbonaceous gray and yellow mudstone and siltstone and subordinate fine-grained quartz-rich sandstone. In the vicinity of Aladdin, coal seams up to 2 m thick occur in this interval. Channel sandstone bodies have average width/thickness ratios of <20; detailed mapping suggests the channels bend sharply with an average radius of curvature

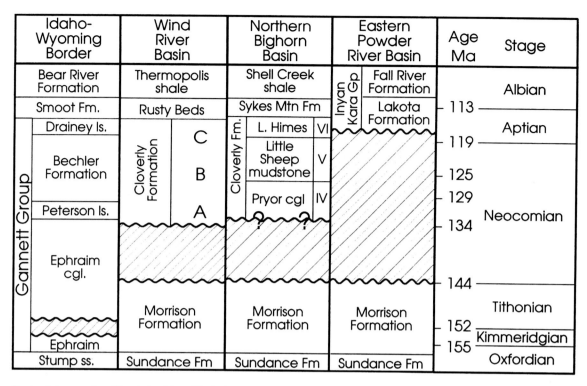

FIG. 4.—Regional time-stratigraphic relationships of Early Cretaceous rocks in Wyoming. Lakota subdivisions are detailed in Figure 3. Early Albian age for middle part of Lakota Formation is based on palynology of Pish (1988). Question marks show uncertainty for the age of the base of the Cloverly Formation in the Wind River Basin. Recent pollen studies and lithostratigraphic correlation to the Bighorn Basin suggest that Cloverly strata may be younger than is shown in this figure. Absolute ages of stage boundaries based on Harland et al. (1982).

of <1 km. Paleocurrent dispersion is relatively high (Fig. 5). On these bases, we infer that L1 strata were deposited in sinuous, mud-dominated fluvial systems.

The contact between the L1 and the overlying L2 Lakota subdivisions is locally an angular unconformity. Earlier workers have incorrectly interpreted this surface as the Lakota–Morrison contact. Near Aladdin, Wyoming (SE1/4, NE1/4, sec. 12, T53N, R61W) and near Sturgis, South Dakota (SW1/4, NE1/4, sec.23, T5N, R5E), the angular L1–L2 contact is evident in outcrop, with the lower L1 strata dipping from 50 to 90° beneath the nearly horizontal L2 sandstone (Fig. 6). In most outcrops, the L1–L2 contact is not angular but is marked by an iron-cemented debris-strewn surface, which commonly caps the uppermost L1 sandstone and may reflect an unconformable contact. Several authors recognize this intraformational unconformity in Lakota exposures in the southern and western Black Hills (Mapel and Pilmore, 1962, 1963a, b, 1964; Mapel and Gott, 1959) where it is also marked by a change in ostracod fauna (Sohn, 1979), a change in plant macrofossil assemblage (Ward, 1899) and a transition in preservation state of fossil wood (i.e., fossil wood in L1 is carbonized, but in L2 it is silicified). In much of the northern Black Hills, the L1 interval is absent, probably because of truncation beneath the unconformity.

Middle Lakota – L2

The middle Lakota (L2) interval is dominantly sandstone in the Aladdin area and commonly forms a prominent cliff. We correlate L2 with the lower part of the Fuson Member as defined for the western Black Hills by Post and Bell (1961). The most common pebbles in the L2 sandstone are black and white pure, spicular, and chalcedonic varieties of chert; but pebbles of silicified limestone containing crinoids and Paleozoic fusilinids are also present. L2 thickness ranges between 10 and 40 m in the Aladdin area, but L2 may be absent in other parts of the northern Black Hills. Where L2 is thick, up to four 10–m–thick sandbodies are stacked vertically, separated by thin sheets of laminated claystone or claystone clast lags. The average sandstone body width thickness ratio is near 100; however, east of Aladdin the edges of these sandbodies overlap and appear from a distance to form nearly continuous sandstone sheets 10 to 20 m thick and over 20 km wide. Sandbodies are straight in map view and can be traced for up to 2.5 km by following paleoflow indicators. Sandbody width thickness ratios, map-view geometry, paucity of fine-grained sediments, abrupt vertical changes in grain size, abundant channel and downstream-accreting bar forms, and low paleocurrent variability (Fig. 5) suggest that L2 sandstone was deposited in straight, sand-dominated, braided-fluvial systems.

The top of L2 is commonly marked by a well-developed paleosol. Pedofacies include pink, orange, red and white, banded and mottled, rooted and silicified surfaces on mudstone. A massive, fine- to medium-grained white sandstone is also present immediately below the upper L2 contact and may reflect intense pedoturbation during deposition.

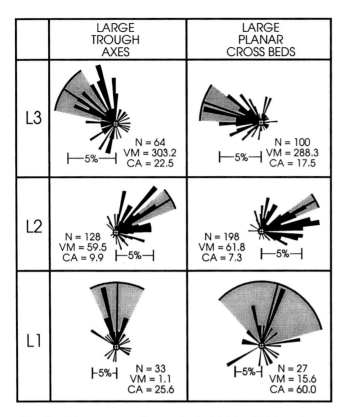

FIG. 5.—Paleocurrent rose diagrams for the L1, L2 and L3 intervals of the Lakota Formation of the Aladdin area. Scale bar for each rose diagram shows relative length of 5% of data. Shaded area shows arc of 95% confidence angle (CA). N – number of data points, VM – vector mean.

Upper Lakota – L3

The upper L3 interval contains the largest clasts within the Lakota Formation. The dominant facies is red, orange or white mudstone locally containing extrabasinal cobble lags. Cobbles up to 15 cm long in diameter include white quartzite, silicified limestone, and pink and red coarse sandstone and pebble conglomerate, the latter apparently derived from either the Cambrian Deadwood Formation or Proterozoic Sioux Quartzite. Sandstone ribbons with lenses of chert-pebble conglomerate are subordinate. Although *Arenicolites* burrows are locally abundant in fine-grained L3 sandstone, other types of trace fossils are conspicuously absent. *Arenicolites* borrows are commonly associated with transitional marine environments (Ekdale et al., 1984). Abundance of a single species where diversity is severely limited is common in stressed environments such as transitional-marine settings. Trace fossils are the basis for our interpretation that L3 strata reflect deposition in a estuary.

PALEOCURRENT ANALYSIS

Paleocurrent data (Fig. 5) were collected from large >1–m–thick trough and planar cross-beds within each Lakota subdivision. According to DeCelles et al. (1983), large-scale trough cross-stratification is the best indicator of the main flow direction of a channel; however, dip directions from large-scale planar cross-stratification generally parallel trough axes in the Lakota sandstone and are also included in this analysis. Although

planar cross-stratification may develop as avalanche faces on laterally migrating bars, dip of most large planar cross-strata does not diverge from the dip of bounding accretion and reactivation surfaces and, therefore, probably formed by downstream migration of bed or bar forms. Some large-scale, planar cross-stratification grade laterally and down dip into inclined higher-order surfaces bounding small-scale cross-beds and may reflect accretion on laterally migrating bars, but this architecture is rarely preserved. We measured dip directions from smaller-scale beds of cross-strata but omitted these data from our analysis, because we believe they reflect flow off emerging bar forms, in subsidiary channels or in local eddies on lateral surfaces of bars. These smaller cross-beds are commonly inclined on higher-order bounding surfaces whose dip direction is oblique (>45°) to flow.

At several localities, dip directions of cross-strata were measured at several points laterally within a single large-scale, planar cross-bed to estimate spatial variability of flow on a bed form. We also measured dip directions in several stacked sets of cross-beds within a coset to estimate sequential variability on a single bed form. In both cases, azimuthal variance was less than 90°. Multiple dip directions measured within a single cross-bed or from several cross-beds within a set were averaged to provide a single data point in our analysis. Measurements from unpaired sets of trough limbs were used to calculate less than 10% of trough-axis data. Instead, most trough axes were measured directly in the field or determined stereographically from two paired limbs on a single trough.

Our analysis indicates that L1 fluvial systems flowed generally northward, L2 systems flowed east-northeastward, and L3 systems flowed westward (Fig. 5). The same the near reversal of flow direction between L2 and L3 is similar to that reported by Mapel and Pillmore (1962) in the undivided Lakota Formation; but we interpret that this reversal reflects the flow direction of straight channels affected by a regional change of gradient, rather than the result of deposition by meandering fluvial systems. This interpretation is supported by paleocurrent data from other parts of the northern Black Hills which show a similar change in flow direction between L2 and L3 (O'Malley, 1994). Also, our detailed mapping of geometry of sandstone bodies in outcrops in the Aladdin area indicate straight channels for both L2 and L3 fluvial systems.

SYNDEPOSITIONAL TECTONISM

Extrabasinal Tectonism

Clast composition and paleocurrent data indicate that both source-area location and regional gradient changed after deposition of the L2 interval. We infer that the L2 rivers originated in the Sevier fold-thrust belt approximately 650 km west-southwest of the study area but that the L3 rivers flowed from sources in the east associated with uplift on the craton, specifically the Transcontinental Arch, at least 2000 km inboard from the edge of the North American plate. Clasts in the L2 interval are similar to those of the Cloverly chert-pebble conglomerate throughout most of Wyoming. In contrast, west-northwest-directed paleocurrents and Proterozoic Sioux Quartzite or Cambrian Deadwood Formation cobbles in the L3 interval, which are not known to be present in correlative strata farther west, demonstrate the later influence of cratonic sources. Earlier, Gott et al.

(1974) inferred that a similar change from an east-dipping to a west-dipping paleoslope occurred between deposition of lower and upper Lakota sandstone based on heavy mineral assemblages and paleocurrent data.

Although early Cretaceous uplift of the Transcontinental Arch has been proposed by several authors (Wilson, 1958; MacKenzie and Ryan, 1962; Gott et al., 1974; DeCelles and Burden, 1992), estimates of the amount of uplift have not been made. Because stream gradient is directly correlated to grain diameter and inversely proportional to channel depth (Heller and Paola, 1989), these parameters can be used to make a general comparison of the gradient of the east and west-flowing rivers. In the Aladdin area, lenses of very coarse-grained sandstone and small (10— diameter) pebble conglomerates occur in L3 channels which have a preserved depth of less than 2 m. Following the method of Heller and Paola (1989), we calculate a slope of about 6.5×10^{-4} for L3 rivers. Using the same calculation, but based on a maximum grain size of 8 mm and channel depth of 10 m as measured in sandstone bodies within the L2 interval, we infer that L2 channels had a gradient of about 1.5×10^{-4}. Because the Aladdin area is nearly equidistant from eastern and western source areas, this difference in gradient indicates that uplift of the Transcontinental Arch supplying L3 clasts resulted in regional slopes greater than those associated with uplift of L2 sources in the fold-thrust belt along the proximal edge of the Rocky Mountain foreland. This implies that extra-basinal structural activity inboard from the cratonic margin of the foreland basin should be a significant factor in modeling of foreland sedimentary processes.

Intrabasinal Tectonism

Several lines of evidence indicate that Lakota deposition and preservation were influenced by subtle structural movement within the foreland basin. These include: (1) alignment and close spatial association of Lakota channels with paleostructural lineaments defined on the basis of independent evidence (Slack, 1981); (2) multistorey stacking of Lakota sandstone bodies in areas of inferred large local subsidence and (3) thickness relationships with underlying strata that indicate preservation of the Lakota was controlled by differential subsidence.

Alignment of Lakota channels with basement-rooted lineaments is best documented by O'Malley (1994) in an area adjacent to our study area to the southwest. Here O'Malley has mapped a major southwest/northeast-trending Lakota (L2) channel in the subsurface paralleling and lying immediately northwest of the Rozet Lineament of Slack (1981; Fig. 7). O'Malley's detailed paleocurrent analysis of L2 sandstone from nearby outcrops confirms the northeasterly flow of the L2 rivers which we propose in this study.

Multistorey stacking of L2 sandstones is best exhibited in outcrop near Aladdin (NE 1/4, sec. 12, T53N, R61W) in association with the angular unconformity described earlier. As many as three amalgamated L2 sandstone bodies totaling 27 m in thickness have been observed 0.5 km east of the angular unconformity. Just 2 km west of the unconformity, L2 sand bodies with an aggregate thickness of <10 m occur. We infer that this multistorey stacking and amalgamation reflect structural trapping of Lakota rivers through time in a zone of relatively large fault-induced subsidence, a phenomenon that has been modeled by Bridge and Leeder (1979).

A

B

FIG. 6.—(A) Photograph of angular unconformity (dashed line) between lower (L1) and middle (L2) Lakota subdivisions near Sturgis, S.D. Lower bar parallels bedding in interbedded carbonaceous mudstone and fine-grained ripple cross-bedded sandstone of the L1. L1 bedding dips 56° to the northeast. The upper bar parallels bedding in the overlying L2 cross-bedded sandstone. L2 bedding dips 4° to the east. Length of the upper bar is 1 m. (B) Close up of the unconformable contact at arrow shown in Figure 5. Distance between marks on staff is in feet.

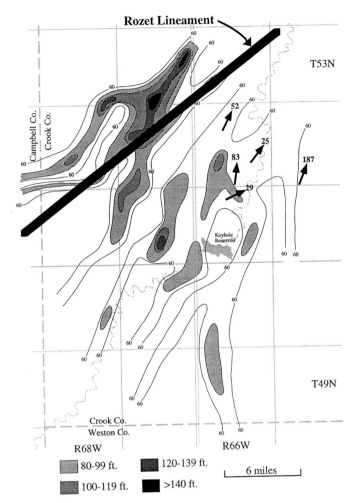

FIG. 7.—Isopach map of the Lakota (L2) sandstone interval in northern Crook County, Wyoming. This area is located about 40 km (25 mi) southwest of our study area. Location of the Rozet Lineament is taken from Slack (1981). Discontinuous nature of sandstone bodies southeast of the lineament reflects local erosion by L3 rivers. Arrows show direction of paleocurrent mean for the L2 sandstone measured in Lakota exposures. Bold numbers denote number of paleocurrent measurements used to determine mean. This is taken from O'Malley (1994).

We propose that localized thickness variations in the Lakota Formation are a function of preservation caused by local differential subsidence and not the result of erosion by Lakota channels. Typically the Lakota Formation is thickest where the underlying Morrison and more rarely the Sundance Formations are thickest (sections 2, 3, 4, 9 and 10 in Fig. 8); conversely, where the Lakota Formation is thinnest, so is the Morrison–Sundance interval (sections 5 and 7 in Fig. 8). If thickened Lakota strata simply reflects incision by Lakota rivers into and through the underlying Morrison Formation, this relationship would not exist. Instead, in order to explain the preservation of the Morrison and Sundance Formations in areas of greatest Lakota thickness, we suggest control by differential subsidence. The most significant structural movement probably occurred prior to deposition of the L2 sandstone to produce the locally angular unconformable relationship between the L1 and L2 intervals observed in outcrop near Aladdin and Sturgis (Fig. 6).

In the subsurface to the northwest, this angular unconformity truncates all of the L1 interval, all of the Morrison Formation and part of the Sundance Formation (section 5 in Fig. 8). The fact that the upper Lakota is thin in this same area suggests that structual relief developed during Lakota deposition.

Nature of Intrabasinal Syndepositional Movement

The influence of recurrent fault movement on the location and migration of early Cretaceous river systems in the Wind River Basin of Wyoming has been documented by Meyers et al. (1992). In the Black Hills, recurrent faulting may have been responsible for development of an angular unconformity separating the upper Lakota from the Sundance Formation, as reported near Devils Tower and east of Newcastle, Wyoming by Mapel and Gott (1959). The mechanical nature of the Early Cretaceous structural deformation that occurred in these areas has not been interpreted.

Spatial variability of the control provided by subsurface data limits our confidence in mapping of paleostructural features in the study area. Detailed uranium exploration maps based on densely spaced drill holes in townships 54 and 55 north, range 67 west were produced by the Homestake Mining Company. These maps suggest that Early Cretaceous deformation was locally intense and characterized by short wavelength folding and faulting along the Little Missouri lineament identified by Slack (1981; Fig. 1). Unfortunately publicly accessible uranium exploration data are limited and oil exploration data only provides sparse control (Fig. 1). Despite these limitations, we have interpreted the Early Cretaceous structural style through consideration of a combination of structural aspects cited below and have applied this interpretation to map structures. We conclude that movement on one such structure north of the Black Hills was associated with transpression along a steeply dipping basement-rooted fault. Based on the thickness of the Lakota, Morrison, and upper part of the Sundance Formations eroded from the upthrown side of this fault, nearly 65 m of stratigraphic throw likely occurred.

Cross section B–B' in Figure 9 is a paleostructural version of section A–A' in Figure 8. It incorporates additional wells (i.e., wells #1, 6 and 8), spaced at their actual relative distance from one another along a line of projection (Fig. 10) and extends down to upper Paleozoic rocks. Like section A–A', B–B' clearly shows erosional truncation of the Lakota L1 interval and the Morrison Formation beneath the Lakota L2 interval in the vicinity of well #5. More importantly, duplication of the Triassic Spearfish Formation in well #5 is unequivocal evidence of a fault cutting at least the upper Paleozoic section. Our interpretation of a southeast-dipping high-angle reverse fault is consistent with the location of the abrupt change in thickness of the Lakota Formation and the presence of repeated section in the Spearfish Formation caused by faulting. Because this fault does not appear to affect thickness or continuity of the Fall River Formation and from evidence cited earlier, we can bracket the age of movement on this structure as post-Lakota L1 and pre-Fall River deposition but more likely pre- or syn-Lakota L2.

A seismic reflection line C–C' (Fig. 10) shot about 40 km

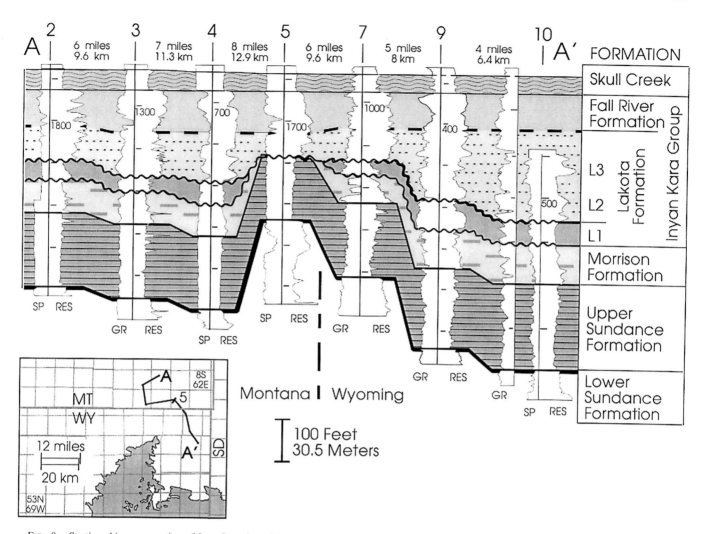

FIG. 8.—Stratigraphic cross section of Late Jurassic and Early Cretaceous strata in the subsurface north of the Black Hills. Datum is top of Fall River Formation. Distance between drill holes is shown at top of cross section. GR– natural gamma-ray radiation, RES– resistivity, SP– spontaneous potential. Informal Sundance Formation subdivisions are those common to industry subsurface correlations. Numbers above logs are keyed to well locations included with Figure 9.

southwest of the location of cross section B–B' (Fig. 10) reveals the existence of a west-dipping basement fault which offsets the Upper Jurassic reflections but not the Fall River reflector (Fig. 11). A line oriented about N75/E separating areas of thick from thin Lakota Formation (Fig. 12) suggests that the fault evident in the seismic line may be the same fault interpreted in cross section B–B' shown in Figure 9. Significantly, areas of thick Lakota rocks occur on opposite sides of the trace of the fault when it is followed along strike. If the thickest Lakota section was preserved on the downthrown side of the fault where subsidence was greatest, this would suggest that the vertical component of motion along this fault reversed direction along strike. Along-strike changes in both direction of dip of a fault plane and direction of oblique slip (schematically shown in Fig. 13) are criteria proposed by Harding (1990) as evidence for transpressional faulting. Additional evidence of transpressional faulting is shown in Figure 11 where the amount and sense of offset of reflectors across the fault change with depth, another feature cited by Harding (1990) as evidence for strike-slip motion.

Timing of Syndepositional Tectonism

On the basis of magnetostratigraphy, fission-track dating and detailed chronostratigraphic correlation, May et al. (1995) suggest that activation of intra-foreland basement faults began to affect early Cretaceous alluvial sedimentation about 134 Ma in west-central Wyoming (i.e., western Wind River Basin) although our preliminary pollen studies now in progress suggest that this date could be significantly younger. By about 116–113 Ma, the C-interval of Cloverly sedimentation in the eastern Wind River Basin was being influenced by intrabasinal movement. Our preliminary correlations (Way et al. 1994) suggest that the Cloverly C-interval likely is slightly older or coeval with the L2 interval of the Lakota Formation in the northern Black Hills, suggesting younger movement along northeast-trending faults in this area, perhaps about 114–110 Ma. Movement on the Transcontinental Arch resulting in a near 120° change in regional paleoslope, and deposition of the L3 interval of the Lakota Formation occurred still later. Palynology of the Lakota Formation in the southern Black Hills (Pish, 1988) and

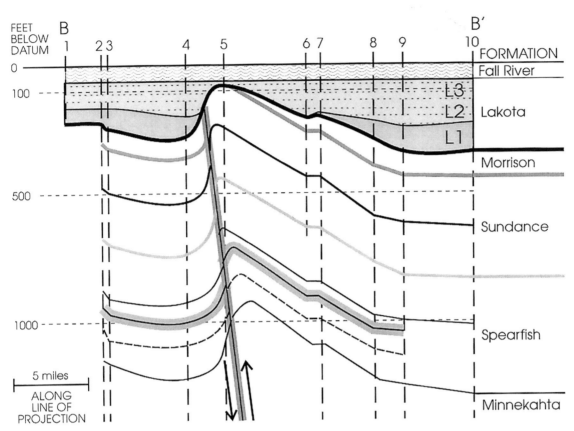

FIG. 9.—Paleostructural cross section showing thickness relationships between Lakota Formation and older rocks at the beginning of Fall River deposition. Scale shows distance between wells along line of projection. Location of line of projection shown in Figure 10. Note fault repeat of marker beds in Triassic Spearfish Formation in well #5. Location of control points: (1) 17–8S–59E, MT, Husky #1 McDowell Govt; (2) 9–8S–60E, MT, Superior Campbell Govt 14–9; (3) 27–8S–59E, MT, Miles Jackson McDowell #1; (4) 23–9S–59E, MT, Union #1 Govt Newton; (5) 17–9S–61E, MT, Continental Govt #1; (6)3–57N–62W, WY, Union Casey Govt #1; (7) 31–58N–61W, WY, Axem Resources; (8) 15–57N–61W, WY, Axem Resources Crow Creek A–15; (9) 28–57N–61W, WY, Jackson #1 Frerichs;(10) 14–56N–61W, WY, Hunt Govt Miller #1.

pollen samples collected by us from the L3 interval suggest an early Albian age for L3 channels (D. Engelhardt, pers. commun., 1996).

Assuming that intrabasinal structural movement was caused by transmission of intraplate stresses through the foreland and into the craton, as first proposed by Meyers et al. (1992) and later supported by the modeling studies of Harry et al. (1995), the above chronology suggests that a stress could have been transmitted nearly 1200 km inboard over approximately 15–25 my. This would explain progressive eastward resurgence of basement structures during early stages of Cretaceous deformation of the western North American plate margin, genetically associated eastward migration of early Cretaceous trunk rivers (Fig. 14) and resultant deposition of Lower Cretaceous conglomerates.

DISCUSSION

Shanley and McCabe (1994) emphasize that for inboard continental settings, climate, basin subsidence and source-area tec-tonism are dominant controls on alluvial sedimentation. Eustatic changes of sea level are subordinate. Comparison of Lakota–Cloverly stratigraphic intervals facilitates preliminary evaluation of the relative effect of each of these controls on the nonmarine and transitional-marine sediment of the distal foreland basin.

A probable climatic change from subhumid to subarid between Lakota L1 and L2 subdivisions in the Black Hills is reflected in floral (Ward, 1899), faunal (Sohn, 1979) and taphonomic (wood fossilization) changes. This climate change coincides with the change in fluvial system morphology from L1 sinuous and mud-dominated systems to straight, braided, and sand-dominated L2 systems.

Unfortunately, eustatic effects of transgression of the Early Cretaceous sea in the Western Interior cannot be isolated from coeval uplift of the Transcontinental Arch. The marine influence recorded by the Lakota L3 sandstone and the Greybull Sandstone of the Bighorn Basin (Kvale and Vondra, 1993) is the only indication of eustatic control on the Lakota–Cloverly

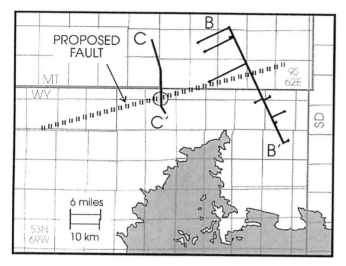

FIG. 10.—Index map showing line of projection for cross section B–B' (FIG. 9), seismic reflection line (C–C') of Ryder et al. (1981), portion of seismic reflection line shown in Figure 11 (shaded circle) and location of proposed transpressional fault (dashed line) discussed in text.

interval. The estuarine depositional setting of the L3 interval and Greybull channels farther west leads us to suspect that backfilling of these channels was related to the first (Skull Creek) transgression of the early Cretaceous seaway, but the influence on backfilling by increased sediment supply resulting from uplift of the Transcontinental Arch cannot be ignored. Although the Skull Creek seaway transgressed over all of Wyoming, paleocurrent data from the known estuarine deposits of the upper Lakota–Cloverly interval indicate that these rocks are associated with an eastern rather than a western source area (Kvale and Vondra, 1993).

Although climate and sea-level change may have affected depositional systems, we suggest that differential subsidence caused by intrabasinal structural deformation was the dominant factor controlling Lakota–Cloverly stratigraphic architecture. In the northern Black Hills, intrabasinal deformation during Lakota deposition created structural and erosional topography resulting in abrupt thickness variations of the Lakota Formation. On structural highs, the lower Lakota rocks are absent because of erosion and the Jurassic–Cretaceous unconformity merges with an unconformity present between L1 and L2 throughout most of the Black Hills.

The basement-rooted transpressional style of intrabasinal deformation documented in this study is significant because this type of faulting has recently been recognized as a common and pervasive feature in the "stable" craton of the western interior (Anna, 1986; Brown and Brown, 1987). Transpressional faults may be common throughout the foreland basin and may be the structural style responsible for tectonic control on many early Cretaceous fluvial systems.

Intrabasinal differential subsidence on a larger scale also may be responsible for the anomolously thick Lakota Formation in the vicinity of the Black Hills. In the classical and simplified models of foreland-basin fill, coarse clastic sediment is represented as a wedge tapering from the foredeep in the direction of the craton. Early Cretaceous alluvial deposits represented by the Gannett Group are thickest (up to 800 m) in the vicinity of the Idaho–Wyoming border, the site of the Early Cretaceous foredeep. In much of central Wyoming, the average thickness of the coeval Cloverly Formation is only about 60 m. However, in many areas of the Black Hills, the aggregate thickness of coarse clastic rocks in the Lakota Formation is as much as 100 m. Bolyard and McGreggor (1966) interpret that the Inyan Kara Group thickness trend is the result of differential subsidence,

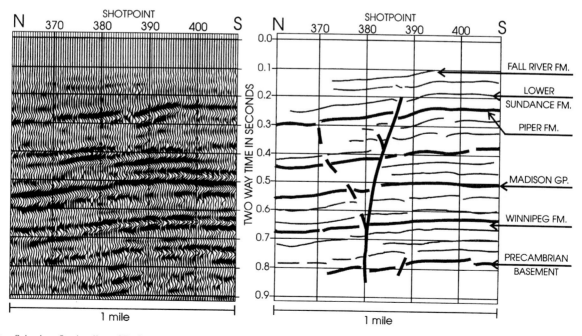

FIG. 11.—Seismic reflection line of Ryder et al. (1981) and structural interpretation showing nearly vertical fault extending to basement. Note the opposite sense of reflector offset evident between 0.5 and 0.4 sec and between 0.2 and 0.3 sec.

DeCELLES, P. G., 1986, Sedimentation in a tectonically partitioned, nonmarine foreland basin: the Lower Cretaceous Kootenai Formation, southwestern Montana: Geological Society of America Bulletin, v. 97, p. 911–931.

DeCELLES, P. G. AND BURDEN, E. T., 1992, Nonmarine sedimentation in the overfilled part of the Jurassic–Cretaceous cordilleran foreland basin: Morrison and Cloverly Formations, central Wyoming, U.S.A.: Basin Research, v. 4, p. 291–313.

DeCELLES, P. G., LANGFORD, R. P., AND SCHWARTZ, R. K., 1983, Two new methods of paleo-current determinaiton from trough cross-stratification: Journal of Sedimentary Petrology, v. 53, p. 629–642.

DOLSON, J. C. AND MULLER, D. S., 1994, Stratigraphic evolution of the lower Cretaceous Dakota Group, Western Interior, U.S.A., in Caputo, M. V, Peterson, J. A., and Franczyk, K. J., eds., Mesozoic Systems of the Rocky Mountain Region, U.S.A.: Denver, Rocky Mountain Section, SEPM, p. 441–456.

DYMAN, T. S., MEREWETHER, E. A., MOLENAAR, C. M., COBBAN, W. A., OBRADOVICH, J. C., WEIMER, R. J., AND BRYANT, W. A., 1994, Stratigraphic transects for Cretaceous rocks, Rocky Mountains and Great Plains regions, in Caputo, M. V, Peterson, J. A., and Franczyk, K. J., eds., Mesozoic Systems of the Rocky Mountain Region, U.S.A.: Denver, Rocky Mountain Section, SEPM, p. 365–392.

EKDALE, A. A., BROMLEY, R. G., AND PEMBERTON, S. G., 1984, Ichnology: Trace Fossils in Sedimentology and Stratigraphy: Tulsa, Society of Economic Paleontologists and Mineralogists Short Course 15, 316 p.

FLEMINGS, P. B. AND JORDAN, T. E., 1990, Stratigraphic modeling of foreland basins: interpreting thrust deformation and lithosphere rheology: Geology, v. 18, p. 430–434.

GOTT, G. B., WOLCOTT, D. E., AND BOWLES, C. G., 1974, Stratigraphy of the Inyan Kara Group and localization of uranium deposits, southern Black Hills, South Dakota and Wyoming: Washington, D.C., United States Geological Survey Professional Paper 763, p. 1–57.

HARDING, T. P., 1990, Identification of wrench faults using subsurface structural data: criteria and pitfalls: American Association of Petroleum Geologists Bulletin, v. 74, p. 1590–1609.

HARLAND, W. B., COX, A., LLEWELLYN P. G., PICKTON, C. A. G., SMITH, A. G., AND WALTERS, R., 1982, A Geologic Time Scale: Cambridge, Cambridge University Press, 131 p.

HARRY, D. L., OLDLOW, J. S., AND SAWYER, D. S., 1995, The growth of orogenic belts and the role of crustal heterogeneities in decollement tectonics: Geological Society of America Bulletin, v. 107, p. 1411–1426.

HAUN, J. D., AND BARLOW, J. A., JR., 1962, Lower Cretaceous stratigraphy of Wyoming, in Enyert, R. L., and Curry, W. H., III, eds., Symposium on Early Cretaceous Rocks of Wyoming and Adjacent Areas: Casper, Wyoming Geological Association, 17th Annual Field Conference Guidebook, p. 15–22.

HELLER, P. L. AND PAOLA, C., 1989, The paradox of Lower Cretaceous gravels and the initiation of thrusting in the Sevier orogenic belt, United States Western Interior: Geological Sociey of America Bulletin, v. 101, p. 864–875.

HOLBROOK, J. M., 1992, Developmental and sequence-stratigraphic analysis of Lower Cretaceous sedimentary systems in the southern part of the United States Western Interior: Interrelationships between eustasy, local tectonics, and depositional environments: Unpublished Ph.D. Thesis, Indiana University, Bloomington, 271 p.

HOLBROOK, J. M., 1993, Evidence for and potential tectonic origin of regional intraplate deformation throughout the Early Cretaceous United States Western Interior: Geological Society of America Abstracts with Programs, v. 25, no. 6, p. A–70.

KVALE, E. P. AND VONDRA, C. F., 1993, Effects of relative sea-level changes and local tectonics on a Lower Cretaceous fluvial to transitional marine sequence, Bighorn Basin, Wyoming, U.S.A. in Marzo, M., Puigdefabregas, C.,eds., Alluvial Sedimentation: Cambridge, Blackwell Scientific Publications, International Association of Sedimentology Special Publication 17, p. 383–399.

MACKENZIE, F. T. AND RYAN, B. J., 1962, Cloverly–Dakota and Fall River paleocurrents in the Wyoming Rockies, in Enyert, R. L., and Curry, W. H., III, eds., Symposium on Early Cretaceous Rocks of Wyoming and Adjacent Areas: Casper, Wyoming Geological Association, 17th Annual Field Conference Guidebook, p. 44–61.

MAPEL, W. J. AND GOTT, G. B., 1959, Diagrammatic restored section of the Inyan Kara Group, Morrison Formation, and Unkpapa Sandstone on the western side of the Black Hills, Wyoming and South Dakota: Washington, D. C., United States Geological Survey Mineral Investigations Field Studies Map MF–218.

MAPEL, W. J. AND PILLMORE, C. L., 1962, Stream directions in the Lakota Formation (Cretaceous) in the northern Black Hills, Wyoming and South Dakota, in Nolan, T. B., ed., Short Papers in the Geologic and Hydrologic Sciences: Washington, D. C., United States Geological Survey Professional Paper 450–B, p.B35–37.

MAPEL, W. J. AND PILLMORE, C. L., 1963a, Geology of the Newcastle area, Weston County, Wyoming: United States Geological Survey Bulletin 1141–N, p. N1–74.

MAPEL, W. J. AND PILLMORE, C. L., 1963b, Geology of the Inyan Kara Mountain quadrangle, Crook and Weston Counties, Wyoming: United States Geological Survey Bulletin 1121–M, p. M1–56.

MAPEL, W. J. AND PILLMORE, C. L., 1964, Geology of the Upton quadrangle, Crook and Weston Counties, Wyoming: United States Geological Survey Bulletin 1181–J, p. J1–46

MAY, M. T., 1992, Intra- and extrabasinal tectonism, climate and intrinsic threshold cycles as possible controls on Early Cretaceous fluvial system architecture, Wind River Basin, Wyoming, in Mullen, C. E., ed., Rediscover the Rockies: Casper, Wyoming Geological Association, 44th Annual Field Conference Guidebook, p. 61–74.

MAY, M. T., FURER, L. C., KVALE, E. P., SUTTNER, L. J., JOHNSON, G. D., AND MEYERS, J. H., 1995, Chronostratigraphy and tectonic significance of Lower Cretaceous conglomerates in the foreland of central Wyoming, in Dorobeck, S. and Ross, G. M., eds., Stratigraphic Evolution of Foreland Basins: Tulsa, SEPM Special Publication 52, p. 97–110.

McGOOKEY, D. P., HAUN, J. D., HALE, L. A., GOODELL, H. G., McCUBBIN, D. G., WEIMER, R. J., AND WULF, G. R., 1972, Cretaceous system, in Mallory, W. W., ed., Geologic Atlas of the Rocky Mountain Region: Denver, Rocky Mountain Association of Geologists, p. 189–228.

MEYERS, J. H., SUTTNER, L. J., FURER, L. C., MAY, M. T., AND SOREGHAN, M. J., 1992, Intrabasinal tectonic control on fluvial sandstone bodies in the Cloverly Formation (early Cretaceous), west-central Wyoming, U.S.A.: Basin Research, v. 4, p. 315–333.

MIALL, A. D., 1981, Alluvial sedimentary basins: tectonic setting and basin architecture, in Miall, A. D., ed., Sedimentation Tectonics in Alluvial Basins, with Examples from North America: Waterloo, Geological Association of Canada Special Publication 23, p. 1–34.

MOBERLY, R., JR., 1960, Morrison, Cloverly, and Sykes Mountain Formations, northern Bighorn Basin, Wyoming and Montana: Geological Society of America Bulletin, v. 71, p. 1137–1176.

O'MALLEY, P. J., 1994, The influence of subtle tectonic movement along basement rooted lineaments on deposition of the Lakota Formation, northeast Wyoming: Unpublished Masters Thesis, Indiana University, Bloomington, 186 p.

OSTROM, J. H., 1970, Stratigraphy and Paleontology of the Cloverly Formation (Lower Cretaceous) of the Bighorn Basin Area, Wyoming and Montana: Peabody Museum of Natural History Bulletin 35, 234 p.

PIPIRINGOS, G. N. AND O'SULLIVAN, R. B., 1978, Stratigraphic sections of some Triassic and Jurassic rocks from Douglas, Wyoming, to Boulder, Colorado: Washington, D. C., United States Geological Survey Oil and Gas Investigations Chart OC–69.

PISH, T. A., 1988, Palynolygy and paleoecology of the Cambria Coal, Weston County, Wyoming: Unpublished Master's Thesis, South Dakota School of Mines and Technology, Rapid City, 66 p.

POST, E. V. AND BELL, H., III, 1961, Chilson member of the Lakota Formation in the Black Hills, South Dakota and Wyoming: Washington, D. C., United States Geological Survey Professional Paper 424–D, p. D173–178.

RICH, F. J., PISH, T. A., AND KNELL, G. W., 1988, Sedimentology, petrography, and paleoecology of the Cambria Coal, Weston County, Wyoming, in Eastern Powder River Basin-Black Hills: Casper, Wyoming Geological Association, 39th Annual Field Conference Guidebook, p. 249–261.

RYDER, R. T., BALCH, A. H., LEE, M. W., AND MILLER, J. J., 1981, Processed and interpreted U.S. Geological Survey seismic reflection profile and vertical seismic profile, Carter County, Montana and Crook County, Wyoming: Washington, D. C., United States Geological Survey Oil and Gas Investigations Chart OC–115.

SCHOON, R. A., 1971, Geology and hydrology of the Dakota Formation in South Dakota: Vermillion, South Dakota Geological Survey, Report of Investigations 104, 55 p.

SCHWARTZ, R. K., 1983, Broken Early Cretacoues foreland basin in southwestern Montana: sedimentation related to tectonism, in Powers, R. G., ed., Geologic Studies of the Cordilleran Thrust Belt: Denver, Rocky Mountain Association of Geologists, p. 159–184.

SHANLEY, K. W. AND McCABE, P. J., 1994, Perspectives on the sequence stratigraphy of continental strata: American Association of Petroleum Geologists Bulletin, v. 78, p. 544–568.

SHURR, G. W., 1981, Cretaceous sea cliffs and structural blocks on the flanks of the Sioux Ridge, South Dakota and Minnesota, *in* Brenner, R. L., Bretz, R. F., Bunker, B. J., Iles, D. L., Ludvigson, G. A., McKay, R. M., Whitley, D. L., Witzke, B. J., eds., Cretaceous Stratigraphy and Sedimentation in Northwest Iowa, Northeast Nebraska, and Southeast South Dakota: Iowa City, Iowa Geological Survey, Guidebook 4, p. 25–41.

SLACK, P. B., 1981, Paleotectonics and hydrocarbon accumulation, Powder River Basin, Wyoming: American Association of Petroleum Geologists Bulletin, v. 65, p. 730–743.

SOHN, I. G., 1979, Nonmarine ostracodes in the Lakota Formation (Lower Cretaceous) from South Dakota and Wyoming: Washington, D. C., United States Geological Survey Professional Paper 1069, 22 p.

WAAGE, K. M., 1959, Regional aspects of Inyan Kara stratigraphy in the Black Hills, South Dakota and Wyoming: Washington, D. C., United States Geological Survey Bulletin 1081–B, p. 11–90.

WARD, L. F., 1899, The Cretaceous formation of the Black Hills as indicated by the fossil plants, *in* Walcott, C. D., ed., United States Geological Survey 19th Annual Report: Washington, D. C., Government Printing Office, p. 521–702.

WAY, J. N., O'MALLEY, P. J., FURER, L. C., SUTTNER, L. J., KVALE, E. P., AND MEYERS, J. H., 1994, Correlations of the Upper Jurassic–Lower Cretaceous nonmarine and transitional rocks in the northern Rocky Mountain Foreland, *in* Caputo, M. V., Peterson, J. A., and Franczyk, K. J., eds., Mesozoic Systems of the Rocky Mountain Region, U.S.A.: Denver, Rocky Mountain Section, SEPM, p. 351–364.

WILSON, J. M., 1958, Stratigraphic relations of nonmarine urassic and pre-thermopolis Lower Cretaceous strata of north-central and northeastern Wyoming, *in* Strickland, J., ed., Powder River Basin: Casper, Wyoming Geological Association, 13th Annual Field Conference Guidebook, p. 77–78.

WILTSCHKO, D. W. AND DORR, J. A., 1983, Timing of deformation in overthrust belt and foreland of Idaho, Wyoming and Utah: American Association of Petroleum Geologists Bulletin, v. 67, p. 1304–1322.

WULF, G. R., 1959, Lower Cretaceous (Albian) in the northern Great Plains: Unpublished Ph.D. Thesis, University of Michigan, Ann Arbor, 350 p.

YINGLING, V. L AND HELLER, P. L., 1992, Timing and record of foreland sedimentation during the initiation of the Sevier orogenic belt in central Utah: Basin Research, v. 4, p. 279–290.

YOUNG, R. G., 1960, Dakota group of Colorado Plateau: American Association of Petroleum Geologists Bulletin, v. 44, p. 156–194.

PART II
ORGANIC DEPOSITS

Bulga Formation) is more complicated. Figure 3 is a 10–km–long section of the amalgamated seam, and its splits along the line A–B in Figure 1. The top of the Archerfield Sandstone, which marks the highest water level at the time, has been chosen as the datum for the cross sections illustrated in Figures 2 and 3.

As mentioned above, the angular discordance between the Archerfield Sandstone and the underlying coal measures weakens with increasing distance from the Muswellbrook Anticline. In the northern part of the section shown in Figure 2, the unconformity becomes aligned with the roof of the Upper Wynn Seam, but the successive erosion of the underlying two tuff bands #2 and #6 (Wub1), illustrated in Figure 3, indicates that the unconformity continues as a ravinement or transgressive surface of erosion. In some bore holes (e.g., Ell 9 and 6000E000), the erosive nature of the contact between the coal and its marine roof is accentuated by a lag of one or two layers of rounded pebbles.

Apart from some irregularity in the vicinity of bore hole 6000E000, the marine interval has a constant thickness of just under 8 m. However, between the bore holes Ell 17 and DDH 18, the ravinement surface rises towards the amalgamated seam by 2.5 m over a lateral distance of approximately 1 km. This relatively gentle rise contrasts sharply with a gain of another

5.5 m recorded for the last 700 m (maximum) so that the split terminates before reaching DDH 18. A correlative surface of the unconformity extends into the amalgamated seam at the base of a 20–cm–thick mudrock marked BB3 in Figure 3. This epiclastic band overlies up to 50 cm of very dull coal (durain) and, while the upward and downward contacts of the BB3 band are gradational in many places, in some bore cores there is evidence of erosion at the base of the band (J. McNamara, pers. commun., 1995).

Coal Properties and Composition

In order to compare coal composition within and outside the amalgamated seam, detailed petrographic and chemical analyses were carried out, in addition to the commercial testing routinely done by the prospecting coal companies, on the relevant cores from two diamond drill holes.

Coal Properties Outside the Amalgamated Seam.—

Some of the analysis results obtained from DDH 5000C500 are displayed in Figure 4. The bore was sunk approximately 5 km down-dip from the split axis (i.e., in an area where the seam splits are fully developed). The illustrated coal characteristics include standard geophysical and lithotype logs, the latter

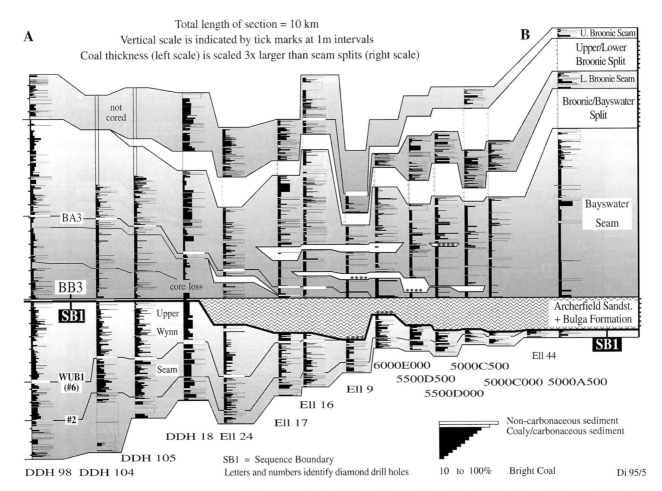

FIG. 3.—Section through the northeastern terminus of the seam splits in the Upper–Wynn–to–Upper–Broonie Seam interval along section line A–B in Figure 1.

fluorescence intensity increases to 0.87 units while mean random vitrinite reflectance drops to its lowest point of 0.66%. This is interpreted as an indication of flooding which occurs at the same level in Figure 4 (i.e., just below the clastic band that in Figure 3 is correlated with the Broonie/Bayswater split).
The Broonie Bench.—The Broonie Seam consists of well bedded bright and dull coal with many thin shale and shaly coal bands. Together with a moderately high-vitrinite fluorescence and a rising sulphur content, this composition indicates frequent base-level fluctuations, presumably because the coal was derived from a predominantly topogenous peat.

Sequence Stratigraphic Interpretation

Together with the angular discordance near the Muswellbrook Anticline, the surface separating the uppermost dull coal of the Upper Wynn from the overlying BB3 band constitutes a No. 1 sequence boundary, *in sensu* Mitchum et al. (1977), marked SB1 in Figures 2 through 5. This sequence boundary caps older rocks, including coal which, at the peat stage, were subjected to exposure and erosion during a lowstand in relative sea level. In the vicinity of the Muswellbrook Anticline, the overlying marine and shoreface sediments of the Bulga Formation and Archerfield Sandstone rest on deeply incised coal measures, but both the angular discordance and the degree of incision decrease towards the amalgamated seam. Within this coal, erosion at the base of the BB3 band was sporadic and generally weak, which agrees with the inferred lagoonal setting.

The High- and Lowstand Systems Tracts Below the Sequence Boundary.—

Since the investigations did not extend below the Upper Wynn Seam, little can be said about the deeper strata, except that they are presumably of highstand origin. This interpretation also holds for the Upper Wynn Seam, which leaves very little room for any sediments that may have been formed as part of the lowstand systems tract (LST) between the unconformity (= sequence boundary) and the base of the marine Bulga Formation.

The only rocks that could possibly be considered to represent LST deposits are the thin pebble layers that are occasionally found on top of the eroded coal in the Upper Wynn Seam (Fig. 3). However, they are more likely to constitute a lag that was left behind from the build-up and destruction of backstepping beaches during the marine transgression. This infers that the marine erosion that affected the Upper Wynn peat outside the amalgamated seam removed all evidence of the lowstand systems tract. The high fluorescence and low reflectance of the vitrinite below the unconformity are testimony of the freshness of the marine-influenced peat at the time of the transgression, which could, wrongly, be taken as an indication that there was no subaerial weathering and no lowstand at all. The up to 40 cm of degraded-peat-derived dull coal underneath the BB3 band in the amalgamated seam is therefore critical for the interpretation because it is the only tangible evidence for a hiatus due to negative accommodation and subaerial weathering under LST conditions. A cartoon depicting the situation near the end of the lowstand systems tract is illustrated in Figure 7A.

The Transgressive Systems Tract.—

The transgressive surface and sequence boundary that overly the Upper Wynn Seam mark the beginning of the transgressive

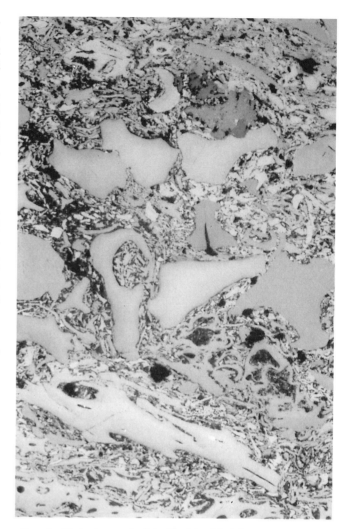

FIG. 6.—Photomicrograph of the inertinite-rich dull coal below the BB3 band in DDH 98. Incident light, oil immersion, antiflex system; the true size of the frame is 0.56 x 0.38 mm.

systems tract (TST) which, outside the amalgamated seam, includes the progradational lower and upper shoreface sediments of the Bulga Formation and the Archerfield Sandstone. Together, they form an upward shoaling lithofacies succession defined as a parasequence by Vail et al. (1977) and van Wagoner et al. (1987). The Bulga Formation and Archerfield Sandstone constitute the most landward projection of the Kulnura Marine Tongue (i.e., they form the only parasequence of the transgressive systems tract that reached this part of the Sydney Basin at the height of the marine transgression). As indicated in Figures 7B and C, the upper contact of the Archerfield Sandstone coincides with the maximum flooding surface. It continues into the amalgamated seam along the top of the BB3 band or, if the assumed allochthonous origin is correct, along the top of the 40 cm of dull coal overlying the BB3 band. Since this band contains the sequence boundary and transgressive surface at its base, the whole of the transgressive systems tract would appear to be little more than 0.5 m thick in the amalgamated seam. However, that depends on the sequence stratigraphic interpretation of the overlying Bayswater Seam.

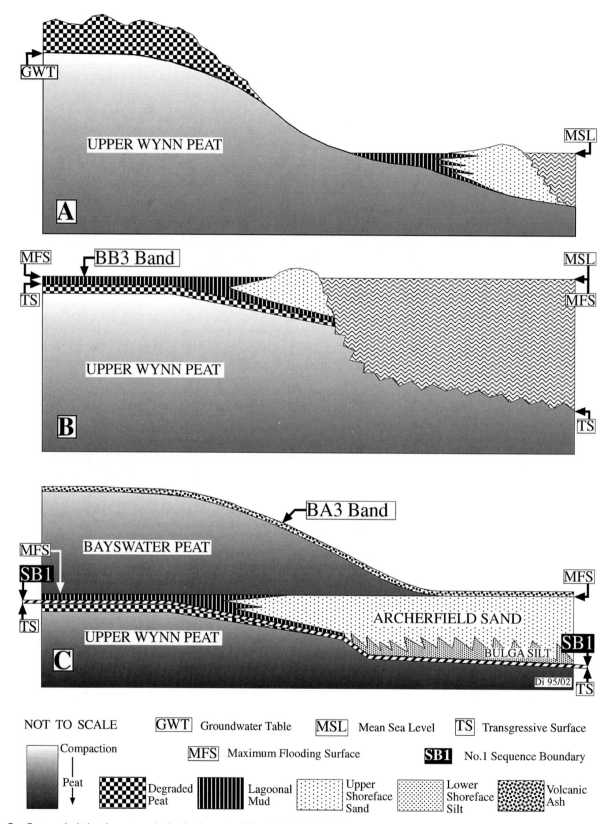

FIG. 7.—Cartoon depicting three stages in the development of the Archerfield–Bulga–Bayswater parasequence with the position of relevant sequence stratigraphic surfaces. A = Oxidative peat degradation due to the lowered groundwater table at the height of the lowstand systems tract. B = Maximum flooding during the transgressive systems tract. C = The development of the Bayswater peat at the time of the BA3 ash fall.

As mentioned in the introduction, progradational coal seams are considered to cap parasequences and are overlain by flooding surfaces or their correlative surfaces at the bases of subsequent parasequences (Van Wagoner et al., 1987, 1990). The Bayswater Seam which overlies the Archerfield Sandstone should therefore be regarded as part of the progradational, upward shoaling "Archerfield–Bulga parasequence" because, apart from some residual lagoonal intercalations, no common flooding surface appears to separate the top of the sandstone from the overlying coal. A well-documented widespread flooding surface occurs at the base of the Broonie/Bayswater split (Figs. 3–5). It is non-marine but probably represents an advance of the sea that did not reach the basin margin. This puts the Bayswater Seam into the top of an expanded "Archerfield–Bulga–Bayswater parasequence" and thus into the top of the transgressive systems tract.

In view of the considerable thickness of up to 14 m of coal, the cause of the fluctuation in vitrinite reflectance and fluorescence illustrated in Figure 5 is relevant to the sequence stratigraphic interpretation of the Bayswater Seam. The vertical trends in optical properties have been interpreted as periods of dryness followed by wet conditions which could indicate parasequence boundaries. In the eutrophic, high-vitrinite, high-ash lower portion of the Upper Wynn (e.g., at 308.5 m in Fig. 5) and upper part of the Bayswater Seams, this interpretation is probably correct. However, the downlap illustrated in Figure 3 of the volcanic ash represented by the BA3 marker band on the Archerfield Sandstone and the likewise downlapping signature of some of the lithotypes in the underlying portion of the Bayswater Seam (i.e., up to the BA3 band) suggest that the variations in optical properties were caused by climatic fluctuations which, from time to time affected the peat surface as the mire followed the basinward progradation of the shoreline.

It can be concluded that the progradational lower portion of the Bayswater Seam is part of an "Archerfield–Bulga–Bayswater parasequence" and, therefore, belongs to the transgressive systems tract. The same may be true for the upper coal section (i.e., above the BA3 marker band), but most of the trend lines illustrated in Figures 4 and 5 indicate an upward tendency towards the establishment of topogenous mire conditions which could be the updip expressions of renewed basinward flooding under highstand conditions. This interpretation infers the existence of at least one more parasequence in the upper part of the Bayswater Seam which is also supported by an apparent separation of the topogenous upper seam portion from the ombrogenous lower portion along a split that is indicated in DDH 118 but appears to develop more fully outside the study area where information is sketchy.

The Highstand Systems Tract.—

Apart from the above-mentioned possibility of the upper portion of the Bayswater Seam being of HST origin, the highstand systems tract includes the two benches of the Broonie Seam. Outside the amalgamated coal, the two benches separate along the deflections shown in the electric logs of Figure 5 at 290.3 m. Their intervening clastics are non-marine, presumably as a result of the loss of the hydrologic connection with sea level due to the basinward migration of the shoreline.

THE MIOCENE COAL MEASURES OF THE NORTH–WEST EUROPEAN TERTIARY BASIN

The sequence stratigraphic interpretation of the Australian example revealed diagnostic criteria for the recognition of a sequence boundary and systems tracts within a coal seam, but the geological circumstances and advanced coalification of the source material did not permit an unequivocal identification of intra-seam parasequences. For this reason, the investigations were extended to incorporate an amalgamated seam of soft brown coal (low-rank lignite) with a clear paleo-environmental record.

The occurrence of brown-coal measures along the southern shores of the North–West European Tertiary Basin is genetically linked to the repeated south- and southeastward directed transgressions of the ancient North Sea across The Netherlands and the north German plains into western Poland (Von Bubnoff, 1948; Lotsch, 1967; Lotsch et al., 1969). Coalfields of considerable economic significance were formed in the Lower Rhine Embayment (= Niederrheinische Bucht) in western Germany and in Lower Lusatia (= Niederlausitz), situated southeast of Berlin. In Figure 8, these two areas are marked A and B, respectively.

Together, the two districts have lignite reserves of 66 000 Mt, of which 80% are contained in the Lower Rhine Embayment (Hager, 1981, 1993; Standke et al., 1993), in spite of a smaller areal extent. This difference in coal content is a function of the thickness of the coal seams which in the Lower Rhine Embayment combine up-dip to form the 100–m–thick Main Seam, while in Lower Lusatia the seams are thinner and only in a few places coalesce to form, for example, up to 14 m of clean coal in the 2nd Lusatian Coal Horizon. This contrasting pattern of Miocene sedimentation resulted from tectonic differences between the two depositional sites.

The Lower Rhine Embayment formed as part of the North Sea rift system which together with differential movements on Variscan (Hercynian) basement structures (Hager, 1993) produced greater tectonic mobility. Although some authors (e.g., Seifert et al., 1993), have stressed the importance of crustal movements for the Tertiary sedimentary history of Lower Lusatia, the depositional pattern of the region and the good correlation of marine transgressions and regressions with sea-level high- and lowstands (Standke et al., 1993; also Fig. 9) suggest a comparatively low level of tectonic influence on coal measure sedimentation apart from mainly epeirogenic crustal movements (Lotsch, 1967). These resulted in a net basin subsidence of several hundred meters and weak deformation (Nowel, 1993). The favorable geological situation combined with the large amount of detailed information that has been obtained as part of the intensive exploration and mining activity in the area for over 150 years renders these low-rank coals an attractive subject for the purpose of conducting a high-resolution sequence-stratigraphic study.

Geological Development of the Lower Lusatian Brown Coal Measures

The coal measures are paralic and were formed in a relatively stable depositional environment on the north German platform where glacio-eustatic sea-level variations appear to have been the main cause of several marine transgressions and regressions

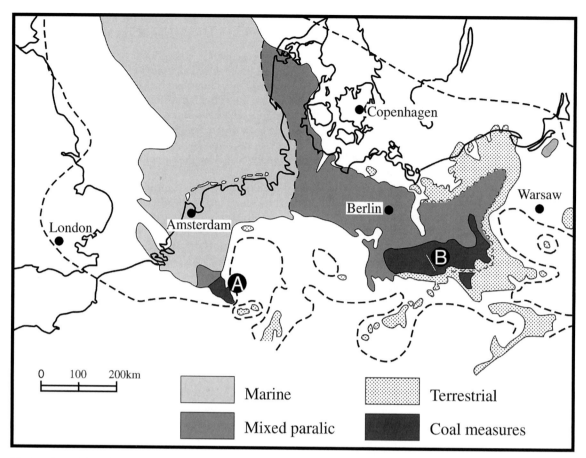

FIG. 8.—The location of the Miocene lignite fields of the Lower Rhine Embayment (A) and Lower Lusatia (B) in relation to the North–West European Tertiary Basin. The dashed line marks the Tertiary basin margin; the shaded area refers to the Miocene deposits. Modified from Vinken (1988) and Standke et al. (1992).

(Standke et al., 1993). The transgressions began to affect Lower Lusatia during the Eocene, and by Late Oligocene time the sea had reached the southern basin margin with the volcanic uplands of the Central German Ridge and the Bohemian Massif to the south. From Early Miocene time onward, an increased supply of terrigenous sediments from the southern uplands reversed the trend and pushed several large clastic fans into the northern basin. One of these fans produced the Spremberg Formation (Fig. 9) at the front of which paralic coal measures were formed as the 4th Lusatian Coal Horizon between the limnic/terrestrial Spremberg Formation to the south and the marine and brackish basin deposits of the Cottbus Formation to the north. The 4th Lusatian Coal Horizon consists of a group of coal seams which are time transgressive and become younger in a northwesterly direction (Lotsch et al., 1969; i.e., they show seaward stacking). In Figure 9 (upper diagram), this group of seams is shown between the maximum flooding surface (mfs) and the sequence boundary (SB) of Cycle TB 1.4, which puts the coals into the highstand systems tract (HST).

The zig-zag pattern between the terrestrial Spremberg and marine Brieske Formation from Nanoplankton Zone 2 to 4 in Figure 9 corresponds to the eustatic rise in sea level followed by a reversal in nanoplankton zone 5. This succession contains several uneconomic coal seams of doubtful sequence-stratigraphic status, only one of which is indicated between the 4th

and 3rd Lusatian Coal Horizons. The latter consists of several backstepping (i.e., landward or retrogradationally stacked), thin coal seams formed in front of the invading sea. This stacking pattern and the biostratigraphic position indicated in Figure 9 place the 3rd Lusatian Coal Horizon in the transgressive systems tract (TST), close to the maximum flooding surface tentatively correlated with TB 2.1.

Maximum southerly advance of the marine Brieske Formation was reached in two relatively brief eustatic cycles of 1–my duration each. They correspond to the TB 2.2 cycle in which the marine Sand G 6.1 represents the maximum landward extent of marine flooding and the TB 2.3 cycle which has been tentatively correlated with the very widespread 2nd Lusatian Coal Horizon. In central and southern Lower Lusatia, this coal horizon begins with the Lower Rider Seam, which is probably situated in the transgressive systems tract of the TB 2.3 cycle. An alternative interpretation has been presented by Rascher and Standke (1991) and Suhr et al. (1992) who put the 3rd Lusatian Coal Horizon into the transgressive systems tract and the 2nd Lusatian Coal Horizon into the highstand systems tract of an enlarged mega-sequence.

The Lower Rider Seam is separated from the HST coal seams of the middle and upper part of the 2nd Lusatian Coal Horizon by the maximum flooding surface of the TB 2.3 cycle which is contained in the G 5 marine sand, a light-colored clean sand-

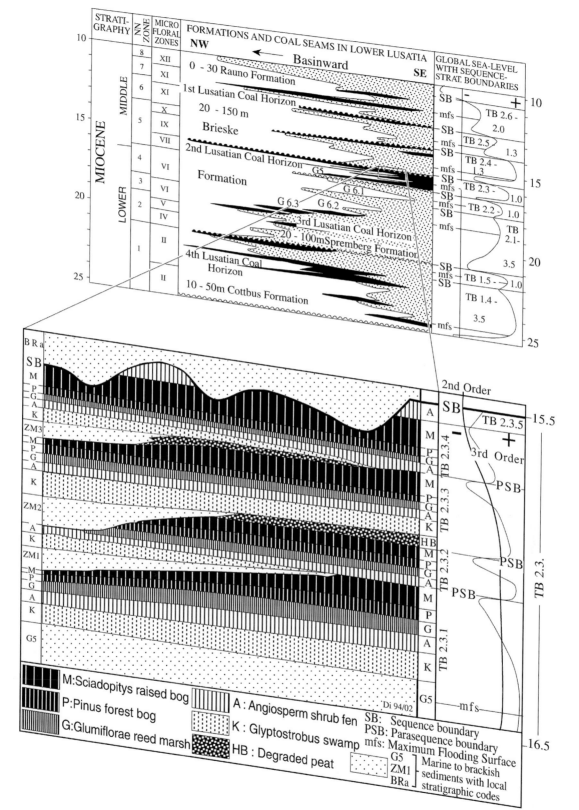

FIG. 9.—Schematic cross-section of part of the Miocene coal measures of Lower Lusatia along the northwest-trending section line at B in Figure 8. Also included are the global sea-level curve and the inferred sequence-stratigraphic boundaries. The lower diagram is an enlarged portion of the 14–m–thick 2nd Lusatian Coal Horizon. Not included is the Lower Rider Seam below the G 5 marine sand. Compiled from Lotsch et al. (1969), Haq et al. (1988), Alexowsky et al. (1989), Brause et al., 1989), Nowel (1993), Suhr et al. (1992), Schneider (1992), Seifert et al. (1993), and Standke et al. (1992, 1993).

stone of shoreface origin. The basinward progradation of the overlying HST coals extended the 2nd Lusatian Coal Horizon as far north as Berlin. Maximum peat accumulation occurred near Senftenberg, close to "B" in Figure 8, where a single seam representing the whole of the coal measures with an original thickness of 40–50 m of peat yielded a maximum thickness of 14 m of brown coal in which several benches can be distinguished. The seam is split northwestward by marine wedges along the bench boundaries so that the upper (younger) benches not only cover successively larger areas than the lower benches but also reach their maximum development more seaward than the underlying benches (Standke et al., 1993).

The HST conditions of the 2nd Lusatian Coal Horizon were terminated by another lowstand which is indicated by an erosional discordance between the deeply incised seam top and the roof rocks. The erosion surface cuts cleanly through the upper coal lithotypes which in the investigated area show no effects of the exposure that might have occurred during or following sea-level fall. In Figure 9, this sequence boundary is marked by the wavy upper boundary of the coal. The roof rocks consist of marine to brackish clays, laminites and clean sands which represent the last large marine transgression across the area. These sediments contain the Upper Rider Seam which may split into two or three benches of coal and/or coaly silt.

The 1st Lusatian Coal Horizon, the youngest of the Tertiary coal measures, has been largely removed, partly by erosion and partly by mining. It is overlain by green, limnic clays (Grüner Flaschenton) of the Rauno Formation, scattered occurrences of thin Pliocene limnic clays and glacigene Pleistocene deposits (Nowel, 1993).

The Composition of the 2nd Lusatian Coal Horizon

Current mining is concentrated on the HST portion of the 2nd Lusatian Coal Horizon between the G 5 sand and the sequence boundary below the Upper Brieske Member (Bra in Fig. 9– lower diagram). The seam consists of soft brown coal with a moisture content of approximately 60%. It has been the subject of detailed investigations (Bönisch et al., 1983; Schneider, 1980, 1988, 1990, 1992; Suhr et al., 1992) and is accessible in several large open cut mines.

As mentioned above, in many parts of the coalfield the 2nd Lusatian Coal Horizon consists of several single coal seams which basinward are separated by marine and brackish silts and sands. The seams are on average between 3 and 5 m thick and contain a single floral cycle each in which Schneider (1992) distinguishes five facies and mire types plus a basal layer of very light colored, reworked, hypautochthonous to partly allochthonous coal. The composition and other characteristics of the coal are summarized in Table 1.

In assessing the analysis results listed in Table 1, it should be realized that they are based on grab samples of 1 to 2 kg taken from each of the major facies. Because of this relatively small sample size, the analysis results are biased towards the groundmass and the enclosed small plant fragments, because mega-xylite (Schneider, 1988), in the form of large tree stumps, trunks, branches and roots, was not sampled. The inclusion of mega-xylite in the samples would have increased the proportions of telohuminite in the K and M facies which are particularly rich in very large wood fragments. A brief description of

the vertical succession from bottom to top is given below, largely after Schneider (1980, 1990, 1992) and Suhr et al. (1992), supplemented by own investigations.

The Reworked Peat (HB Facies).—

When present, the HB facies occurs at the base of a floral and lithotype cycle where it may trace erosional contacts between the stacked coal benches of the amalgamated seam. This dense, unbedded, cream-colored band, an example of which is illustrated in Figure 10, consists of a concentration of comparatively stable macerals such as liptinite (mainly sporinite and liptodetrinite) and inertinite (mainly fusinitic inertodetrinite and some corposclerotinite), set in a matrix of finely fragmented humodetrinite.

The HB facies has been interpreted (Schneider, 1992) as the product of degradation, flooding and partial erosion of the underlying former peat (usually M facies), and its redistribution in backbarrier lagoons. Because of its reworked nature, the ash content of the HB facies is higher than that of any other lithotype. Table 1 lists and ash content of 10.4% (db) which agrees with Seifert et al. (1993) who note that the ash figures for the HB facies are generally in excess of 10%.

It is of interest to note that the petrographic and chemical composition given by George (1982) for the "pale lithotypes" from the brown coals of the Gippsland Basin in Australia is almost identical to the HB facies in Lower Lusatia, which also agrees well with the pale "Baipao" (= light in color and weight) coal described by Qi Ming et al. (1994) from the Yunnan Province in China.

The Conifer Facies (K Facies).—

This facies is represented mainly by *Glyptostrobus europaea* whose well-developed pneumatophores indicate limno-telmatic conditions, possibly with some variation in water depth (Schneider, 1992). Associated are various angiosperms and a Nyssa–Taxodium assemblage near the swamp margin. This horizon is reddish-brown when fresh, but darkens rapidly on exposure due to its high content of humus gels. Another characteristic is the high concentration of large tree stumps (mainly *Sequoia*) in growth position in the upper part of this horizon which in such proliferation is repeated only in the M facies. However, the root/shoot ratio is lower in the K facies due to the large contribution made to the peat by downed trunks, stems, branches and other above-ground phyterals that accumulated in the water-logged limnotelmatic environment.

It is significant that the K facies is well-developed only where it is underlain by a clayey or silty floor sediment or by the HB facies, which suggests that the *Glyptostrobus* swamp required eutrophic conditions. It may even have tolerated weak brackish conditions which is indicated by the presence of planktonic nanofossils. In the eastern part of the coalfield, the nanofossils extend from the underlying marine clastic wedge into the K facies of Bench 1 (R. Bönisch and W. Schneider, pers. commun., 1993).

Other properties which support the notion of limnotelmatic conditions for the K facies are the high-ash contents of over 15% on a dry basis (Seifert et al. 1993), although the analysis results listed in Table 1 yielded only 5.2% (db). The low tissue preservation indicated by the large proportion of humodetrinite in Table 1 (74.1%) appears to be also related to the eutrophic

TABLE 1.—SUMMARY OF THE CHARACTERISTICS OF THE SIX COAL FACIES DISTINGUISHED BY SCHNEIDER (1992) IN THE 2ND LUSATIAN SEAM HORIZON. THE M-FACIES HAS BEEN DIVIDED INTO Ml AND Ms IN REFERENCE TO THE MORE LANDWARD AND THE MORE SEAWARD POSITION OF THE SAMPLE LOCALITIES

	HB Pale Band	K Conifer Facies	A Angiosperm Facies	G Glumiflorae Facies	P Pinus Facies	Ml Marcoduria	Ms Facies
Mire Type*	Allochthonous peat	Glyptostrobus swamp	Shrub fen	Reed marsh	Forest bog	Sciadopitys raised bog	
Nutrient Supply*	n.a.	eutrophic	mesotrophic	oligotrophic	oligotrophic	ombrotrophic	
Lithotypes	very light	dark	medium light	dark	medium light	medium light	
Root/Shoot Ratio*	nil	low	low	low ⟶	variable	high	
Accommodation Pot.	high					low	
Moisture (ad)	**8.2**	15.8	14.6	15.2	**31.4**	16.4	28.6
Ash (db)	**10.4**	5.2	4.6	4.6	**3.7**	5.0	6.6
V.M. (daf)	72.5	56.2	57.9	56.6	56.2	56.6	64.2
Tot. S. (daf)	**0.47**	0.59	0.53	1.13	1.34	1.78	**3.08**
Pyr. S. (daf)	0.03	0.01	0.01	0.01	0.01	0.02	**0.06**
Org. S. (daf)	0.44	0.58	0.52	1.01	1.24	1.70	**2.56**
C (daf)	**71.9**	66.7	67.1	66.5	66.7	**65.1**	66.7
H (daf)	**7.76**	5.03	5.24	5.05	4.88	**4.61**	4.78
N (daf)	0.49	0.67	0.67	0.59	0.65	0.61	0.56
O (daf)	**19.85**	27.60	26.99	27.86	27.77	**29.68**	27.96
Atom. H/C	**1.30**	0.91	0.94	0.91	0.88	**0.85**	0.86
Rr Huminite	n.a.	0.18	0.18	0.17	**0.19**	**0.15**	0.16
Fluor. I. Hum.	n.a.	11.1	**12.3**	7.1	**8.3**	12.0	11.2
Humotelinite (mf)	1.8	19.2	11.3	**41.9**	39.3	30.7	24.9
Humodetrinite (mf)	**67.6**	**74.1**	66.7	38.2	41.5	47.1	67.2
Humocollinite (mf)	0.8	3.8	5.4	10.0	10.7	**13.4**	13.0
Inertinite (mf)	**6.3**	0.4	2.5	0.9	0.6	0.8	0.6
Liptinite (mf)	**23.5**	2.4	**1.4**	8.7	7.9	8.5	2.6

*after Schneider (1990, 1992)

origin of this facies which provided the low-acid environment necessary for a high level of bacterial activity. However, as mentioned above, these analysis results refer to groundmass samples only from which large wood fragments have been excluded.

The Angiosperm Facies (A Facies).—

The low degree of tissue preservation and large proportion of fine detritus give this coal a poorly banded appearance with a light to medium brown color when dry. The ash content is reduced to 4.6% (db) which suggests a decreasing contact of the peat with nutrient-laden waters, possibly due to an increasing peat thickness. Under such conditions, plant roots cannot reach the seat earth any more causing a change from eutrophic to oligotrophic conditions. The A facies signals a shift to sedentary (autochthonous) peat accumulation on peat islands (hammocks) in which some laurophyllous shrubs of the K facies persist. Trees, too, are restricted to the peat islands, which are surrounded by open water in which fine-grained humic detritus collects. This interpretation is supported by the high humodetrinite content listed in Table 1.

The Glumiflorae Facies (G Facies).—

Various water plants and laurophyllous shrubs are represented in this highly banded, dark brown coal which Seifert et al. (1993) consider to have been formed in deeper water than the other facies. However, the first appearance of *Sphagnum* moss, presumably on higher ground, suggests the beginning of raised mire conditions (Schneider, 1992). In Table 1, this development is indicated by the drop in humodetrinite and a further decrease in ash contents.

The Pinus Facies (P Facies).—

This facies is characteristic of the spreading of a forested raised-mire environment in which *Pinus spinosa* constitutes the main plant in association with *Cryptomeria* and various shrubs (Schneider, 1992). The ash content has been reduced further to below 4%, and there is good tissue preservation in the form of pine bark and needles (Table 1). The humotelinite content is therefore higher than in the underlying facies, and this feature is also shared with the overlying M facies. Another characteristic maceral of this facies is resinite which occurs either *in situ* within pine needles or as isolated and dispersed lumps. In the coal face both are conspicuous by their light color against the dark background of the coal.

In fresh exposures, the coal has a reddish color but it darkens quickly on account of the high humus colloid content. The latter appears to have been percolated through the peat from the overlying M facies.

The Marcoduria Facies (M Facies).—

Because of its large content of dispersed tissue known as *Marcoduria inopinata*, the M facies is considered to be of ombrotrophic, raised-mire origin (Schneider, 1992). The *Marcoduria* tissue is well adapted to peaty, anaerobic soil conditions on account of the large air-filled, intercellular spaces (aerenchyma) which after death fill with capillary water and thus sustain raised mire conditions. *Marcoduria inopinata* has been derived from the primary roots of taxodiaceous conifers, such as *Sciadopitys tertiaria* whose needles are also frequently found in the Marcoduria coal together with fibrous xylite. In common with other ombrogenous peats, the M facies originated mainly as a sedentary root peat (Schneider, 1992).

The low pH of the oligotrophic conditions in the ombrogenous setting not only caused a low diversity of plant species in the raised mire but also suppressed tissue-destroying microbial activity except for fungi which can tolerate relatively acid conditions. For this reason, the medium- to light-brown M facies contains even more tree stumps than the K facies at the base of the floral cycle, and its tissue preservation is relatively high, in

FIG. 10.—Photograph of the light-colored HB band (centre of frame) separating the upper two benches of the 2nd Lusatian Coal Horizon at Welzow Süd open cut. Lithotypes are indicated by the facies codes listed in Table 1. The coal face is 6 m high.

spite of the exclusion of mega-xylite from the sample. There is a basinward decline in the proportion of humotelinite from Sample Ml to Ms because of a shift towards humodetrinite. Presumably this trend is related to bacterial attack as part of the increased marine influence in Ms which is demonstrated by the concomitant trend in the sulphur distribution listed in Table 1. The increase in ash content may also be related to the partial destruction of organic matter, although percolating sea water may also have added inorganic material to the peat.

Sequence–Stratigraphic Interpretation

Peat formation began in the 2nd Lusatian Coal Horizon with the Lower Rider Seam above a poorly-defined sequence boundary. The predominance of the limnotelmatic K facies (Suhr et al., 1992) plus a large contribution to the peat by the eutrophic *Alnus–Liquidambar* facies (F facies of Schneider, 1980) suggest that this comparatively thin seam formed in a predominantly topogenous setting as part of the transgressive systems tract.

The maximum landward extent of marine flooding in the 2nd Lusatian Coal Horizon occurs within the G–5 marine bed (Fig. 9) between the Lower Rider Seam and the overlying coal. Only in a relatively small landward position is the Lower Rider Seam connected to the base of the amalgamated seam. The latter is situated in the early highstand systems tract which is suggested by the seaward stacking of interseam wedges.

When fully developed, each of the three to four benches of the amalgamated seam consists of the superposition of up to six coal facies with distinct botanical, petrographic and technological characteristics. They identify an upward gradation from limnotelmatic conditions at the beginning of each cycle to raised-mire conditions near the top. The abrupt upward changes from the raised mire (P or M facies) to the limnotelmatic K facies are the up-dip correlatives of the flooding surfaces that mark parasequence boundaries. A cartoon depicting this situation is illustrated in Figure 11C between the 3rd and 4th parasequence (PS 3 and 4). The boundaries may be accentuated by some erosion of the upper portions of the ombrogen-

ous M facies, or by the intercalation of the light-colored, high-ash HB facies, as indicated between PS 2 and 3 in Figure 11C.

Because of the lack of outcrops and sparse bore hole coverage, the landward termination of the composite seam against the lacustrine clays and fluvial sands of the Rietschen Member is poorly understood. Down-dip, the beginning of the seam splits, referred to as ZM 1 to 3 in Figure 9 (lower diagram) is mostly indicated by the HB facies which grades first into coaly and then bioturbated silts. This is not unlike the intermingling of the dull coal with the BB3 band in the Australian example. Further basinward, very clean but, initially, thin washover and foreshore sands are introduced between the silts and the underlying M facies. With increasing thickness, first upper then lower shoreface sands and sand-laminated, bioturbated silts begin to dominate the seam splits. In reverse order, this means that up-dip the Lusatian parasequences lose their constituent components from the bottom upwards so that in the amalgamated seam only the coal is left. A schematic illustration of this trend is given in Figure 11 by the change in the composition of the columnar sections A to C.

As the parasequences thicken basinward and become more marine, the former coal benches change into parasequence-capping coal seams. Down-dip, the coals display a conspicuous decrease in the proportion of raised-mire coal (M facies) which brings about a concomitant change in the microscopic and chemical composition of the coal. According to Schneider (1980), there is a strong decline in the root/shoot ratio due to the decrease in the root tissue *Marcoduria inopinata* which is expressed petrographically by a shift in the composition of the tissued components (i.e., away from the sedentary rhizo-textite to the sedimentary xylo-textite, both microlithotype varieties after Sonntag et al., 1965). The trend is paralleled by a basinward increase in the over-all ash content from mostly less than 7% (db = dry basis) in the amalgamated seam with a relatively high proportion of ombrogenous coal to more than 11% (db) in the split portion of the seam horizon where the eutrophic influence was higher (Standke et al., 1993). As mentioned above, there is also a tendency for the sulphur content to increase basinward.

The contact of the upper coal with its roof sediments illustrated in Figure 9 (lower diagram) represents a Type–1 sequence boundary which is indicated by widespread and substantial erosion followed by another marine transgression.

SUMMARY AND CONCLUSIONS

The two amalgamated coals discussed in this paper split laterally into single seams which cap parasequences dominated by clastic sediments. Although both coals differ in detail due to differences in rank, age and geological setting, they share a number of properties with high diagnostic value for onshore sequence-stratigraphic analysis. These diagnostic properties can be identified by standard coal-analytical methods (e.g., proximate and elemental analysis, in particular ash and sulphur content, lithotype logs, maceral counts, vitrinite reflectance and fluorescence intensity measurements in conjunction with geophysical bore logs and paleobotanical investigations).

Sequence Stratigraphic Setting

In the 2nd Lusatian Coal Horizon, which has been tentatively correlated with Sequence TB 2.3 of Haq et al. (1988), the max-

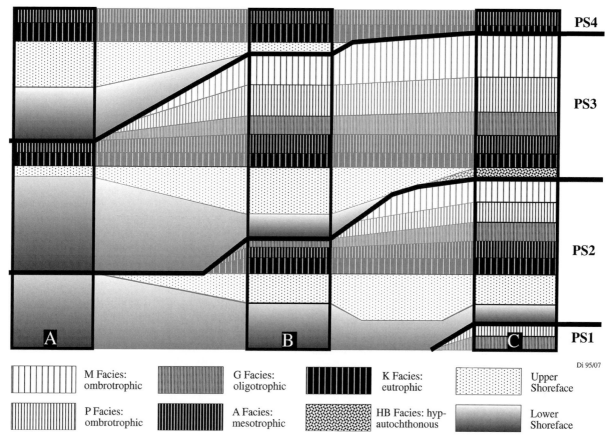

Di 95/07

FIG. 11.—Cartoon illustrating the observed and inferred orders of superposition in three columnar sections A (seaward) to C (landward). PS 1 to 4 identify parasequences of which only PS 3 is complete in all three columns.

imum landward marine flooding horizon is contained in the G 5 marine sand which presumably also contains the maximum flooding surface. This is likely to put the Lower Rider Seam and the first coal seam above the G 5 marine sand into the transgressive systems tract while the remainder of the horizon, which contains the bulk of the coal, belongs to the highstand systems tract. This interpretation is supported by differences in coal composition and the seaward stacking of the coal seams that split away from the amalgamated coal.

Peat accumulation in the 2nd Lusatian Coal Horizon was terminated by a fall in relative sea level that led to considerable incision under lowstand systems tract conditions and the formation of a sequence boundary along the top of the coal. This contrasts with the example from the Sydney Basin where a sequence boundary passes through the middle of the amalgamated coal seam so that the transgressive surface, the maximum landward marine flooding and the maximum flooding surface are all concentrated within one parasequence comprising the Bulga Formation, the Archerfield Sandstone and the Bayswater Seam. This parasequence represents the most landward extension of a group of marine horizons known as the Kulnura Marine Tongue which reaches its maximum thickness along the present coastline of the Sydney Basin.

The Archerfield–Bulga–Bayswater parasequence constitutes the whole of the transgressive systems tract near the northeastern basin margin, where it is situated above a Type 1 sequence boundary. This boundary consists of an unconformity between the marine sediments of the Archerfield–Bulga–Bayswater parasequence above, and partially eroded Wittingham Coal Measures underneath. These coal measures are part of a highstand systems tract. They were incised during the subsequent lowstand which resulted in the formation of the unconformity (= sequence boundary). Another highstand occurs above the Broonie/Bayswater split. Within the studied interval, it is represented by two parasequences, but neither its upper nor lower boundaries were defined with sufficient certainty.

Parasequence Anatomy

The parasequences represented by the Bayswater/Broonie split and the Upper/Lower Broonie split are dominated by alluvial sediments for which no equivalents were found in the Lower Lusatian coal measures. The latter contain marine sediments which are very similar to the Bulga Formation and Archerfield Sandstone in the Sydney Basin. The Bayswater Seam and the Lusatian seams are therefore similar in the way they separate down-dip from the amalgamated coal body and become the topmost members of upward shoaling parasequences which range from lower shoreface silts to upper shoreface sands. In the Lusatian coals, the trend towards terrestrialisation is continued in each of the coal seams by a succession of facies or lithotypes representing an upward development from eutrophic, limnotelmatic to oligotrophic, raised-mire conditions. The latter decline in proportion down-dip where limnotelmatic conditions become more common.

Erosion below the Lusatian flooding surfaces at the parasequence boundaries is relatively mild. Seaward of the shorelines (i.e., where marine sediments are intercalated with the coal), the marine flooding surfaces are slightly undulating due to the erosive removal of some of the underlying coal. Landward of the shorelines (i.e., within the amalgamated seam), the conjoined coal splits appear as benches with their lithotype and floral successions intact. The parasequence boundaries are therefore recognized as contacts between the underlying M facies, representing ombrotrophic, raised-mire conditions and the overlying K facies of eutrophic, limnotelmatic origin. These contacts which mark a break in peat accumulation between two growth cycles, may be straight or slightly undulating which is due to some tidal (?) erosion of the underlying coal.

The preservation of the stable residual peat components represented by the HB facies landward of the shoreline and its absence further seaward appear to be a function of contrasting environmental energy. While peat degradation at the change from one parasequence cycle to another was probably widespread across the mire, it was the maximum landward extent of the strandline belonging to the new parasequence that divided the underlying peat surface into offshore and onshore portions. The offshore portions of the flooded peat were subjected to marine erosion and dispersal of the degraded peat down to the lower limits of the degradation. Because of its interlocking fabric the underlying fresh woody peat (M or P facies) would have been considerably more resistant so that weathering and erosion surfaces might share a similar position.

In the onshore region, the peat surface would also have been subjected to degradation, erosion and dispersal but the lower energy conditions and geographical constraints prevented complete elimination of the degraded peat. Instead, its reworked stable components were preserved partly in situ and partly redistributed in back-barrier lagoons.

Coal Lithotype Organisation

Irrespective of the substantial differences in their degrees of coalification, the Lusatian and Hunter Valley coal lithotypes that originated in similar depositional environments share a common signature. Assuming an increase in coalification to the rank of bituminous coal, a typical Lusatian coal seam would begin at the base with a high-ash, thin layer of dull coal, possibly with a sapropelic tendency, representing the HB facies. This would give way to well-bedded, bright lithotypes and still high, but decreasing ash content in the lower portion of the seam where the eutrophic limnotelmatic K facies with its high proportion of wood-derived telovitrinite set in a matrix of detrovitrinite is dominant.

Further upward, the coal would lose its distinct stratification and become more massive. While the bright lithotypes in the lower part of the seams are the remnants of fallen logs, branches and twigs with a high shoot/root ratio, any bright coal occurring in the upper raised mire consist primarily of roots and other subsurface material of high root/shoot ratio. Many bituminous coals follow this general pattern which has been associated with a genetic trend from eutrophic limnotelmatic to ombrotrophic

raised-mire conditions (Smith 1961, 1962, 1964, 1968; Shibaoka and Smyth 1975; Smyth and Cook 1976; Littke 1985, 1987; Fulton 1987, Grady et al., 1993). As a generalization, this interpretation has not remained unchallenged (for a discussion see Diessel, 1992a), but the upward change from predominantly bright to dull coal in the Upper Wynn Seam and, to a lesser extent, in the lower portion of the Bayswater Seam agrees with the notion of a change from eutrophic to oligotrophic peatforming conditions. Its main support is seen in the decline in the number of epiclastic stone bands and the reduction in rawcoal ash content from more than 20% in the lower eutrophic portion to less than 10% and in some samples down to 5% (db) in the upper oligotrophic part (Fig. 5).

It can be concluded that the parasequence-capping coals tend to continue the upward shoaling trend, beginning with eutrophic (sp. minerotrophic) limnotelmatic peat formation and ending with oligotrophic (sp. ombrotrophic) conditions.

Coal Lithotypes with Sequence Stratigraphic Significance

Two coal lithotypes were found to have a particularly high diagnostic value for the identification of significant sequence stratigraphic surfaces in coal seams. One is the inertodetrinite- and liptinite-rich degraded-peat-derived dull coal that occurs below the sequence boundary at the base of the BB3 band in the amalgamated seam from the Hunter Valley. The other lithotype is the HB facies which occurs above the parasequence boundaries in the amalgamated 2nd Lusatian Coal Horizon.

The Dull Coal Below the Sequence Boundary at Dartbrook.—

Apart from a small amount of highly reflecting pyrofusinite and semifusinite, the maceral content of the dull coal below the sequence boundary and BB3 band at Dartbrook is dominated by low-reflecting inertinite, mostly in the form of inertodetrinite. While it is possible for some of the low-reflecting inertinite to have been formed by incomplete combustion (McGinnes et al., 1971; Scott, 1989; Scott and Jones, 1991, 1994; Diessel, 1992a), its large proportion suggests an additional source, probably, in the form of fungal attack as discussed by Teichmüller (1982), Prokopovich (1985), Grady et al. (1993; Moore and Shearer (1993), and Moore et al. (1996). For example, white- and brown-rot fungi extract carbohydrates (e.g., polysaccharides; Blanchette and Abad, 1988) from the cell tissue leaving behind carbon-enriched remnants of the cell walls (Adaskaveg et al., 1991) which appear as inertinite in the resultant coal. The fungi require moist and aerobic conditions but cannot operate under a complete water cover (Moore, 1989). The possible presence of mainly fungus-derived inertinite in the dull coal below the BB3 band agrees well with the notion of a lowered groundwater table.

A small but important component of the dull coal are vitrinitized rootlets. Mostly, they are less than 1 mm thick, and many are not aligned with stratification but penetrate obliquely into the attritus of inertodetrinite and sporinite. Cross-cutting roots are rarely observed in coals because the compaction of the fresh peat during coalification causes them to become more or less aligned with bedding. However, as pointed out by Prokopovich (1985), and Shearer et al. (1994), a peat that is subaerially exposed due to a lowering of the groundwater table, dewaters and compacts under its own weight. A similar mechanism of auto-compaction may have operated in the exposed upper peat layer during the lowstand systems tract thus preventing the alignment of the roots sunk into the pre-compacted peat. The fact that these rootlets were vitrinitized (some with a high-micrinite content and poor preservation) and not subjected to the same degradation as their host peat, is further testimony to their late-stage emplacement.

The Dull Coal Lithotypes in the Upper Wynn and Bayswater Seams.—

In coals formed from autochthonous peat, the proportion of inertinite is loosely related to the duration of subaerial exposure of the plant litter in the oxidizing or aerobic zone (acrotelm) on its passage to the reducing or anaerobic zone (catatelm) below the groundwater table. High concentrations of inertinite indicate therefore less than optimum peat-forming conditions due to a low groundwater table, or because of a low sedimentation rate. In either case, there exists an imbalance between the relatively high rate of plant production and the low accommodation potential of the mire to which the peat and subsequent coal responds as indicated in Table 2.

While the occurrence of inertinite-rich, degraded-peat-derived dull coal below a sequence boundary appears to be a natural consequence of a trend towards negative accommodation at the height of the lowstand systems tract, lesser base-level shifts, as for example associated with parasequence cycles and/or smaller-scale climatic shifts, may produce similar results. For example, the downlap of inertinite-rich lithotypes in the lower ombrotrophic part of the Bayswater Seam, including the tuffaceous BA3 marker band illustrated in Figure 3, on the Archerfield Sandstone suggests that they mark successive positions of the peat surface as the mire followed the basinward progradation of the shoreline. Rather then indicating regional base-level related stages of desiccation and flooding, the compositional changes appear to constitute the response of the peat to small-scale climatic changes which accelerated or retarded peat growth. According to Esterle and Ferm (1994), the margins of peat deposits are likely to show compositional variations because of their lower peat accumulation rates and more frequent groundwater table fluctuations.

The Lusatian HB Band Above the Parasequence Boundary.—

While the dull coal below the sequence boundary at Dartbrook is akin to the "oxidised organic partings" of Shearer et al. (1994), the HB facies shows similarities to their "organic, non-oxidised, degradative partings". The vegetable matter of both lithotypes is highly degraded but, in contrast to the prolonged subaerial exposure above the groundwater table in the first example, the aerobic influence on the HB facies was shorter. Although the inertinite content in the HB band is significantly higher than in the other Lusatian lithotypes, it remains well below that of the dull coal underneath the BB3 band and may even be allochthonous (i.e., the result of alluvial or aeolian concentration).

The 6.3% inertinite in the HB facies consists mainly of high-reflecting pyrofusinite and semifusinite plus some fungal spores but with little evidence of low-reflecting, fungus-derived inertinite. With increasing coalification, there would be some relative increase in the inertinite proportion due to the devolatilization and condensation of the more volatile humic substances, but the dominance of the humodetrinite and liptinite macerals would not be affected. Such a maceral composition is the product of a largely non-oxidative and non-fungal destruction of cell tissue leading to a concentration of humodetrinite and relatively stable plant residues (liptinite and inertinite) which is likely to take place in a subaqueous, near-neutral pH environment (Shearer et al., 1994). In their discussion of the compositionally similar gyttja facies in the Holocene peat deposits of the Fraser River delta, British Columbia, Styan and Bustin (1983a, b) also refer to the intensity of the destruction of cell tissue in ponded water and the paucity of fungal hyphae, which leave aerobic bacteria as the most likely agents of the degradation. The bacterial influence on the formation of the fine humic detritus and its intimate mixing with fine liptodetrinite and subaqueous origin are in keeping with the high atomic H/C value and huminite fluorescence intensity of the HB facies.

Because the HB facies appears to be more the product of the degradation of a pre-existing peat rather than of the accumulation of new peat, there is a genetic similarity to the dull coal below the BB3 band at Dartbrook. However, there is an important difference between the two lithotypes in their spatial relationship to the flooding surface. The dull coal has to be situated below the flooding surface (see position of TS in Fig. 7), because the compositional emphasis is on peat exposure and aerobic tissue destruction. In contrast, the HB band has been put above the flooding surface (see position of the parasequence boundary between PS 2 and 3 in Fig. 11C) because it was formed as part of the subaqueous bacterial degradation and reworking of the peat surface after the flooding of the mire.

TABLE 2.—THE RESPONSE OF COAL PROPERTIES TO BALANCED AND UNBALANCED ACCOMMODATION

Accommodation/ Accumulation Ratio:	Balanced	High	Low
Mire type	topogenous or ombrogenous	topogenous, limno-telmatic	ombrogenous (raised)
Root/shoot ratio	variable	low	high
Cell tissue preservation	high	variable, low when marine influenced	variable, low when strongly oxidised
Ash content	generally low	high, adventitious	high, inherent
Main maceral groups	high vitrinite (telocollinite)	medium to high vitrinite (desmocollinite)	high inertinite
Atomic H/C ratio	medium to high	high to very high	low to very low
Microfabric	laminated	strongly laminated (sedimentary)	weakly laminated (sedentary)

The Applicability of Reflectance and Fluorescence
Determinations to Sequence Stratigraphic Problems

The final conclusion that can be drawn from the investigations concerns the applicability of optical coal properties to sequence-stratigraphic problems. Ever since it was found that vitrinite reflectance is consistently suppressed and its fluorescence intensity is raised in coals that contain alginite (i.e., the remnants of algae, and large amounts of other liptinite macerals; Taylor, 1966; Correia and Connan, 1974; Jones and Edison, 1978; Hutton and Cook, 1980; Hutton et al., 1980; Goodarzi, 1985; Kalkreuth, 1982; Kalkreuth and Macauley, 1984, 1987; Wolf and Wolff-Fischer, 1984; Correa da Silva, 1989; Hagemann et al., 1989; Teichmüller, 1989; Diessel, 1990, 1992b; Barker, 1991; Raymond and Murchison, 1991; Rathbone and Davis, 1993), vitrinite reflectance has ceased to be a unique rank indicator. The association of suppressed vitrinite reflectance and high-liptinite concentrations need not be a causal relationship but a manifestation of unique paleo-environmental conditions (Bostick and Foster, 1975; Newman and Newman, 1982; Wenger and Baker, 1987; Veld and Fermont, 1990; Wilkins et al., 1992; Veld, 1995).

Coals whose peat precursors were subjected for a limited time to mild oxic depositional conditions contain high-reflecting, slightly subhydrous vitrinite (Wenger and Baker, 1987; Fang and Jianyu, 1992). Conversely, coals which were subjected to anoxic conditions and marine flooding at the peat stage, like the seams in the 2nd Lusatian Coal Horizon and the Upper Wynn Seam in the Sydney Basin, tend to be slightly perhydrous (Teichmüller, 1987, 1989; Veld and Fermont, 1990; Fang and Jianyu, 1992), and record the degree of dilution of the acid peat water very sensitively by lowered vitrinite reflectance and raised fluorescence intensity (Price and Barker, 1985; Diessel, 1990; Wilkins et al., 1992, Newman, 1992; Stevenson, 1993; Quick, 1994). The acquired differences in optical properties between isometamorphic vitrinites are largely retained during physico-chemical coalification (Jones and Edison, 1978; Wenger and Baker, 1987) but are strongest in the rank range from sub-bituminous to medium volatile bituminous coal (Diessel, 1992a, Fig 5.28).

The vertical trends in optical properties shown by the coals in this study are consistent with the results of previous investigations which indicate that coals covered by marine roof sediments display a strong upward decrease in vitrinite reflectance and a concomitant increase in vitrinite fluorescence intensity (Diessel, 1992b; Stevenson, 1993; George et al., 1994). Coal seams with brackish or lacustrine roof sediments show the same trends in optical properties but at lower intensity levels (Diessel, 1992a, Figs. 8.11, 8.15).

The trends in optical properties appear to be similar in coals with syngenetic (syn-depositional) and epigenetic (post-depositional) marine influence, provided the underlying peat was fresh and reactive enough to respond to the raised pH and chemical compounds introduced by the downward percolating seawater. This point is highlighted by the Upper Wynn Seam below the sequence boundary by the difference between the low fluorescence and high reflectance recorded in the degraded-peat-derived vitrinite within the amalgamated seam and the high fluorescence and low reflectance measured in the fresh-peat-derived vitrinite outside the amalgamated seam where the ma-

rine erosion had removed the weathered peat. The magnitude of the contrast in the optical properties is significant even though the root-derived vitrinite in the dull coal of the amalgamated seam largely escaped the full duration and severity of the peat degradation that led to the formation of the surrounding inertinite.

ACKNOWLEDGMENTS

The research for this paper was greatly assisted by several colleagues and organisations. R. Doyle of Dartbrook Coal Pty. Limited, J. Dwyer and B. Preston of Coal & Allied Operations Pty. Limited, and R. Bönisch and W. Schneider of Lusitzer Braunkohle AG (Laubag) are thanked for their help in the field and for facilitating the acquisition of samples, bore logs, analysis results and general background material.

Many of the petrographic analyses were carried out by L. Gammidge of The University of Newcastle, N.S.W., which is gratefully acknowledged.

G. Standke and P. Suhr of Sächsisches Landesamt für Umwelt und Geologie are thanked for helpful discussions about the geology of the Lower Lusatian coal deposits, while R. Kerr and R. Boyd of The University of Newcastle, N.S.W. provided counsel in stratigraphic and sequence-stratigraphic matters, respectively.

The two reviewers of the manuscript, K. Bohacs of Exxon Production Research and M. Bustin of The University of Britisch Columbia are thanked for their thoughtful suggestions which considerably improved the quality of the paper.

The research for this paper is part of an ARC-supported investigation into the timing and effects of marine influence on coal. The Australian Research Council is thanked for providing financial assistance.

REFERENCES

ADASKAVEG, J. E., BLANCHETTE, R. A., AND GILBERTSON, R. L, 1991, Decay of date palm wood by white-rot and brown-rot fungi: Canadian Journal of Botany, v. 69, p. 615–629.

AITKEN, J. F., 1994, Coal in a sequence stratigraphic framework: Geoscientist, v. 4, no. 5, p. 9–12.

ALEXOWSKY, W., STANDKE, G., AND SUHR, P., 1989, Beitrag zur weiteren lithostratigraphischen Untergliederung des Tertiärprofils in der Niederlausitz: Geoprofil, v. 1, p. 57–62.

BAILEY, J. G., 1981, The geology of the Broke–Mt. Thorley area: Unpublished B.Sc. Honours Thesis, The University of Newcastle, Newcastle, N.S.W., 320 p.

BARKER, C. E., 1991, An update on the suppression of vitrinite reflectance: The Society of Organic Petrology, News Letter 8, p. 8–11.

BEAUBOUEF, R. T., MCLAUGHLIN, P. P., BOHACS, K. M., DEVLIN, W. J., AND SUTER, J. R., 1995, Sequence stratigraphy of coal-bearing strata, Upper Cretaceous, Washakie Basin, SW Wyoming: Tulsa, Annual Convention of the American Association of Petroleum Geologists, Abstracts, p. 7A.

BEAUMONT, E. C., SHOEMAKER, P. W., AND KOTTLOWSKI, F. E., 1971, Stratidynamics of coal deposition in southern Rocky Mountain region, U. S. A., in Beaumont, E. C., Shoemaker, P.W., and Kottlowski , F. E., eds., Strippable Low-Sulfur Coal resources of the San Juan Basin in New Mexico and Colorado: Santa Fe, New Mexico Bureau of Mines and Mineral Resources Memoir 25, p. 175–185.

BECKETT, J., HAMILTON, D. S., AND WEBER, C. R., 1983, Permian and Triassic stratigraphy and sedimentation in the Gunnedah–Narrabri–Coonabarabran region: New South Wales Geological Survey, Quarterly Notes, v. 51, p. 1–16.

BELYAEV, S. S., YY-LEIN, A., AND IVANOV, M. V., 1981, Role of methane-producing and sulphate-reducing bacteria in the destruction of organic matter, in Trudinger, P. A. and Walter, M. R., eds., Biochemistry of Ancient and Modern Environments: Berlin, Springer-Verlag, p. 235–242.

BERNER, R. A., 1970, Sedimentary pyrite formation: American Journal of Science, v. 268, p. 1–23.

BERNER, R. A., 1985, Sulphate reduction, organic matter decomposition and pyrite formation: London, Royal Society Philosophical Transactions, v. A315, p. 25–38.

BERNER, R. A., BALDWIN, T., AND HOLDERN, G. R., Jr., 1979, Authigenic iron sulfides as paleosalinity indicators: Journal of Sedimentary Petrology, v. 49, p. 1345–1350.

BLANCHETTE, R. A. AND ABAD, A. R., 1988, Ultrastructural localization of hemicellulose in birch wood (*Petula papyrifera*) decayed by brown and white rot fungi: Holzforschung, v. 42, p. 393–398.

BOHACS, K. M. AND SUTER, J., 1995, Sequence stratigraphic distribution of coaly rocks: Fundamental controls and paralic examples: Houston, Annual Convention of the American Association of Petroleum Geologists, Abstracts, p. 10A.

BÖNISCH, R., GRUNERT, K., AND SCHNEIDER, W., 1983, Neue Erkenntnisse zum Stand der kohlengeologischen Erkundung im Förderraum Lausitz: Zeitschrift für angewandte Geologie, v. 29, p. 469–475.

BOSTICK, N. H. AND FOSTER, J. N., 1975, Comparison of vitrinite reflectance in coal seams and in kerogen of sandstones, shales and limestones in the same part of a sedimentary section, *in* Alpern, B., ed., Pétrographie Organique et Potential Pétrolier: Paris, Centre National de la Recherche Scientifique, p. 14–25.

BOYD, R. AND DIESSEL, C., 1995, The effects of accommodation, base level and rates of peat accumulation on coal measure architecture and composition: Houston, Annual Convention of the American Association of Petroleum Geologists, Abstracts, p. 12A.

BRAKEL, A. T., 1984, Correlation of the Permian coal measures sequences of eastern Australia, *in* Permian Coals of Eastern Australia, v. 1: Canberra, National Energy Research, Development and Demonstration Council, Report 78/2617.

BRAUSE, H., RASCHER, J., AND SEIFERT, A., 1989, Transgressionsgeschichte und Kohlenqualitäten im Miozän der Lausitz: Geoprofil, v. 1, p. 18–30.

BRITTEN, R. A., 1972, A review of the stratigraphy of the Singleton Coal Measures and its significance to coal geology and mining in the Hunter Valley region of New South Wales: Newcastle, Conference of the Australasian Institute of Mining and Metallurgy, Proceedings, p. 11–22.

BRITTEN, R. A. AND HANLON, F. N., 1975, North Western Coalfield, *in* Knight, C. L., ed., Economic Geology of Australia and Papua New Guinea, v. 2 Coal: Melbourne, The Australasian Institute of Mining and Metallurgy, Monograph 6, p. 236–44.

BRUENIG, E. F., 1990, Oligotrophic forested wetlands in Borneo, *in* Lugo, A. E., Brinson, M., and Brown, S., eds., Ecosystems of the World: Forested Wetlands: Amsterdam, Elsevier, p. 299–344.

CASAGRANDE, D. J., 1987, Sulphur in peat and coal, *in* Scott, A. C., ed., Coal and Coal-bearing Strata: Recent Advances: Boulder, Geological Society of America, Special Publication, v. 32, p. 87–105.

CLYMO, R. S., 1983, Peat, *in* Gore, A. J. P., ed., Ecosystems of the World: Mires: Swamp, Bog, Fen and Moor: New York, Elsevier, p. 159–224.

CLYMO, R. S., 1984, The limits of peat and bog growth: Philosophical Transactions of the Royal Society of London, v. B303, p. 605–654.

CLYMO, R. S., 1987, Rainwater-fed peat as a precursor of coal, *in* Scott, A. C., ed., Coal and Coal-bearing Strata: Recent Advances: Boulder, Geological Society of America Special Publication 32, p.17–23.

CORREA DA SILVA, Z. C., 1989, The rank evaluation of South Brazilian Gondwana coals on the basis of different chemical and physical parameters, *in* Lyons P. C. and Alpern, B., eds., Coal: Classification, Coalification, Mineralogy, Trace-Element Chemistry, and Oil and Gas Potential: International Journal of Coal Geology, v. 13, p. 21–39.

CORREIA, M. AND CONNAN, J., 1974, Analyses physio-chimiques et observations microscopiques de la matière organique de schistes bitumineux, *in* Tissot, B. and Bienner, F., eds., Advances in Organic Geochemistry 1973, Paris, Editions Technip, p. 153–161.

CROSS, T. A., 1988, Controls on coal distribution in transgressive-regressive cycles, Upper Cretaceous, Western Interior, U.S.A., *in* Wilgus, C. K., Hastings, B. S., Kendall, C. G. St. C., Posamentier, H. V., Ross, C. A., and Van Wagoner, J. C., eds., Sea-Level Changes: An Integrated Approach: The Society of Economic Paleontologists and Mineralogists, Special Publication 42, p. 371–380.

DEGENS, E. T., 1958, Geochemische Untersuchungen zur Faziesbestimmung im Ruhrkarbon und im Saarkarbon: Glückauf, v. 94, p. 513–520.

DIESSEL, C. F. K., 1965, Correlation of macro- and micropetrography of some New South Wales coals: Melbourne, Eighth Commonwealth Mining and Metallurgy Congress, Proceedings 6, p. 669–677.

DIESSEL, C. F. K., 1990, Marine influence on coal seams, *in* Advances in the Study of the Sydney Basin: Newcastle, 24th Newcastle Symposium, Proceedings, p. 33–40.

DIESSEL, C. F. K., 1992a, Coal-Bearing Depositional Systems: Berlin, Springer-Verlag, 721 p.

DIESSEL, C. F. K., 1992b, The problem of syn- versus post-depositional marine influence on coal composition, *in* Advances in the Study of the Sydney Basin: Newcastle, 26th Newcastle Symposium, Proceedings, p. 154–163.

DIESSEL, C. F. K. AND STODDART, F. G., 1986, Geological guide to BHP Saxonvale Mine. Excursion 2, *in* Advances in the Study of the Sydney Basin: Newcastle, 20th Newcastle Symposium, Proceedings, p. 109–129.

DIESSEL, C. F. K., BOYD R., AND WARBROOKE, P. R., 1989, The Waratah Sandstone: a wave/tide dominated barrier beach, *in* Boyd, R. L. and MacKenzie, G. A., eds., Advances in the Study of the Sydney Basin: Newcastle, 23rd Newcastle Symposium, Proceedings, p. 109–115.

DIESSEL, C. F. K., BECKETT, J., AND WEBER, C., 1995, On the anatomy of an angular unconformity in the Wittingham Coal Measures near Muswellbrook, *in* Boyd, R. L. and MacKenzie, G. A., eds., Advances in the Study of the Sydney Basin: Newcastle, 23rd Newcastle Symposium, Proceedings, p. 133–139.

ESTERLE, J. S. AND FERM, J. C., 1994, Spatial variability in modern tropical peat deposits from Sarawak, Malaysia and Sumatra, Indonesia: analogues for coal: International Journal of Coal Geology, v. 25, p. 1–42.

FANG, H. AND JIANYU, C., 1992, The cause and mechanism of vitrinite reflectance anomalies: Journal of Petroleum Geology, v. 15, p. 419–434.

FULTON, I. M., 1987, Genesis of the Warwickshire Thick Coal: a group of long-residence histosols, *in* Scott, A. C., ed., Coal and Coal-bearing Strata: Recent Advances: Boulder, Geological Society of America Special Publication 32, p. 201–218.

GEORGE, A. M., 1982, Latrobe Valley brown coal-lithotypes: macerals: coal properties, *in* Mallett C. W., ed., Coal Resources–Origin, Exploration and Utilization in Australia: Melbourne, Coal Group, Geological Society of Australia Symposium, Australian Coal Geology, v. 4, p. 111–130.

GEORGE, S. C., SMITH, J. W., AND JARDINE, D. R., 1994, Vitrinite reflectance suppression in coal due to a marine transgression: Case study of the organic geochemistry of the Greta Seam, Sydney Basin: Journal of the Australian Petroleum Exploration Association, v. 22, p. 241–255.

GIVEN, P. H. AND MILLER, R. N., 1985, Distribution of forms of sulfur in peats from saline environments in the Florida Everglades: International Journal of Coal Geology, v. 5, p. 397–409.

GLUSKOTER, H. J., 1977, Inorganic sulfur in coal: Energy Sources, v. 2, p. 125–131.

GOLDHABER, M. B. AND KAPLAN, I. R., 1974, The sulfur cycle, *in* Goldberg, E. D., ed., Ideas and Observations on Progress in the Study of the Seas: New York, Wiley, p. 569–655.

GOODARZI, F., 1985, Organic petrology of Hat Creek coal deposit No. 1, British Columbia: International Journal of Coal Geology, v. 5, p. 377–396.

GRADY, W. C., EBLE, C. F., AND NEUZIL, S.G., 1993, Brown coal maceral distribution in a modern domed tropical Indonesian peat and a comparison with maceral distributions in Middle Pennsylvanian-age Appalachian bituminous coal beds, *in* Cobb, J. C. and Cecil, C. B., eds., Modern and Ancient Coal-Forming Environments: Boulder, Geological Society of America, Special Paper 286, p. 63–82.

HAGEMANN, H. W., OTTENJANN, K., PÜTTMANN, W., WOLF, M., AND WOLFF-FISCHER, E., 1989, Optische und chemische Eigenschaften von Steinkohlen: Erdöl und Kohle–Erdgas–Petrochemie vereinigt mit Brennstoff–Chemie, v. 42, p. 99–110.

HAGER, H., 1981, Das Tertiär des Rheinischen Braunkohlenreviers, Ergebnisse und Probleme: Fortschritte der Geolologie von Rheinland und Westfalen, v. 29, p. 529–563.

HAGER, H., 1993, The origin of the Tertiary lignite deposits in the Lower Rhine region, Germany: International Journal of Coal Geology, v. 23, p. 251–262.

HAMILTON, D. S., 1985, Deltaic depositional systems, coal distribution and quality, and petroleum, Permian Gunnedah Basin, New South Wales: Sedimentary Geology, v. 45, p. 35–75.

HAQ, B. U., HARDENBOL, J., AND VAIL, P. R., 1988, Mesozoic and Cenozoic chronostratigraphy and cycles of sea-level change, *in* Wilgus, C. K., Hastings, B. S., Kendall, C. G.St. C., Posamentier, H. V., Ross, C. A., and Van Wagoner, J. C., eds., Sea-Level Changes: An Integrated Approach: Tulsa, Society of Economic Paleontologists and Mineralogists Special Publication 42, p. 71–108.

HERBERT, C. AND HELBY, R., 1980, A Guide to the Sydney Basin: Sydney, Geological Survey of New South Wales, Bulletin 26, 603 p.

HUNT, J. W., BRAKEL, A. T., AND SMYTH, M., 1986, Origin and distribution of the Bayswater Seam and correlatives in the Permian Sydney and Gunnedah Basins, Australia: Australian Journal of Coal Geology, v. 6, p. 59–75.

HUTTON, A. C. AND COOK, A. C., 1980, Influence of alginite on the reflectance of vitrinite from Joadja, NSW, and some other coals and oil shales containing alginite: Fuel, v. 59, p. 711–716.

HUTTON, A. C., KANSTLER, A. J., COOK, A. C., AND MCKIRDY, D. M., 1980, Organic matter in oil shales: Journal of the Australian Petroleum Exploration Association, v. 20, p. 44–68.

JONES, R. W. AND EDISON, T. A., 1978, Microscopic observations of kerogen related to geochemical parameters with emphasis on thermal maturation, in Oltz, D. F., ed., Symposium in Geochemistry: Low Temperature Metamorphism of Kerogen and Clay Minerals: Los Angeles, Pacific Section of the Society of Economic Paleontologists and Mineralogists, Proceedings, p. 1–12.

KALKREUTH, W., 1982, Rank and petrographic composition of selected Jurassic-Lower Cretaceous coals of British Columbia, Canada: Bulletin of Canadian Petroleum Geology, v. 30, p. 112–139.

KALKREUTH, W., AND MACAULEY, G., 1984, Organic petrology of selected oil shale samples from the Lower Carboniferous Albert Formation, New Brunswick, Canada: Bulletin of Canadian Petroleum Geology, v. 32, p. 38–51.

KALKREUTH, W., AND MACAULEY, G., 1987, Organic petrology and geochemical (rock eval) studies on oil shales and coals from the Pictou and Antigonish areas, Nova Scotia, Canada: Bulletin of Canadian Petroleum Geology, v. 35, p. 263–295.

LECKIE, D. A., 1994, Canterbury Plains (New Zealand) and sequence stratigraphy: The role of tectonism and wave climate-eustacy is not so important: Denver, Annual Convention of the American Association of Petroleum Geologists, Abstracts, p. 194.

LIN, R. AND DAVIS, A., 1988, A fluorogeochemical model for coal macerals: Organic Geochemistry, v. 12, p. 363–374.

LIN, R., DAVIS, A., AND DERBYSHIRE, F. J., 1986, Vitrinite secondary fluorescence: Its chemistry and relation to the development of a mobile phase and thermoplasticity in coal: International Journal of Coal Geology, v. 9, p. 87–108.

LITTKE, R., 1985, Flözaufbau in den Dorstener, Horster und Essener Schichten der Bohrung Wulfener Heide 1 (nördliches Ruhrgebiet): Fortschritte der Geologie von Rheinland und Westfalen, v. 33, p. 129–159.

LITTKE, R., 1987, Petrology and genesis of Upper Carboniferous seams from the Ruhr region, West Germany: International Journal of Coal Geology, v. 7, p. 147–184.

LOTSCH, D., 1967, Zur Paläogeographie des Tertiärs im Gebiet der DDR: Berichte der deutschen Gesellschaft für geologische Wissenschaften, Reihe A–Geologie und Paläontologie, v. 12, p. 369–374.

LOTSCH, D., KRUTZSCH, W., MAI, D., KIESEL, Y., AND LAZAR, E., 1969, Stratigraphisches Korrelationschema für das Tertiär der Deutschen Demokratischen Republik, in Abhandlungen: Berlin, Zentrales Geologisches Institute, v. 12, 438 p.

LOVE, L. G., 1957, Micro-organisms and the presence of syngenetic pyrite: Geological Society of London, Quarterly Journal. v. 113, p. 429–440.

MCCABE, P. J., 1993, Sequence stratigraphy of coal-bearing strata: New Orleans, American Association of Petroleum Geologists, Shortcourse, Unpublished Course Notes, 81 p.

MCCABE, P. J. AND SHANLEY, K. W., 1992, Organic control on shoreface stacking patterns: Bogged down in the mire: Geology, v. 20, p. 741–744.

MCCABE, P. J. AND SHANLEY, K. W., 1994, The role of tectonism, eustasy, and climate in determining the location and geometry of coal deposits: Denver, Annual Convention of the American Association of Petroleum Geologists, Abstracts, p. 209.

MCGINNES, E. A., KANDEEL, S. A., AND SZOPA, P. S., 1971, Some structural changes observed in the transformation from wood to charcoal: Wood and Fibre, v. 3, p. 77–83.

MCHUGH, E. A., 1984, The geology of the Jerry's Plains area: Unpublished Diploma of Science, The University of Newcastle, Newcastle, N.S.W., 87 p.

MITCHUM, R. M., Jr., VAIL, P. R., AND THOMPSON, S., III., 1977, Seismic stratigraphy and changes of sea level, Part 2: The depositional sequence as a basic unit for stratigraphic analysis, in Payton C. E., ed., Seismic Stratigraphy: Applications to Hydrocarbon Exploration: Tulsa, American Association of Petroleum Geologists, Memoir 26, p. 53–62.

MOORE, T. A., 1989, The ecology of peat-forming processes: a review: International Journal of Coal Geology, v. 12, p. 89–103.

MOORE, T. A. AND SHEARER, J. C., 1993, Processes and possible analogues in the formation of Wyoming's coal deposits, in Snoke, A. W., Steidtmann, J. R., and Roberts, S. M., eds., Geology of Wyoming: Cheyenne, Geological Survey of Wyoming Memoir 5, p. 874–896.

MOORE, T. A., SHEARER, J. C., AND MILLER, S. L., 1996, Fungal origin of oxidised plant material in the Palangkaraya peat deposit, Kalimantan Tengah, Indonesia: Implications for 'inertinite' formation in coal: International Journal of Coal Geology, v. 30, p. 1–23.

NEWMAN, J., 1992, Isorank variations in vitrinite chemistry– 1. Vitrinite reflectance: University Park, Nineth Annual Meeting, The Society of Organic Petrology, Abstracts and Program, p. 44–47.

NEWMAN, J. AND NEWMAN, N. A., 1982, Reflectance anomalies in Pike River coals: evidence of variability in vitrinite type, with implications for maturation studies and "Suggate rank": New Zealand Journal of Geology and Geophysics, v. 25, p. 233–243.

NOWEL, W., 1993, Geologische ¨Ubersichtskarte des Niederlausitzer Braunkohlen-Reviers: Senftenberg, Lausitzer Bergbau A.G. (LAUBAG), Presse und ¨Offentlichkeitsarbeit.

PRICE, F. T. AND SHIEH, Y. N., 1979, The distribution and isotopic composition of sulfur in coal from the Illinois Basin: Economic Geology, v. 74, p.1445–1461.

PRICE, L. C. AND BARKER, C. E., 1985, Suppression of vitrinite reflectance in amorphous rich kerogen– a major unrecognised problem: Journal of Petroleum Geology, v. 8, p. 59–84.

PROKOPOVICH, N. P., 1985, Subsidence of peat in California and Florida: Bulletin of the Association of Engineering Geologists, v. 22, p. 395–420.

QI MING, RONG XILIN, TANG DAZHONG, XIA JIAN AND WOLF, M., 1994, Petrographic and geochemical characterization of pale and dark brown coal from Yunnan Province, China: International Journal of Coal Geology, v. 25, p. 65–92.

QUICK, J. C., 1994, Iso-rank variation of vitrinite reflectance and fluorescence intensity, in Mukhopadhyay, P.K. and Dow, W.G., eds., Vitrinite Reflectance as a Maturity Parameter. Applications and Limitations: American Chemical Society, Symposium Series, v. 570, p. 64–75.

QUICK, J. C., DAVIS A., AND LIN, R., 1988, Recognition of reactive maceral types by combined fluorescence and reflectance microscopy: Toronto, American Institute of Metallurgy, Ironmaking Conference, Proceedings, p. 331–337.

RASCHER, J. AND STANDKE, G., 1991, Paläogeographie und Kohlebildung am Beispiel eines paralischen "Kohlenmoores" im Untermiozän der Lausitz: Freiberg, Mitteilung der geologischen Landesuntersuchung, v. 758, p. 73–76.

RATHBONE, R. F. AND DAVIS, A., 1993, The effects of depositional environment on vitrinite fluorescence intensity: Organic Geochemistry, v. 20, p. 177–186.

RAYMOND, A. C. AND MURCHISON, D. G., 1991, Influence of exinitic macerals on the reflectance of vitrinite in Carboniferous sediments of the Midland Valley of Scotland: Fuel, v. 70, p. 155–161.

RYER, T. A., 1984, Transgressive-regressive cycles and the occurrence of coal in some Upper Cretaceous strata of Utah, U.S.A., in Rahmani, R. A. and Flores, R. M., eds., Sedimentology of Coal and Coal-Bearing Sequences: Oxford, International Association of Sedimentologists Special Publication 7, p. 217–227.

SCHNEIDER, W., 1980, Mikropaläobotanische Faziesanalyse in der Weichbraunkohle: Neue Bergbautechnik, v. 10, p. 670–675.

SCHNEIDER, W., 1988, Die Bedeutung des Xylitgehaltes der Braunkohle im Rahmen der Erkundung von Kohlequalitäten: Zeitschrift für Angewandte Geologie, v. 34, p. 105–108.

SCHNEIDER, W., 1990, Die neue Deutung von Marcoduria inopinata Weyland 1957 und ihre kohlengeologische Konsequenz: Zeitschrift für Geologische Wissenschaften, v. 18, p. 911–918.

SCHNEIDER, W., 1992, Floral successions in Miocene swamps and bogs of Central Europe: Zeitschrift für Geologische Wissenschaften, v. 20, p. 555–570.

SCOTT, A. C., 1989, Observations on the nature and origin of fusain: International Journal of Coal Geology, v. 12, p. 443–475.

SCOTT, A. C. AND JONES, T. P., 1991, Microscoipical observations of recent and fossil charcoal: Microscopy and Analysis, v. 25, p. 13–15.

SCOTT, A. C. AND JONES, T. P., 1994, The nature and influence of fire in Carboniferous ecosystems. Palaeogeography, Palaeoclimatology, Palaeoecology, v. 106, p. 91–112.

SEIFERT, A., BRAUSE, H., AND J. RASCHER, 1993, Geology of the Niederlausitz lignite district, Germany: International Journal of Coal Geology, v. 23, p. 263–289.

SHEARER, J. C., STAUB, J. R., AND MOORE, T. A., 1994, The conundrum of coal bed thickness: A theory of stacked mire sequences: The Journal of Geology, v. 102, p. 611– 617.

SHIBAOKA, M. AND SMYTH, M., 1975, Coal petrology and the formation of coal seams in some Australian sedimentary basins: Economic Geology, v. 70, p. 1463–1473.

SINNINGE DAMSTÉ, J. S., RIJPSTRA, W. I. C., DE LEEUW, J. W., AND SCHENCK, P. A., 1988, Origin of organic sulphur compounds and sulphur-containing high molecular weight substances in sediments and immature oils: Organic Geochemistry, v. 13, p. 593–606.

SINNINGE DAMSTÉ, J. S., EGLINGTON, T. I, DE LEEUW, J. W., AND SCHENCK, P. A., 1989, Organic sulphur in macromolecular sedimentary organic matter. 1. Structure and origin of sulphur-containing moities in kerogen, asphaltenes and coal as revealed by flash pyrolysis: Geochimica et Cosmochimica Acta, v. 53, p. 873–889.

SMITH, A. H. V., 1961, Palaeoecology of Carboniferous peat bogs: Nature, v. 189, p. 744–745.

SMITH, A. H. V., 1962, The palaeoecology of Carboniferous peats based on the miospores and petrography of bituminous coals: Yorkshire Geological Society, Proceedings 33, p. 423–474.

SMITH, A. H. V., 1964, Zur Petrologie und Palynologie der Kohlenflöze des Karbons und ihrer Begleitschichten: Fortschritte der Geologie von Rheinland und Westfalen, v. 12, p. 285–302.

SMITH, A. H. V., 1968, Seam profiles and seam characters, in Murchison, D. G. and Westoll, T.S., eds., Coal and Coal-Bearing Strata: London, Oliver and Boyd, p. 31–40.

SMYTH, M. AND COOK, A. C., 1976, Sequence in Australian coal seams: Mathematical Geology, v. 8, p. 529–547.

SONNTAG, E., TZSCHOPPE, E., AND CHRISTOPH, H.-J., 1965, Beitrag zur mikropetrographischen Nomenklatur und Analyse der Weichbraunkohle: Zeitschrift für Angewandte Geologie, v. 11, p. 647–658.

STANDKE, G., SUHR, P., STRAUSS, C., AND RASCHER, J., 1992, Meeresspiegelschwankungen im Miozän von Ostdeutschland: Geoprofil, v. 4, p. 43–48.

STANDKE, G., RASCHER, J., AND STRAUSS, C., 1993, Relative sea-level fluctuations and brown coal formation around the Early-Middle Miocene boundary in the Lusation Brown Coal District: Geologische Rundschau, v. 82, p. 295–305.

STEVENSON, D. K., 1993, The determination of marine influences on the Greta and Pelton Seams by the use of selected palaeo-environmental indicators, in Advances in the Study of the Sydney Basin: Newcastle, 27th Newcastle Symposium, Proceedings, p. 197–204.

STYAN, W. B. AND BUSTIN, R. M., 1983a, Petrography of some Fraser River Delta peat deposits: coal maceral and microlithotype precursors in temperate-climate peats: International Journal of Coal Geology, v. 2, p. 321–370.

STYAN, W. B. AND BUSTIN, R. M., 1983b, Sedimentology of Fraser River Delta peat deposits: a modern analogue for some deltaic coals: International Journal of Coal Geology, v. 3, p. 101–143.

SUHR, P., SCHNEIDER, W., AND LANGE, J.-M., 1992, Facies relationships and depositional environments of the Lausitzer (Lusatic) Tertiary: Jena, 13th Regional Meeting of the International Association of Sedimentologists, Excursion Guidebook, p. 229– 260.

SWEENEY, R. E. AND KAPLAN, I. R., 1973, Pyrite framboid formation: Laboratory synthesis and marine sediments: Economic Geology, v. 68, p. 618–634.

TADROS, N. Z., ed., 1993, The Gunnedah Basin, N.S.W: Sydney, Geological Survey of New South Wales, Memoir 12, 649 p.

TAYLOR, G. H., 1966, The electron microscopy of vitrinites, in Gould, R. F., ed., Coal Science: Washington D.C., Advances in Chemistry Series , v. 55, p. 274–283.

TAYLOR, G. H. AND LIU, S. Y., 1987, Biodegradation in coal and other organic-rich rocks: Fuel, v. 66, p. 1269–1273.

TEICHMÜLLER, M., 1982, Origin of the petrographic constituents of coal, in Stach, E., Mackowsky, M.-Th., Teichmüller, M., Taylor, G. H., Chandra D., and Teichmüller, R., eds., Stach's Textbook of Coal Petrology: Berlin, Gebrüder Borntraeger, p. 219–294.

TEICHMÜLLER, M., 1987, Recent advances in coalification studies and their application to geology. In, in Scott, A. C., ed., Coal and Coal-bearing Strata: Recent Advances: Boulder, Geological Society of America Special Publication 32, p. 127–169.

TEICHMÜLLER, M., 1989, The genesis of coal from the viewpoint of coal petrology, in Lyons P. C. and Alpern, B., eds., Peat and Coal: Origin, Facies and Depositional Models: International Journal of Coal Geology, v. 12, p. 1–87.

TYLER, R. AND HAMILTON, D. S., 1995, Genetic Sequence stratigraphy of the intermontane Paleocene Fort Union Formation, Greater Green River Basin, southwest Wyoming and northwest Colorado: Houston, Annual Convention of the American Association of Petroleum Geologists, Abstracts, p. 98A.

UREN R. E., 1983, Facies and sedimentary environments of the Wittingham Coal Measures near Muswellbrook, in Advances in the Study of the Sydney Basin: Newcastle, 17th Newcastle Symposium, Abstrscts, p. 30–33.

UREN, R. E., 1985, The Margin of a marine incursion in the Wittingham Coal Measures, Upper Hunter Valley, in Advances in the Study of the Sydney Basin: Newcastle, 19th Newcastle Symposium, Proceedings, p. 45–48.

VAIL, P. R., 1987, Seismic stratigraphy interpretation procedure, in Bally, A. W., ed., Atlas of Seismic Stratigraphy: Tulsa, American Association of Petroleum Geologists, Studies in Geology, v. 27, p. 1– 10.

VAIL, P. R., MITCHUM, R. M., JR., TODD, R. G., WIDMIER, J. M., THOMSON, S. III, SANGREE, J. B., BUBB, J. N., AND HATLELID, W. G., 1977, Seismic stratigraphy and global changes of sea level, in Payton, C. E., ed., Seismic Stratigraphy: Applications to Hydrocarbon Exploration: Tulsa, American Association of Petroleum Geologists Memoir 26, p. 49– 212.

VAN WAGONER, J. C., MITCHUM, R. M., JR., POSAMENTIER, H. W., AND VAIL, P. R., 1987, Key definitions of sequence stratigraphy Part 2, in Bally, A. W., ed., Atlas of Seismic Stratigraphy: Tulsa, American Association of Petroleum Geologists Studies in Geology 27, p.11–14.

VAN WAGONER, J. C., MITCHUM, R.M., JR., CAMPION, K. M., AND RAHMANIAN, V. D., 1990, Siliclastic sequence stratigraphy in well logs, cores and outcrops: concepts for high resolution correlation of time and facies: Tulsa, American Association of Petroleum Geologists Methods in Exploration Series 7, 55 p.

VELD, H., 1995, Organic petrology of the Westphalian of the Netherlands: Den Haag, CIP–Gegevens Koninklijke Bibliotheek, 191 p.

VELD, H. AND FERMONT, W. J. J., 1990, The effect of a marine transgression on vitrinite reflectance values. Medelingen rijks Geologische Dienst, Nieuwe serie, v. 45, p. 151–169.

VINKEN, R., 1988, The Nortwest European Tertiary Basin: Geologisches Jahrbuch, Reihe A, v. 100, p. 1–608.

VON BUBNOFF, S., 1948, Einführing in die Erdgeschichte, Teil 2: Halle, Mitteldeutsche Druckerei und Verlagsanstalt, 771 p.

WENGER, L. M. AND BAKER, D. R., 1987, Variations in vitrinite reflectance with organic facies—Examples from Pennsylvanian cyclothems of the Midcontinent, USA: Organic Geochemistry, v. 11, p. 411–416.

WILKINS, R. W. T., WILMSHURST, J. R., RUSSELL, N. J., HLADKY, G., ELLACOTT, M. V., AND BUCKINGHAM, C. P., 1992, Fluorescence alteration and the suppression of vitrinite reflectance: Organic Geochemistry, v. 18, p. 629–640.

WOLF, M. AND WOLFF-FISCHER, E., 1984, Alginit in Humuskohlen karbonischen Alters und sein Einfluss auf die optischen Eigenschaften des begleitenden Vitrinits: Glückauf-Forschungshefte, v. 45, p. 243–246.

PART III
LACUSTRINE DEPOSITS

COARSE-GRAINED GILBERT DELTAS: FACIES, SEQUENCE STRATIGRAPHY AND RELATIONSHIPS TO PLEISTOCENE CLIMATE AT THE EASTERN MARGIN OF LAKE BONNEVILLE, NORTHERN UTAH

MARK R. MILLIGAN

Utah Geological Survey, P.O. Box 146100, Salt Lake City, Utah 84114–6100

AND

MARJORIE A. CHAN

Department of Geology and Geophysics, University of Utah, Salt Lake City, UT 84112–1183

ABSTRACT: The geomorphology of coarse-grained Pleistocene deltas was described over one hundred years ago by G. K. Gilbert. These deltas are exposed in gravel pit faces in northern Utah, at the eastern margin of ancient Lake Bonneville. Detailed sedimentologic and sequence stratigraphic studies indicate that the Gilbert deltas are the result of the complex interplay of climate and glaciation, tectonics and lake-level history.

The formation of Gilbert-type deltas requires coarse-grained sediment supply from steep canyons draining directly into a standing body of water. Active uplift along the Wasatch fault zone created the necessary condition of steep canyons. Uplift along the Wasatch fault zone began about 15 Ma. Offset on the fault continues at a rate of about 0.76–1.5 mm/yr (average slip rate for last 15 ka in study areas). The combination of climatic conditions and basin physiography brought the shoreline of Lake Bonneville within close proximity of the Wasatch Front. The lake-level highstand occurred about 15 ka. Glaciation reached a maximum about 19 ka.

An established lake-level history provides a framework for defining sequence stratigraphy in the shoreline deposits. Two end member coarse-grained delta systems are distinguished: (1) topset-dominated deltas at the Bonneville level (maximum high at ≈15 ka) comprising the transgressive/highstand systems tract and (2) foreset-dominated deltas at the Provo level (≈14.5–12 ka) comprising the lowstand systems tract. Where valley glaciers extended to the canyon mouths (close to the Bonneville shoreline), extensive horizontal subaerial outwash gravels were deposited at the Bonneville level. Where valley glaciation was limited to the mid to upper reaches of the canyon, horizontally stratified subaqueous gravel was deposited to form a Bonneville transgressive delta. A catastrophic 100–m drop in the lake level (to Provo shoreline) was followed by a climate-driven regression that caused reworking of Bonneville-level deltaic sediments and the development of large foresets in the Provo-level deltas in the lowstand systems tract.

INTRODUCTION

Coarse-grained (gravel) deltas at basin margins are sensitive recorders of tectonic, climatic and base-level conditions. Pleistocene Lake Bonneville of Utah (Fig. 1) has a well-established lake-level history and presents an excellent area in which to study the relationships of allocyclic factors. In addition, this is the classic study area of G. K. Gilbert's (1885, 1890) landmark work on coarse-grained deltas. Although Gilbert's work has

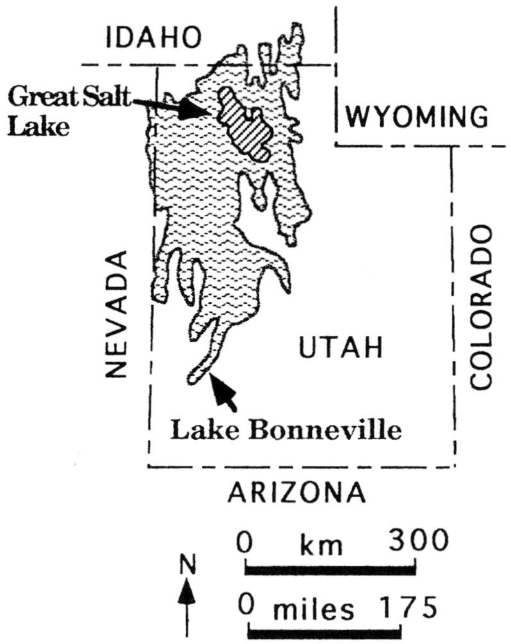

FIG. 1.—Location of Pleistocene Lake Bonneville at its highest (Bonneville) shoreline (modified from Currey, 1989).

often been applied to many modern and ancient examples, there are few sedimentologic studies on these seminal localities. Gravel pit exposures at American Fork, Big Cottonwood and Brigham City (Fig. 2) along the eastern margin of Lake Bonneville present an exceptional opportunity to evaluate the original Gilbert deltas in a modern context of basin analysis.

This study focuses on description of facies and depositional styles of Gilbert deltas to interpret the influence of tectonics and climate. Much of Lake Bonneville history is defined by radiocarbon age dating of remnant lake shoreline features which are used to construct a lake hydrograph (Fig. 3), summarized by Currey and Oviatt (1985). The initial uplift along the Wasatch mountains provided the necessary condition of steep canyons for the development of coarse-grained deltas. However, over the relatively short time span that Lake Bonneville existed (approximately 20,000 yr), the magnitude of subsequent tectonic uplift was overprinted (and relatively minor) in comparison to larger lake-level fluctuations. The sequence stratigraphy is interpreted largely as a function of lake level and implied influence of climate. Two delta examples are examined at the highest Bonneville level (mouths of American Fork and Big Cottonwood Canyons) and two examples are studied at the Provo level (mouth of Big Cottonwood Canyon and Brigham City locality). The recognition of facies and geometries has application to understanding problems of groundwater flow and contaminant migration in analogous facies along the populated Wasatch Front.

Previous Work

There is an enormous wealth of published literature on Lake Bonneville deposits, the cornerstone of which is G. K. Gilbert's (1890) classic work on Lake Bonneville and his earlier associated paper (Gilbert, 1885) on topographic features. Subsequent works on Lake Bonneville contain detailed geologic map-

Relative Role of Eustasy, Climate, and Tectonism in Continental Rocks, SEPM Special Publication No. 59

ping and stratigraphic studies (Hunt et al., 1953; Williams, 1962; Bissell, 1963; Morrison, 1965a, b, c, 1966). More recently, various studies have amply established and documented the Lake Bonneville stratigraphy, chronology, climate history, paleolimnology, paleohydrology and geomorphology. An exhaustive list of previous work on Lake Bonneville is not presented here, but important works include: Currey (1980, 1982, 1990, 1994a, b), Currey et al. (1984), Scott et al. (1983), Currey and Oviatt (1985), Oviatt (1987), Oviatt et al. (1987, 1992, 1994), McCoy (1987) and Machette (1988a). Aspects of basin isostatic rebound with removal of Lake Bonneville water, are discussed by Gilbert (1890), Crittenden (1963), Currey (1982), Currey et al. (1983) and Bills and May (1987). Recent surficial geology maps (Machette, 1989; Nelson and Personius, 1990; Personius and Scott, 1992) depict the distribution of Bonneville deposits along the Wasatch Front of Utah.

Delta Terminology

To date, there is no general agreement on terminology that should be applied to Pleistocene Lake Bonneville deltaic deposits. Terms that have been used to described coarse-grained deltas (including the deltas Gilbert, 1890 described as well as other delta systems) are:

1. Gilbert-type Delta = "coarse-grained fan delta with steeply dipping foresets" (Fouch and Dean, 1982).
2. Fan Delta = "an alluvial fan that is deposited directly into a standing body of water" ... "composed of a subaerial component, which is an alluvial fan, and a subaqueous component" (Mcpherson et al., 1987).
3. "Classic" Gilbert-type delta = shallow water delta with a Gilbert-type profile (steeply dipping foresets; Postma, 1990).

In this paper, we use the term 'Gilbert deltas' to describe coarse-grained (gravel) deltaic deposits originally described by G. K. Gilbert (1885, 1890). This distinguishes the original, true Gilbert deltas as being different from other deltas that exhibit similar features. In addition, it should be noted that Gilbert's original description includes both steeply dipping foreset-dominated deltas, as well as horizontal topset-dominated deltas. We do not propose different names for the topset vs. foreset deltas since both set types (of facies) could conceivably exist simultaneously. In the definition of fan deltas, there is a subaerial component. However, the Gilbert-delta deposits of this study appear to be largely subaqueous, and it is difficult to distinguish subaerial alluvial fan portions (either due to non-deposition or erosion).

GEOLOGIC SETTING

Wasatch Fault Zone

All three study localities (Fig. 2) are in the immediate vicinity of the Wasatch fault zone, which lies along the foot of the Wasatch Mountains. The Wasatch fault zone is a long (about 343 km), active, normal fault comprised of ten seismically independent segments (e.g., Machette et al., 1991). Uplift along the Wasatch fault zone began about 15 Ma and has produced about 2,000 m of erosional relief (Madsen and Currey, 1979). Slip rates at American Fork are about 1.0–1.5 mm/yr for the last 15

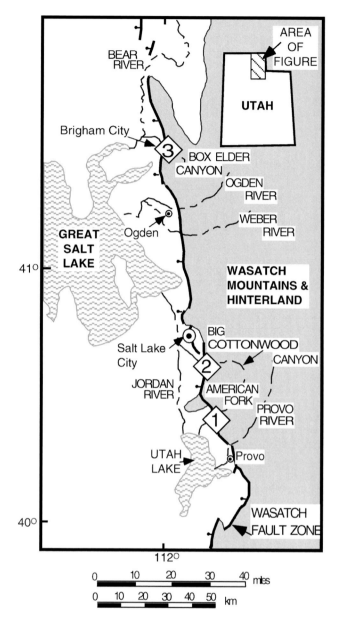

FIG. 2.—Study localities of the eastern margin of Lake Bonneville, exposed along the Wasatch Front, UT: (1) American Fork delta, (2) Big Cottonwood delta and (3) Brigham City delta. Bar and ball on downthrown side of Wasatch fault zone.

ka (Machette, 1988b). The Gilbert delta at the Big Cottonwood locality lies along the Salt Lake City fault segment. Slip rates for the Salt Lake City fault segment are about 0.76 mm/yr for the last 20 ka (based on offset across the left lateral moraine at Bells Canyon, just south of Big Cottonwood Canyon; Schwartz and Lund, 1988). Fault scarps on the Provo-level delta at Brigham City exhibit 10–16 m of surface offset (Personius and Scott, 1992). This offset occurred after deposition of the Provo-level deltaics (less than 13.5 ka) and thus suggests a minimum slip rate of 1.2 mm/yr.

Lake Bonneville

Pleistocene Lake Bonneville existed (30–10 ka) during marine oxygen isotope stage 2 and occupied a topographically-closed basin at the eastern margin of the Great Basin in the Basin and Range physiographic province (Fig. 1). At its maximum height of 1552 m above sea level, the lake covered an area of approximately 51,281 km² (Currey et al., 1984) making it the largest lake in the Pleistocene Great Basin (Grayson, 1993).

Various factors controlled lake level during the Bonneville lake cycle. Lowering of the lake was largely controlled by outlet elevation and evaporation. Input to the lake is attributed to precipitation (pluvial character), glacial melting and groundwater and stream contributions. Lake Bonneville began to rise about 30 ka and continued to transgress until the Stansbury oscillation (22–20 ka) where it formed the Stansbury shoreline (Fig. 3; Currey and Oviatt, 1985). Following the Stansbury oscillation, the lake continued to transgress to its highest level (1552 m) where it formed the Bonneville shoreline shortly after 15.3 ka (Currey and Oviatt, 1985). At the Bonneville shoreline, the lake overflowed intermittently near Red Rock Pass into the Snake River drainage basin of southeastern Idaho (Currey et al., 1984). During the open system conditions that occurred at the Bonneville and Provo shorelines, outlet elevation controlled lake level. Overflow waters breached the relatively unconsolidated sediments controlling outlet elevation (near Red Rock Pass), scoured a channel down to bedrock and released the catastrophic Bonneville Flood. This lowered lake level 100 m to the Provo shoreline, where the lake overflowed intermittently until the Provo regression (14–10 ka; Currey and Oviatt, 1985). The lake remained a closed basin throughout the rest of its history, leaving the present remnant Great Salt Lake.

Sediments deposited during the Bonneville Lake cycle are known as the Bonneville Alloformation (McCoy, 1987). The sediments include: lacustrine clays and marls with deltaic, beach and other shoreline silts, sands and gravels. The Fielding geosol (≈40–25 ka), where present, separates these Bonneville deposits from the underlying lacustral and interlacustral (e.g., alluvium, loess, eolian sand and colluvium) sediments of the Cutler Dam Alloformation (Oviatt et al., 1987). The Bonneville and Cutler Dam Allofomations are then underlain by older lacustral and interlacustral deposits of the Pokes Point (≈200 ka; Scott et al., 1983) and Little Valley (≈150–130 ka; McCoy, 1987) Alloformations. The bedrock underlying this sequence of valley fill is up to 1500 m deep in the Salt Lake Valley (Radkins, 1990).

Glaciation

Glacial melting after the Pinedale glacial maximum in the Lake Bonneville area (18–26 ka; Madsen and Currey, 1979) contributed to the overall lake-level rise and clastic sediment input (glacial outwash) into the lake. Evidence for glaciation (moraines and striations) of Pinedale age is present in Big Cottonwood Canyon and American Fork Canyons (Fig. 2; Atwood, 1909; Hintze, 1988). However, Box Elder Canyon (delta at Brigham City) shows no evidence of glaciation (Currey, pers. commun., 1993). Glaciers in American Fork Canyon did not advance low enough to reach the Bonneville shoreline. Glaciers in Big Cottonwood Canyon came close to the elevation of the Bonneville shoreline, but predated the Bonneville shoreline (Atwood, 1909; Madsen and Currey, 1979; Currey and Oviatt, 1985). Thus, the terminus of the Big Cottonwood glacier might

have been at a higher elevation than the Bonneville shoreline. Pinedale-age glaciation in the central Wasatch Mountains (based on Little Cottonwood and Bells glaciers, located just south of Big Cottonwood Canyon of Fig. 3) reached its maxima after 26 ka and began to retreat between 20 and 18 ka (Madsen and Currey, 1979; Currey and Oviatt, 1985; Fig. 3). Deglaciation of the middle and upper reaches of canyons (based on the Little Cottonwood glacier) probably began about 13 ka, although the canyon head probably approached complete deglaciation by about 8 ka (Madsen and Currey, 1979).

Paleoclimate

Inferences of paleoclimate are largely based on interpretations of glacial conditions and the established lake hydrograph. McCoy's (1981) snow melt model of the Little Cottonwood glacier and rates of amino-acid racemization (temperature as the rate-controlling factor for the racemization reaction in fossil mollusks of a given genus) and differences between glacial and present snowline altitudes (Porter et al., 1983) suggests a glacial maxima climate that was 12 to 16°C colder than present mean annual temperatures. Lake Bonneville was not at its zenith (maximum high) until 5,000 years or more after the Pinedale glacial maximum (26 and 18 ka). In other words, the glacial maxima was achieved while Lake Bonneville increased from 35 to 80% of its highest lake level (Bonneville highstand), thus suggesting that the full-glacial climate was cold and dry (Scott, 1988). If Great Salt Lake (a remnant of Lake Bonneville) were to experience its present precipitation conditions at a mean annual temperature 7°C lower than the present, the Great Salt Lake would fill to the Pleistocene shorelines of Lake Bonneville (McCoy, 1981).

Scott (1988) has suggested that this scenario of a colder late Pleistocene climate implies that precipitation (during full glaciation) must have been substantially less than today. He further suggests that because the maximum Bonneville highstand corresponds with the retreat of glaciation (≈18–14.5 ka), the climate must have become relatively wetter (more precipitation) and warmer during the lake-level maximum high (compared to the full glacial climate). The increase in precipitation during the Bonneville highstand must have been great enough to offset increased lake evaporation (due to increased temperature).

Other proxy data (e.g., frost wedges, lower tree line elevations and pollen records) also suggests cold full glacial conditions at ≈18 ka (Thompson et al., 1993). In addition, hydrogen isotope ratios from plant cellulose in packrat middens on the western margin of Lake Bonneville suggest a major warming and/or shift in the source of precipitation toward more southerly origins between full glacial and 14 ka (Siegal, 1983; Long et al., 1990; Thompson et al., 1993).

An alternative idea that Scott (1988) does not discuss is the importance of water surface temperature. Although increased mean air temperature could have induced glacial melting, cold melt waters may have increased peri-surface condensation (even during dry periods) to contribute to the water budget and decrease the overall effect of evaporation (T. Marjanac, pers. commun., 1994). However, this peri-surface condensation is probably minor compared to larger water input from glacial meltwaters. It should be noted that although glacial meltwater may have played a role in the final transgression of Lake Bon-

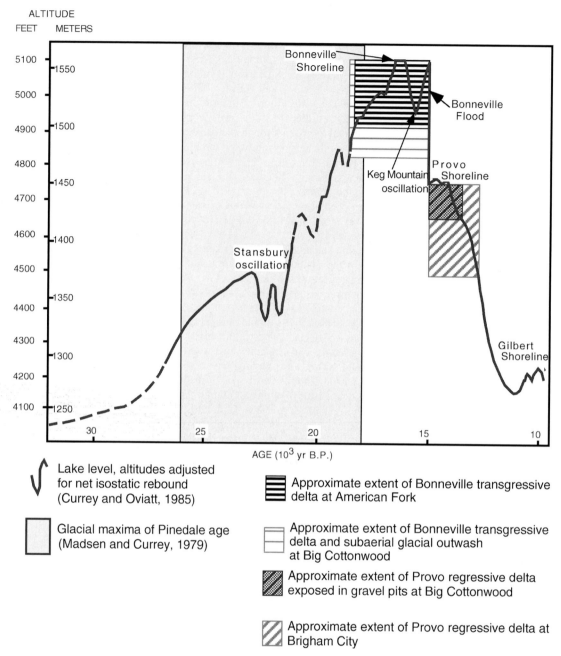

FIG. 3.—Hydrograph of Lake Bonneville (modified from Currey and Oviatt, 1985), with Pinedale glacial maxima (Madsen and Currey, 1979) and Gilbert delta elevations. Age span of Bonneville Lake cycle is within marine oxygen isotope stage 2. Elevations/altitudes are adjusted for isostatic rebound and faulting.

neville, ice-volume estimates only equal a maximum of about 5% of the total Bonneville-level lake water volume (Gilbert, 1890; Scott, 1988).

FACIES DESCRIPTIONS AND INTERPRETATIONS

Four major sedimentary facies are identified in the Gilbert deltas, based on field observations, measured stratigraphic sections and the use of photomosaics. These facies (Table 1, Fig. 4) include delta coarse-grained topset (TS), coarse-grained foreset (FS), fine-grained bottomset (BS) and associated shoreline deposits. The TS and FS facies distinguish two important end-

member Gilbert deltas: topset-dominated and foreset-dominated, respectively. These two end-member deltas also characterize different systems tracts in the sequence stratigraphy (discussed later). Associated shoreline deposits can be further subdivided into: delta front/beach fines (DF), gravel barrier (GB) and back barrier (BBF). A more detailed examination of lithofacies, architectural elements (characterized by their geometries, compositions and environments) and bounding surfaces (after Miall, 1988a, 1988b, 1991) was used to examine the temporal and spatial variability of the deltaic deposits, although these discussions are not presented here.

TABLE 1.—LITHOFACIES ASSEMBLAGES FOR AMERICAN FORK, BIG COTTONWOOD AND BRIGHAM CITY DELTAS

Lithofacies	Description	Geometry and Relations	Depositional Processes
Coarse-Grained Topset (TS)	Coarse pebble to cobble gravel locally bouldery, clast to sandy matrix supported, moderate to very poorly sorted, horizontal bedding	Sheet geometry: overlies FS or DF, at Big Cottonwood locality grades upward to FS	Traction carpets driven by unconfined stream flow, wave action?
Coarse-Grained Foreset (FS)	Pebble to cobble gravel, clast supported, moderate to poorly sorted, some localities show interbedded silty sand, mollusks (*Lymnaea/Amnicola*), openwork gravel lenses, sand lenses, steeply dipping bedding (25–35°)	Sheet to wedge-like geometry; overlies and grades into (basinward) BS, underlies TS	Slipface processes (sediment gravity flow)
Fine-Grained Bottomset (BS)	Very fine to fine sand and clayey silt, dropstones, mollusks (*Lymnaea/Amnicola*), horizontal laminations, asymmetric and symmetric ripples, soft sediment deformation, low-angle to subhorizontal, interbedded gravel near FS contact	Sheet geometry: overlain by FS, grades basinward to lake bottom deposits, interfingers shoreward with FS	Density currents, suspension settling, wave action
Associated Shoreline Deposits			
Delta Front/Beach Fines (DF)	Coarse to very coarse sand with oblate pebbles and sandy clay, cross-bedding, wave ripples, discontinuous to continous beds	Sheet geometry; grades above and below to TS	Littoral drift, wave action
Gravel Barrier (GB)	Coarse pebble and cobble gravel, clast to sandy matrix supported, moderate to poorly sorted, horizontal bedding	Bar geometry; overlain by BBF and FS, base not exposed	Wave action
Back Barrier Fines (BBF)	Very fine silty sand, mollusks (*Lymnaea/Amnicola*), steeply dipping bedding (up to 27° toward shore)	Wedge geometry; draped over the shoreward side of GB	Suspension settling

Coarse-grained Topset (TS)

Coarse-grained topset deposits consist of clast to sandy matrix-supported pebble to cobble gravel with sand lenses. Beds (m-scale thicknesses) are distinguished by grain size variations, but internally show no obvious grading. These topset deposits occur on the transgressive rise to, or at the Bonneville-level shoreline, with two different expressions shown at the Big Cottonwood and the American Fork localities.

At the mouth of Big Cottonwood Canyon, the topsets (> 60 m thick) are composed of clast- to matrix-supported cobble gravel with sand lenses (Fig. 5) in a sheet geometry. The matrix is largely silt to very coarse-grained sand (5–8% clay). At this locality, the topsets are interpreted as horizontal gravel outwash (Morrison, 1965b; Van Horn, 1972; Scott, 1981; Scott et al., 1982; Personius and Scott, 1992). They grade upwards into gravel foresets (≈1 m) and lacustrine sands and silts (≈5 m) (Scott, 1981; Scott, pers. commun., 1994). Both the gravel foresets and fine-grained lacustrine deposits are now poorly exposed and thus difficult to study.

At the American Fork locality, the topsets (> 55 m thick) consist of horizontal clast-supported pebble and cobble gravel with lenses of silty sand. This topset facies has a sheet-like geometry with an intervening 9–m section of delta front/beach fines (TS–DF–TS sequence; Figs. 4, 6). Coarse-grained delta progradation into shallow water commonly creates topset beds with no break in slope and little change in facies between the subaerial and subaqueous parts of the system (Nemec et al., 1984; Postma, 1990; Orton and Reading, 1993). Thus, the subaerial and subaqueous topsets can be difficult to distinguish. However, the overall geomorphic expression and fan-shape of the American Fork deposits implies that these topsets are largely deltaic (subaqueous), although some portions were likely deposited subaerially as glacial outwash (Machette,

1988b). On the south portion of the delta there are local variations in the topsets, which include silty clay, rippled sand (southwest transport direction) and pebble gravel (locally deposited in subaerial channels).

The coarse-grained topsets are generally inferred to represent bedload transport in planar sheets under high-energy (sediment gravity) flow conditions. Some workers present various flow processes and depositional mechanics for subaerial and subaqueous horizontally stratified, fan delta gravels (e.g., Nemec et al., 1984). In these Bonneville deposits, the topsets probably represent a range of mass flow processes with the Big Cottonwood example being most strongly influenced by glacial outwash.

Coarse-grained Foresets (FS)

Coarse-grained foresets (FS) are characterized by steeply dipping (25 to 35°) gravel with sand lenses and interbedded silty sand and pebble gravel. Some foresets contain allochthonous freshwater gastropod shells (*Lymnaea* and *Amnicola*). This foreset facies exhibits a sheet to wedge geometry. Big Cottonwood pits (Fig. 7) expose about 30 m of foreset beds that overlie and grade into bottomsets (discussed later). Two Brigham City pits expose about 80 m of foreset beds (Fig. 8). Ground-penetrating radar at the upper pit shows at least another 32 m of foreset beds beneath the 70–m pit face exposure; however, it is not known whether the subsurface foreset beds were deposited during the Provo regression (Smith and Jol, 1992). Limited exposures of foreset beds (< 2 m) are exposed in the northwestern portion of the Bonneville-level delta at American Fork. Foresets presumably compose the Provo-level delta at American Fork, however no exposures currently exist.

The foresets are clast supported in a matrix of poorly sorted fine- to very coarse-grained sand with less than 1% clay. The

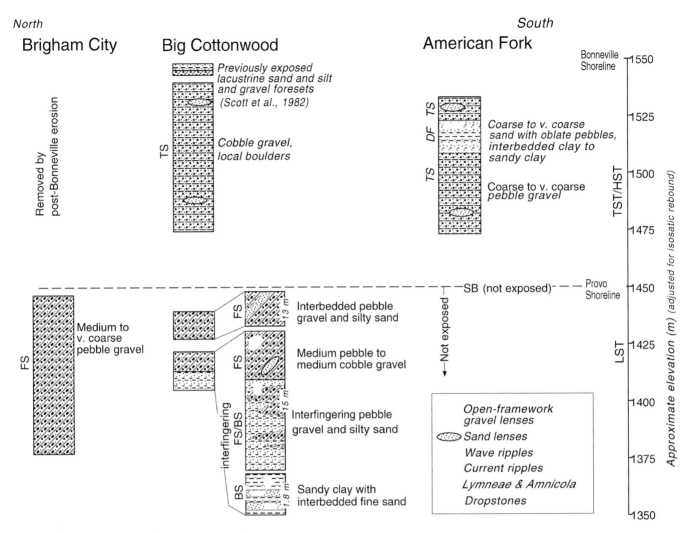

FIG. 4.—Representative stratigraphic columns for Gilbert deltas at Brigham City, Big Cottonwood, and American Fork localities. Provo-level column at Big Cottonwood locality is enlarged at right to show detail. A sequence boundary (SB) at the Provo shoreline separates the lowstand systems tract (LST) and transgressive/highstand systems tract (TST/HST).

FIG. 5.—Horizontal outwash topsets (TS) characterized by poorly sorted clast- to sandy-matrix supported cobble gravel with sand lenses (Bonneville-level gravel pit exposure at Big Cottonwood locality).

moderate to poorly sorted clasts range from pebbles to cobbles and show no preferred orientation. Exposures at Big Cottonwood Canyon commonly show inverse grading, but exposures at Brigham City show no grading. The steep, generally westward dip (away from canyon mouth sources) of the beds suggests deposition in deltaic foresets due to mass transport processes on the slipface. The mass transport process may be one or some combination of sediment gravity flows. The lack of clay content suggests cohesionless flow and the random clast orientation indicates that fluid turbulence played an important role during deposition (Hwang and Chough, 1990).

For the non-graded foresets, various workers have suggested high-density turbidity current origins (e.g., Lowe, 1982). However, others suggest high-density, non-cohesive, sandy debris mechanisms for the random fabric in clast-supported conglomerates (e.g., Surlyk, 1984). For inversely graded gravels, other workers suggest many gravity flow mechanisms (Hwang and Chough, 1990): gravely high-density turbidity currents (Aalto, 1976; Walker, 1978; Lowe, 1982; Massari, 1984); density-modified grain flows (Middleton and Hampton, 1976; Lowe, 1979)

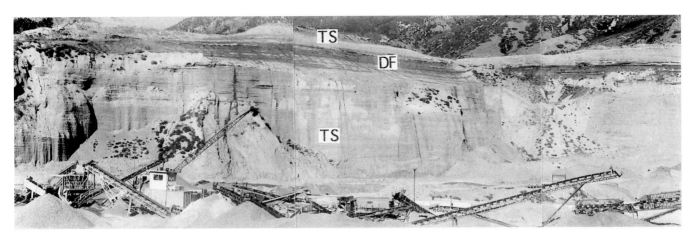

FIG. 6.—Extensive Bonneville-level gravel pit exposure of coarse-grained topset (TS), fine-grained delta front (DF), coarse-grained topset (TS) sequence (north side of the American Fork delta). Amphitheater view is approximately east-west at the left, and north-south at the right.

FIG. 7.—Sequence of interfingering bottomsets and foresets (BS/FS). The sequence is overlain by bottomsets (BS), which in turn are overlain by foresets (FS). View is southwest (left) to northeast (right) in a gravel pit exposure in the Big Cottonwood Provo-level delta.

and cohesionless debris flows (Curry, 1966; Winn and Dott, 1977; Lowe, 1982; Postma, 1986).

Open-framework gravel lenses (i.e., no matrix) are found in some localized areas at Big Cottonwood Canyon. These lenses are coarser than surrounding sediments and show random grain orientation with moderate to poor sorting. The lenses are similar to flow slide deposits of Colella et al. (1987), resulting from resedimentation by relatively high-viscosity debris flows.

Fine-grained Bottomsets (BS)

Fine-grained bottomsets (BS) are exposed in the lowermost Provo level at the mouth of Big Cottonwood Canyon (Fig. 7). Bottomsets are dominantly characterized by moderately sorted silty sand. Where the bottomsets interfinger with the gravely

foreset beds, abundant granule- and pebble-rich lenses are present. Basinward of the interfingering (distances of meters to tens of meters), more clayey silt beds are found. Other common features in the bottomsets include: laminar to thin bedding, asymmetric current ripples, localized symmetric wave ripples, soft sediment deformation and freshwater gastropods (*Lymnaea* and *Amnicola*). Isolated outsized clasts (up to 5 cm in diameter) are interpreted as dropstones and occur in sandier bottomset beds (at distances of tens of meters from the foresets).

The bottomsets have a sheet geometry (>10 m thick) that is overlain by foresets and grades basinward to lake bottom deposits (not discussed in this paper). The contact between the bottomsets and overlying foresets is both sharp and interfingering (Figs. 4, 7). The laminar to thin bedding, asymmetric ripples and proximity to foresets suggests deposition due to

FIG. 8.—Well-developed steeply dipping gravel foresets of Provo-level Brigham City delta (lowstand systems tract). Pit face strikes northwest (left).

density currents (Postma, 1990). The clayey silt beds found further from the foresets imply some suspension settling. Localized symmetric ripples may denote wave action (above wave base) or periodic storm activity. Dropstones suggest the presence of floating ice. However, as previously mentioned, Pinedale age glaciation in the Big Cottonwood area predates the Bonneville shoreline. Thus, glaciers are not thought to have calved into the lake (Atwood, 1909; Madsen and Currey, 1979; Currey and Oviatt, 1985; Hintze, 1988).

Associated Shoreline Deposits

Delta front/beach fines (DF).—

This shoreline deposit is characterized by sandy clay and coarse-grained to very coarse-grained sand with granules and oblate pebbles (Fig 9). Sedimentary structures include wave ripples and tabular cross-bedding. The cross-bedding suggests a southerly flow direction (parallel to the shoreline) and is likely to have been created by littoral currents. The presence of oblate pebbles and symmetric ripples suggest shallow water, wave-influenced deposition in a delta front/beach environment. This facies is exposed only in the upper portion of the gravel pit on the north side of the American Fork locality (Fig. 6).

This facies consists of a sheet (approximately 9 m thick) which grades above and below to horizontal topsets. Current exposures show that the sheet geometry is at least 600 m wide. If the sheet is continuous across the fan-shaped delta, its radius would be approximately 1.6 km. The occurrence of this fine-grained beach facies amidst coarse-grained delta topsets may be attributed to the Keg Mountain oscillation (Machette, 1988b). The drop in lake level during this oscillation (Fig. 3) probably caused the American Fork River to incise a channel through the delta, thus transferring the river deposition westward into the basin. This river channel (until filled) would have cut off the coarse-grained sediment supply (during the oscillation rise and final transgression to the Bonneville shoreline, Fig. 3) allowing the accumulation of the finer grained beach and delta front sands.

FIG. 9.—Delta front/beach fines (DF) overlying coarse-grained topsets (TS) (north side of the American Fork delta). One-meter scale shows 10–cm divisions.

Gravel barrier (GB).—

The gravel barrier (GB) consists of alternating horizontal, thin-bedded, moderate to poorly sorted, coarse pebble gravel and thick-bedded sandy medium cobble gravel (Fig. 10). The coarse pebble gravel beds are clast supported, and the sandy medium cobble beds range from clast to matrix supported, with very fine- to coarse-grained sand comprising the matrix of both. Foresets (FS) and steeply dipping silty sand (BBF described below) overlie this gravel. The overlying silty sand forms a relatively sharp planar contact that dips 27° to the east (toward the canyon mouth). The steep shoreward dip of this contact suggests that the horizontal gravel formed a barrier to incoming wave action. Thus, this style of horizontal gravel is inferred to have been reworked by wave action. This facies is found in Provo-level gravel pit faces at the Big Cottonwood locality where limited exposures suggest a bar-shaped geometry at least 5.5 m thick, 12 m wide and 60 m long.

Back barrier fines (BBF).—

Very fine- to fine-grained silty sand draped over the shoreward side of a gravel barrier distinguish this facies. About 1.5 m of this facies is exposed in Provo-level deposits at the Big Cottonwood locality (Fig. 10). The facies has a wedge-shaped geometry overlain by gravel pit spoil (where exposed) and underlain by a relatively sharp shoreward dipping contact (described above) with the gravel barrier. The matrix of the underlying gravels (at its upper surface) is filled with very fine-grained sand suggesting the contact is not due to slumping. The units steep shoreward dip (strike 40°, dip 27° east) and fine-grained nature suggest suspension settling in a lower energy environment behind the barrier (GB).

SEQUENCE STRATIGRAPHY

Sequence stratigraphy uses a chronostratigraphic framework to study repetitive genetically related strata bounded by surfaces of erosion or non-deposition or their correlated conformities (Van Wagoner et al., 1988, 1990). These concepts and principles can be applied to lacustrine systems (e.g., Shanley and McCabe, 1994 Oviatt et al., 1994). For the Bonneville deltaic deposits, the lake hydrograph provides a chronostratigraphic framework for defining transgressive (TST), highstand (HST) and lowstand (LST) systems tracts (Vail, 1987; Van Wagoner et al., 1988, 1990).

Two end-member Gilbert-delta systems are identified: *topset-dominated* and *foreset-dominated* (Table 2). Each of these delta systems characterize different systems tracts related to the known lake hydrograph. Topset-dominated delta systems, characterized by subaerial and subaqueous horizontally stratified (topset) gravel (facies TS), were deposited during or immediately prior to the Bonneville transgression (Fig. 11). TST deposits probably formed where accommodation space allowed for the preservation and thick accumulation of the delta topsets. During the HST, one would expect to see more progradation of topsets into foresets. However, at the Big Cottonwood locality where topset thicknesses are greatest (60 + m), only about 1 m of foresets are exposed. Less than 2 m of foresets are exposed at the American Fork locality. Thus, no distinction is made between the TST and HST and both are grouped as TST/HST of the Bonneville lake cycle (Fig. 12).

FIG. 10.—Gravel barrier (GB) with back barrier fines (BBF) draped over east side (at left). Back barrier fines dip to the east (toward the mountain front) (strike 40°, dip 27° east). Upper meter consists of gravel pit spoil (graded material). Exposure is in the Provo-level delta at the Big Cottonwood Canyon locality.

TABLE 2.—PRINCIPAL CHARACTERISTICS FOR AMERICAN FORK, BIG COTTONWOOD AND BRIGHAM CITY DELTAS

Delta type	Lithofacies	Lake Level	Systems Tract	Locations
Topset Dominated	TS (abundant) FS (limited) DF	Bonneville	TST/HST	Big Cottonwood American Fork
Foreset Dominated	FS (abundant) BS GB BBF	Provo	LST	Big Cottonwood Brigham City

For facies (also see Table 1): TS = coarse-grained topsets, FS = coarse-grained foresets, BS = fine-grained bottomsets, DF = delta front/beach fines, GB = gravel barrier, and BBF = back-barrier fines. For sequence stratigraphy systems tracts: TST/HST = transgressive/highstand systems tract and LST = lowstand systems tract.

Well-developed steeply dipping foresets (facies FS) characterize the second type of delta systems (classical Gilbert delta), deposited as the LST during the Provo regression (Fig. 11). Fine-grained bottomsets (facies BS) are also found in the LST at Big Cottonwood locality. Although present exposures show no topsets in the LST (due to quarrying and urbanization), other workers reported topsets at the Provo Level (unreported thickness of topsets in Provo-level delta at Big Cottonwood Canyon locality; Scott et al., 1982, and less than 2 m of topsets in Provo-level Dry Creek delta near American Fork; Fouch and Dean, 1982).

The LST consists of regressive Provo-level deltaic sediments separated from the underlying TST/HST by a sequence boundary (Fig. 12). Prior to the Provo regression, the catastrophic Bonneville flood dropped lake level about 100 m to the Provo shoreline leaving Bonneville-level deltaic sediments high and dry. Reworking of these freshly exposed Bonneville-level deltaic systems provided the abundant coarse-grained sediment found in the LST near Big Cottonwood Canyon and Brigham City.

Sequence stratigraphy of the Pleistocene Lake Bonneville deposits is considerably different from most sequence stratigraphy interpretations in the following respects:

1. Most interpretations of sequence stratigraphy (including the original concepts) are based on marine deposits and observations of stratal surfaces and successions of strata. Interpretations of sea-level change (the unknown) are deduced from the physical criteria of stratal surfaces and sequences. The Bonneville deposits of this study are continental in or-

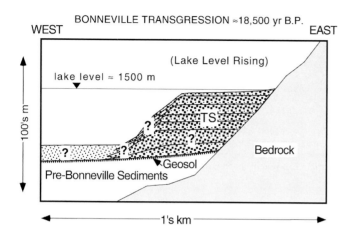

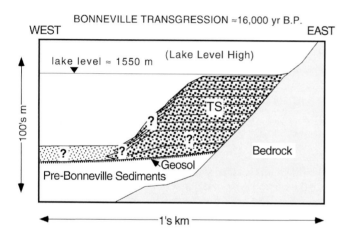

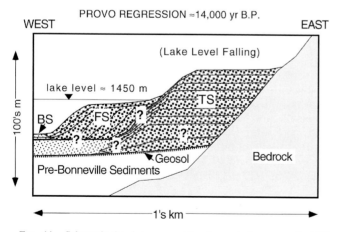

FIG. 11.—Schematic showing proposed development of coarse-grained Gilbert deltas found along the eastern margin of Pleistocene Lake Bonneville. Figure is generalized and does not show the following variations: (1) fine-grained delta front/beach (DF) sediments associated with the Keg Mountain oscillation, (2) post-Bonneville erosion that removed topsets (TS) at Brigham City and (3) no exposed Provo-level foresets (FS) at American Fork locality. For description of topsets (TS), foresets (FS), and bottomsets (BS) see Table 1. TS largely form the Tst/Hst and FS largely form the Lst. '?' indicates areas or contacts that are not exposed.

igin and in many ways are poorly exposed, where important stratal surfaces in the unconsolidated sediments are simply not observable. However, the deposits are well dated (based on radiocarbon dates and shoreline elevation positions) and there is a well-established hydrograph (Fig. 3) which clearly shows the lake-level change. Although individual surfaces or unconformities are commonly not exposed, there are still packages of sediments that are definitively tied to the hydrograph. Hence, we can relate different sediment packages (systems tracts) to the lake-level curve more convincingly (and on a smaller time scale) than one might ever find in the many studies of marine sequence stratigraphy.

2. In these Bonneville deposits, sequence boundaries do not coincide with lake cycles. The placement of the sequence boundary (SB) at the base of the LST (here at the Provo level) divides the Bonneville Lake cycle into two separate sequences. The older Cutler Dam Lake cycle (Oviatt et al., 1987) includes a LST that belongs to the same sequence as the transgressive rise (TST) of Lake Bonneville. Thus, the beginning of the Bonneville Lake cycle starts with the Bonneville transgression (TST) of one sequence and ends after the Provo regression (LST). Boundaries between lake cycles (commonly marked by geosols or interlacustral sediments) may be more evident or more prominent than the correlative conformity of the sequence boundary.

DISCUSSION

Allocyclic controls of both climate and tectonism contributed to the formation of Pleistocene Lake Bonneville Gilbert deltas (topset- and foreset-dominated). Active uplift along the Wasatch fault zone created the steep drainages of the three study localities along the eastern margin of Lake Bonneville. River gradients and drainage basin areas for the three study localities were estimated from 1:100,000 topographic maps. The river gradients of American Fork, Big Cottonwood and Brigham City are approximately 0.06, 0.06 and 0.09 respectively. Their drainages are all relatively small (American Fork ≈160 km², Big Cottonwood ≈135 km² and Brigham City ≈98 km²) and sourced within the Wasatch Range that supplied coarse-grained sediments (gravel) to the Gilbert deltas. Other deltas with headwaters outside of the Wasatch Range, such as the Bear and Weber river deltas (not discussed in this paper), originated in the Uinta Mountains and had low-gradient (0.003–0.01) drainages that developed fine-grained (silt and sand) delta systems (Lemons, 1994).

Basin subsidence, over the short time span represented in the exposed deposits (≈6,000 yr), is minor (<9 m based on slip rates) relative to lake-level fluctuations (≈175 m total elevation difference). Thus, the stratal patterns and facies distributions were strongly influenced by the lake-level history and climate. The accommodation space for the style of delta sedimentation (topset- and foreset-dominated) was largely controlled by the lake depth.

Climate affected Lake Bonneville and its Gilbert deltas in two ways:

1. The general lake level rise/fall (Bonneville Deep Lake Cycle) was due to an interplay of evaporation, precipitation and possibly effects of deglaciation on water surface temperature.

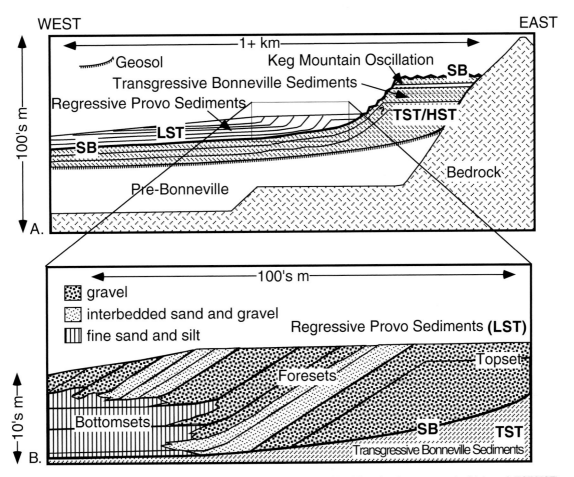

FIG. 12.—Schematic representation of American Fork, Big Cottonwood, and Brigham City deltas showing transgressive/highstand (TST/HST) systems tract, lowstand (LST) systems tract, and sequence boundary (SB).

2. Pinedale Glaciation (supply of sediment from deglaciation) provided coarse-grained glacial sediment to the deltas at the American Fork and Big Cottonwood localities.

Melting of the Big Cottonwood Canyon glacier probably contributed to sedimentation rates that outpaced the rate of transgression and produced thick, extensive horizontal subaerial glacial outwash topsets. This glacial outwash is capped by a thin layer of deltaic deposits (in gradational contact with the glacial outwash) at the Bonneville shoreline, possibly due to a reduction in sediment supply as the valley glacier retreated further up canyon.

CONCLUSIONS

Two coarse-grained Gilbert-delta systems are recognized in the Pleistocene shoreline deposits at the eastern margin of Lake Bonneville. Topset-dominated deltas (composed of topset, foreset and delta front/beach fine facies) were deposited as the lake level rose to the Bonneville shoreline (≈30–15 ka). These topset-dominated deltas comprise the TST/HST. Foreset-dominated deltas (composed of foreset, bottomset, gravel barrier and back-barrier fine facies) were deposited during the catastrophic lake-level drop (≈15 ka) and the subsequent regression (post-Provo shoreline ≈14–10 ka). These foreset-dominated deltas comprise the LST.

Active uplift along the Wasatch fault zone created the steep canyons responsible for the coarse-grained nature of the deposits. However, uplift rates (< 1.5 mm/yr) and implied basin subsidence rates are relatively small compared to lake-level fluctuations (> 300-m rise and fall in entire lake cycle ≈20,000 yr). Climate-induced deglaciation (≈20–8 ka) also contributed sediment to the deltas. Thus, climatic changes that induced lake-level fluctuations were largely responsible for the observed facies distributions and stratal patterns.

The sequence stratigraphy of the Lake Bonneville deposits varies from traditional marine stratigraphy in two prominent ways: (1) Sequence boundaries are based on the established lake-level hydrograph rather than the physical stratal surfaces and sequences that are poorly exposed in the unconsolidated Lake Bonneville deposits. (2) Sequence boundaries do not coincide with lake cycles boundaries. Lake cycle boundaries are commonly marked by geosols or interlacustral sediments that are more conspicuous than the correlative conformities of sequence boundaries.

Finally, this Pleistocene example provides an unique opportunity to evaluate allocyclic controls and sequence stratigraphy in the context of the established tectonic setting and lake hydrograph. This example can contribute to a better understanding of non-marine sequence stratigraphy and may be a useful analog to other ancient coarse-grained delta systems.

ACKNOWLEDGMENTS

This project was funded by NSF grant EAR 9303519. Reviewers J. M. Boyles and T. Marjanac, and editor K. W. Shanley provided many helpful comments and suggestions on the original draft of this paper. We thank the sand and gravel operations for access to their pits: Westroc Inc. and Geneva Rock Products Inc. (American Fork locality); Geneva Rock Products Inc. and Harper Sand and Gravel Inc. (Provo-level Big Cottonwood locality); A. J. Dean and Sons Ready Mix Concrete Co. and Concrete Products Co. (Bonneville-level Big Cottonwood locality); and Fife Rock Products Co. and Brigham Sand and Gravel (Brigham City locality). We are grateful to Don Currey (Department of Geography, University of Utah) who oriented us to the Pleistocene stratigraphy of the Bonneville basin and David Lemons for his field assistance and valuable feedback.

REFERENCES

AALTO, K. R., 1976, Sedimentology of a melange: Franciscan of Trinidad, California: Journal of Sedimentary Petrology, v. 46, p. 913–929.

ATWOOD, W. W., 1909, Glaciation of the Uinta and Wasatch Mountains: Washington, D. C., United States Geological Survey Professional Paper 61, 96 p.

BILLS, B. G. AND MAY, G. M. 1987, Lake Bonneville – Constraints on lithospheric thickness and upper mantle viscosity from isostatic warping of Bonneville, Provo, and Gilbert stage shorelines: Journal of Geophysical Research, v. 92, p. 11493–11508.

BISSELL, H., 1963, Lake Bonneville geology of southern Utah Valley, Utah: Washington, D. C., United States Geological Survey Professional Paper 257–B, p. 101–130.

COLELLA, A., DEBOER, P. L., AND NIO, S. D., 1987, Sedimentology of a marine Pleistocene Gilbert-type fan-delta complex in the Crati Basin, Calabria, southern Italy: Sedimentology, v. 34, p. 721–736.

CRITTENDEN, M. D., JR., 1963, New data on the isostatic deformation of Lake Bonneville: Washington, D. C., United States Geological Survey Professional Paper 454–E, 31 p.

CURREY, D. R., 1980, Coastal geomorphology of Great Salt Lake and vicinity, in Gwynn, J. W., ed., Great Salt Lake – A Scientific, Historical, and Economic Overview: Salt Lake City, Utah Geological and Mineralogical Survey Bulletin 116, p. 69–82.

CURREY, D. R., 1982, Lake Bonneville – Selected features of relevance to neotectonic analysis: Washington, D. C., United States Geological Survey, Open–File Report 82–1070, 30 p.

CURREY, D. R., 1990, Quaternary palaeolakes in the evolution of semidesert basins, with special emphasis on Lake Bonneville and the Great Basin U.S.A.: Palaeogeography, Palaeoclimatology, Palaeoecology, v. 76, p. 189–214.

CURREY, D. R., 1994a, Hemiarid lake basins: Hydrographic patterns, in Abrahams, A. D. and Parons, A. J., eds., Geomorphology of Desert Environments: London, Chapman and Hall, p. 403–421.

CURREY, D. R., 1994b, Hemiarid lake basins: Geomorphic patterns, in Abrahams, A. D. and Parons, A. J., eds., Geomorphology of Desert Environments: London, Chapman and Hall, p. 422–444.

CURREY, D. R., ATWOOD, G., AND MABEY, D. R., 1984, Major levels of Great Salt Lake and Lake Bonneville: Utah Geological and Mineral Survey Map 73.

CURREY, D. R. AND OVIATT, C. G., 1985, Durations, average rates, and probable causes of Lake Bonneville expansion, still-stands, and contractions during the last deep-lake cycle, 32,000 to 10,000 years ago, in Kay, P. A. and Diaz, H. F., eds., Problems of and Prospects for Predicting Great Salt Lake Levels – Proceedings of a NOAA Conference held March 26–28, 1985: Salt Lake City, Center for Public Affairs and Administration, University of Utah, p. 9–24.

CURREY, D. R., OVIATT, C. G., AND PLYLER, G. B., 1983, Lake Bonneville stratigraphy, geomorphology, and isostatic deformation in west-central Utah, in Gurgel, K. D., ed., Geologic Excursions in Neotectonics and Engineering Geology in Utah, Guidebook – Part IV: Salt Lake City, Utah Geological and Mineralogical Survey Special Studies 62, p. 63–82.

CURRY, R. R., 1966, Observation of alpine mudflows in the Ten Mile Range, Central Colorado: Geologic Society of America Bulletin, v. 77, p. 771–776.

FOUCH, T. D. AND DEAN, W. E., 1982, Lacustrine and associated clastic depositional environments, in Scholle, P. A. and Spearing, D., eds., Sandstone Depositional Environments: Tulsa, American Association of Petroleum Geologists Memoir 31, p. 87–114.

GRAYSON, D. K., 1993, The Desert's Past: A Natural History of the Great Basin: Washington D. C., Smithsonian Institution Press, 355 p.

GILBERT, G. K., 1885, The topographic features of lake shores: Washington, D. C., United States Geological Survey 5th Annual Report, p. 69–123.

GILBERT, G. K., 1890, Lake Bonneville: Washington, D. C., United States Geological Survey Monograph 1, 438 p.

HINTZE, L. F., 1988, Geologic History of Utah: Provo, Department of Geology, Brigham Young University, 202 p.

HUNT, C. B., VARNES, J. D., AND THOMAS, H. E., 1953, Lake Bonneville geology of northern Utah Valley, Utah: Washington, D. C., United States Geological Survey Professional Paper 257–A, 99 p.

HWANG, I. G. AND CHOUGH, S. K., 1990, The Miocene Chunbuk Formation, southeastern Korea: marine Gilbert-type fan-delta system, in Colella, A. and Prior, D. B., eds., Coarse Grained Deltas: Oxford, International Association of Sedimentologists Special Publication No. 10, Blackwell Scientific Publications, p. 235–254.

LEMONS, D. R., 1994, Stratigraphic heterogeneity in fine-grained lacustrine deltas in Pleistocene Lake Bonneville, northern Utah and southern Idaho (abstr.): American Association of Petroleum Geologists 1994 Annual Convention, p.195–196.

LONG, A., WARNEKE, L. A., BETANCOURT, J. L., AND THOMPSON, R. S., 1990, Deuterium variations in plant cellulose from fossil packrat middens, in Betancourt, J. L., VanDevender, T. R., and Martin, P. S., eds., Packrat Middens – The Last 40,000 Years of Biotic Change: Tucson, University of Arizona Press, p. 380–396.

LOWE, D. R., 1979, Sediment gravity flows: their classification and some problems of application to natural flows and deposits, in Doyle, L. J., and Pilkey, O. H., eds., Geology of continental slopes: Tulsa, Society of Economic Paleontologist and Mineralogist, Special Publication 27, p. 75–82.

LOWE, D. R., 1982, Sediment gravity flows: II. Depositional models with special reference to the deposits of high-density turbidity currents: Journal of Sedimentary Petrology, v. 52, p. 279–297.

MACHETTE, M. N., 1988a, ed., In the Footsteps of G. K. Gilbert – Lake Bonneville and Neotectonics of the Eastern Basin and Range Province: Denver, Geological Society of America Field Trip Guidebook 12, Utah Geological and Mineral Survey Miscellaneous Publication 88–1, 120 p.

MACHETTE, M. N., 1988b, American Fork Canyon, Utah: Holocene faulting, the Bonneville fan-delta complex, and evidence for the Keg Mountain oscillation, in Machette, M. N., ed., In the Footsteps of G. K. Gilbert – Lake Bonneville and Neotectonics of the Eastern Basin and Range Province: Denver, Geological Society of America Field Trip Guidebook 12, Utah Geological and Mineral Survey Miscellaneous Publication 88–1, p. 89–96.

MACHETTE, M. N., 1989, Preliminary surficial geologic map of the Wasatch fault zone, eastern part of Utah Valley, Utah County and parts of Salt Lake and Juab Counties, Utah: Washington, D. C., United States Geological Survey Miscellaneous Field Studies Map MF–2109.

MACHETTE, M. N., PERSONIUS, S. F., NELSON, A. R., SCHWARTZ, D. P., AND LUND, W. R., 1991, The Wasatch fault zone, Utah: Segmentation and history of Holocene earthquakes, in Hancock, P. L., Yeats, R. S., and Sanderson, D. J., eds., Characteristics of active faults: Journal of Structural Geology, v. 13, p. 137–150.

MADSEN, D. B. AND CURREY, D. R., 1979, Late Quaternary glacial and vegetation changes, Little Cottonwood Canyon area, Wasatch Mountains, Utah: Quaternary Research, v. 12, p. 254–270.

MASSARI, F., 1984, Resedimented conglomerates of a Miocene fan-delta complex, southern Alps, Italy, in Koster, E. H. and Steel, R. J., eds., Sedimentology of Gravels and Conglomerates: Calgary, Canadian Society of Petroleum Geologists Memoir 10, p. 259–278.

MCCOY, W. D., 1981, Quaternary aminostratigraphy of the Bonneville and Lahontan basins, western U.S., with paleoclimatic implications: Unpublished Ph.D Dissertation, University of Colorado, Boulder, 603 p.

MCCOY, W. D., 1987, Quaternary aminostratigraphy of the Bonneville basin, western United States: Geological Society of America Bulletin, v. 98, no. 1, p. 99–112.

MCPHERSON, J. G., SHANMUGAM, G. AND MOIOLA, R., 1987, Fan-deltas and braid deltas: Varieties of coarse-grained deltas: Geological Society of America Bulletin, v. 99, p. 331–340.

MIALL, A. D., 1988a, Reservoir heterogeneities in fluvial sandstones: Lessons from outcrop studies: American Association of Petroleum Geologists Bulletin, v. 72, p. 682–697.

MIALL, A. D., 1988b, Facies architecture in clastic sedimentary basin, *in* Klein-spehn, K. and Paola, C., eds., New Perspectives in Basin Analysis: New York, Springer Verlag, p. 63–81.

MIALL, A. D., 1991, Hierarchies of architectural units in terrigenous clastic rocks, and their relationship to sedimentation rate, *in* Miall, A. D. and Tyler, N., eds., The Three-dimensional Facies Architecture of Terrigenous Clastic Sediments and its Implications for Hydrocarbon Discovery and Recovery: SEPM (Society for Sedimentary Geology) Concepts in Sedimentology and Paleontology 3, p. 6–12.

MIDDLETON, G. V. AND HAMPTON, M. A., 1976, Subaqueous sediment transport and deposition by sediment gravity flows, *in* Stanley, D. J. and Swift, D. J. P., eds., Marine Sediment Transport and Environmental Management: New York, Wiley and Sons, p. 197–218.

MORRISON, R. B., 1965a, New evidence on Lake Bonneville stratigraphy and history from southern Promontory Point, Utah: Washington, D. C., United States Geological Survey Professional Paper 525–C, p. C110–119.

MORRISON, R. B., 1965b, Lake Bonneville Quaternary stratigraphy of eastern Jordan Valley, south of Salt Lake City, Utah: Washington, D. C., United States Geological Survey Professional Paper 477, 80 p.

MORRISON, R. B., 1965c, Quaternary geology of the Great Basin, *in* Wright, H. E. and Frey, D. G. eds., The Quaternary of the United States: Princeton, Princeton University Press, p. 265–285.

MORRISON, R. B., 1966, Predecessors of Great Salt Lake, *in* Stokes, W. L., ed., The Great Salt Lake: Salt Lake City, Utah Geological Society Guidebook to the Geology of Utah, no. 20, p. 77–104.

NELSON, A. R. AND PERSONIUS, S. F., 1990, Preliminary surficial geologic map of the Weber segment, Wasatch fault zone, Davis, Weber, and Box Elder Counties, Utah: Washington, D. C., United States Geological Survey Miscellaneous Field Investigation Series Map I–2132.

NEMEC, W., STEEL, R. J., POREBSKI, S. J., AND SPINNANGER, A., 1984, Domba Conglomerate, Devonian, Norway: process and lateral variability in a mass flow-dominated lacustrine fan-delta, *in* Koster, E. H. and Steel, R.J., eds., Sedimentology of Gravels and Conglomerates: Calgary, Canadian Society of Petroleum Geologists Memoir 10, p. 295–320.

ORTON, G. J. AND READING, H. G., 1993, Variability of deltaic processes in terms of sediment supply, with particular emphasis on grain size: Sedimentology, v. 40, p. 475–512.

OVIATT, C. G., 1987, Lake Bonneville stratigraphy at the Old River Bed, Utah: American Journal of Science, v. 287, p. 383–398.

OVIATT, C. G., CURREY, D. R., AND SACK, D., 1992, Radiocarbon chronology of Lake Bonneville, Eastern Great Basin, USA: Palaeogeography, Palaeo-climatology, Palaeoecology, v. 99, p. 225–241.

OVIATT, C. G., MCCOY, W. D., AND NASH, W. P., 1994, Sequence stratigraphy of lacustrine deposits: A Quaternary example from the Bonneville basin, Utah: Geological Society of America Bulletin, v. 106, p. 133–144.

OVIATT, C. G., MCCOY, W. D., AND REIDER, R. G., 1987, Evidence for shallow early or middle Wisconsin age lake in the Bonneville Basin, Utah: Quaternary Research, v. 27, p. 248–262.

PERSONIUS, S. F. AND SCOTT, W. E., 1992, Surficial geologic map of the Salt Lake City segment and parts of adjacent segments of the Wasatch fault zone, Davis, Salt Lake and Utah Counties, Utah: Washington, D. C., United States Geological Survey Miscellaneous Field Investigation Series Map I–2106.

PORTER, S. C., PIERCE, K. L., AND HAMILTON, T. D., 1983, Late Wisconsin mountain glaciation in the western United States, *in* Porter, S. C., ed., Late Quaternary Environments of the United States, Vol. 1: The Late Pleistocene: Minneapolis, University of Minnesota Press, p. 71–114.

POSTMA, G., 1986, A classification for sediment gravity flows based on flow characteristics during deposition: Geology, v. 14, p. 291–294.

POSTMA, G., 1990, Depositional architecture and facies of river and fan-deltas: a synthesis, *in* Colella, A. and Prior, D. B., eds., Coarse Grained Deltas: Oxford, International Association of Sedimentologists Special Publication 10, Blackwell Scientific Publications, p. 13–27.

RADKINS, H. C., 1990, Bedrock topography of the Salt Lake Valley, Utah, from constrained inversion of gravity data: Unpublished M.S. Thesis, University of Utah, Salt Lake City, 59 p.

SCOTT, W. E., 1981, Field-trip guide to the Quaternary stratigraphy and faulting in the area north of the mouth of Big Cottonwood Canyon, Salt Lake County,

Utah: Washington, D. C., United States Geological Survey Open File Report 81–773, 12 p.

SCOTT, W. E., 1988, Temporal relations of lacustrine and glacial events at Little Cottonwood and Bells Canyons, Utah, *in* Machette, M. N., ed., In the Foot-steps of G.K. Gilbert – Lake Bonneville and Neotectonics of the Eastern Basin and Range Province: Denver, Geological Society of America Field Trip Guidebook 12, Utah Geological and Mineral Survey Miscellaneous Publication 88–1, p. 78–81.

SCOTT, W. E., MCCOY, W. D., SHROBA, R. R., AND RUBIN, M., 1983, Reinter-pretation of the exposed record of the last two cycles of Lake Bonneville, Western U.S.: Quaternary Research, v. 20, p. 2612–285.

SCOTT, W. E., SHROBA, R. R., AND MCCOY, W. D., 1982, Guidebook for the 1982 Friends of the Pleistocene, Rocky Mountain Cell, field trip to Little Valley and Jordan Valley, Utah: Washington, D. C., United States Geological Survey Open File Report 82–845, p. 32–37.

SCHWARTZ, D. P. AND LUND, W. R., 1988, Paleoseismicity and earthquake re-currence at Little Cottonwood canyon, Wasatch fault zone, Utah, *in* Mach-ette, M. N., ed., In the Footsteps of G. K. Gilbert – Lake Bonneville and Neotectonics of the Eastern Basin and Range Province: Denver, Geological Society of America Field Trip Guidebook 12, Utah Geological and Mineral Survey Miscellaneous Publication 88–1, p.82–85.

SHANLEY, K. W. AND MCCABE, P. J., 1994, Perspectives on the sequence stra-tigraphy of continental strata: American Association of Petroleum Geologists Bulletin, v. 78, p. 544–568.

SIEGAL, R. D., 1983, Paleoclimatic significance of D/H and $^{13}C/^{12}C$ ratios in Peistocene and Holocene wood: Unpublished M.S. Thesis, University of Arizona, Tucson, 105 p.

SMITH, D. G. AND JOL, H. M., 1992, Ground-penetrating radar investigation of a Lake Bonneville delta, Provo level, Brigham City, Utah: Geology, v. 20, p. 1083–1086.

SURLYK, F., 1984, Fan-delta to submarine fan conglomerates of the Vogian-Valanginian Wollastone Forland Group, East Greenland, *in* Koster, E. H. and Steel, R. J., eds., Sedimentology of Gravels and Conglomerates: Calgary, Canadian Society of Petroleum Geologists Memoir 10, p. 359–382.

THOMPSON, R. S., WHITLOCK, C., BARTLEIN, P. J., HARRISON, S. P., AND SPAULDING, W. G., 1993, Climatic changes in the western United States since 18,000 yr B.P., *in* Wright, H. E., Jr., Kutzbach, J. E., Webb, T., III, Ruddiman, W. F., Street-Perrott, F. A., and Bartlein, P. J., eds., Global Cli-mates Since the Last Glacial Maximum: Minneapolis, University of Min-nesota Press, p. 468–513.

VAIL, P. R., 1987, Seismic stratigraphy interpretation using sequence stratig-raphy, part 1, seismic stratigraphy interpretation procedure, *in* Bally, A. W., ed., Atlas of Seismic Stratigraphy: Tulsa, American Association of Petro-leum Geologists 27, v. 1, p. 1–10.

VAN HORN, R., 1972, Surficial geologic map of the Sugar House Quadrangle, Salt Lake County, Utah: Washington, D. C., United States Geological Survey Miscellaneous Field Investigation Series Map I–766–A.

VAN WAGONER, J. C., POSAMENTIER, H. W., MITCHUM, R. M., VAIL, P. R., SARG, J. F., LOUITT, T. S., AND HARDENBOL, J., 1988, An overview of the fundamentals of sequence stratigraphy and key definitions, *in* Wilgus, C. K., Hastings, B. S., Kendall, C. G. St. C., Posamentier, H. W., Ross, C. A., and Van Wagoner, J. C., eds., Sea-level changes: an integrated approach: SEPM Special Publication No. 42, p. 39–45.

VAN WAGONER, J. C., MITCHUM, R. M., CAMPION, K. M., AND RAHMANIAN, V. D., 1990, Siliciclastic Sequence Stratigraphy in Well Logs, Core, and Outcrops: Concepts for High-Resolution Correlation of Time and Facies: Tulsa, American Association of Petroleum Geologists Methods in Explora-tion Series 7, 55 p.

WALKER, R. G., 1978, Deep-water sandstone facies and ancient submarine fans: models for exploration for stratigraphic traps: American Association of Petroleum Geology Bulletin, v. 62, p. 932–966.

WILLIAMS, J. S., 1962, Lake Bonneville: Geology of southern Cache Valley, Utah: Washington, D. C., United States Geological Survey Professional Pa-per 257–C, 152 p.

WINN, R. D. AND DOTT, R. H., JR., 1977, Large-scale traction-produced struc-tures in deep-water fan-channel conglomerates in southern Chile: Geology, v. 5, p. 41–44.

CARBONATE LAKES IN CLOSED BASINS: SENSITIVE INDICATORS OF CLIMATE AND TECTONICS: AN EXAMPLE FROM THE GETTYSBURG BASIN (TRIASSIC), PENNSYLVANIA, USA

CAROL B. DE WET
Department of Geosciences, Franklin and Marshall College, Lancaster, PA 17604
DANIEL A. YOCUM
Department of Geology and Geophysics, University of Wisconsin, Madison, WI 53706
AND
CLAUDIA I. MORA
Department of Geological Sciences, University of Tennessee, Knoxville, TN 37996

ABSTRACT: The New Oxford Formation, consisting of alluvial fan conglomerates, fluvial sandstones, overbank mudstones and carbonate lacustrine strata, extends along the southern margin of the Triassic Gettysburg Basin. Vertisols, calcretes and alpha-type microfabrics within carbonate nodules indicate that the paleoclimate was semi-arid throughout Late Triassic time; however, the presence of lake deposits suggests a period of increased humidity. Alternatively, tectonically-driven subsidence within the rift basin could have affected fluvial drainage patterns, causing lakes to form in topographic lows. The sensitivity of carbonate precipitation in lakes to ground and surface water chemistry and to the amount of detrital clastic input permits discrimination between the role of tectonics and climate in governing lake formation. Two distinct carbonate lakes, 30 km apart, formed in different structural settings along the Gettysburg Basin's southern margin. The lacustrine sediments near Rheems, PA were deposited in a shallow, oligotrophic, palustrine setting associated with perennial flood-plain saturation. In contrast, lacustrine deposits from near Thomasville, PA formed in a deep, episodically stratified or meromictic lake. The sediments record initial flooding over a coarse-grained basal conglomerate, followed by an upward deepening trend, that reversed to a shallowing-upward trend, ending in subaerial exposure and fluvial deposition. Together, lacustrine and associated facies indicate that fault-controlled, localized differential subsidence was responsible for creating the accommodation space required for lacustrine deposition and that the climate remained semi-arid throughout late Carnian-early Norian time.

INTRODUCTION

Controversy continues within the geologic community over the importance of local tectonics, regional climate and/or global eustasy in determining sedimentary successions (Burton et al., 1987; Haq et al., 1987; Drummond and Wilkinson, 1993; Baum et al., 1994; Ross et al., 1994). This study was designed to essentially eliminate one of the three variables and then determine if one or the other variable could be the controlling factor driving sedimentation and accumulation in a given location. We examined Triassic lacustrine rocks within a landlocked basin, effectively ruling out eustasy as a depositional factor (Smoot, 1985). The object was then to distinguish between the role of tectonics and climate in sedimentation. That information could then be re-applied to marine sequences of similar age to constrain their depositional mechanism(s). We studied carbonate lake sediments because of their sensitivity to both climatic and tectonic forces. The presence of a *carbonate lake* implies low siliciclastic input, otherwise the carbonate component would be overwhelmed and would not be preserved. Low detrital clastic input is generally associated with low relief and/or tectonic quiescence (Flores and Pillmore, 1987). Climates which are either very humid or very arid will not yield carbonate lake deposition because the dissolved load in incoming streams and groundwater will be too low, but a semi-arid climate is ideal for transporting ions in incoming water and allowing for their concentration within lacustrine waters (Cecil, 1990). However, tectonically-induced basin subsidence can cause water ponding, initiating lake development (Blair and Bilodeau, 1988), and increased humidity has also been interpreted as the reason lakes form (Barron, 1990). Detailed sedimentological analysis of as many factors as possible is required to differentiate between these various possibilities.

Triassic-Jurassic Rift Basin Geology

The Gettysburg Basin is one of a series of northeast-trending rift basins which extend 2000 km from Newfoundland to Flor-

ida along the eastern edge of North America (Fig. 1) (Manspeizer, 1988). Rift basin development in North America and western Africa occurred in an extensional regime during the breakup of Pangea. Triassic rifting reactivated Alleghenian-Variscan thrust faults, which had formed during orogenic thrusting as the Gondwana, Laurentian and Baltic cratons collided during the Paleozoic (Rast, 1988). The thrust faults were reactivated as listric and planar normal faults, producing elongate grabens and half-grabens (Ando et al., 1984; Ratcliffe and Burton, 1985; Manspeizer, 1988).

The Gettysburg Basin is a half-graben located in southeastern Pennsylvania and western Maryland (Fig. 1). The basin contains strata that dip approximately 25–30° NW towards a northeast-trending border fault. The eastern edge of the basin is partially fault-bounded (Root, 1988; Manspeizer, 1988) and partially onlapping unconformably over Cambro-Ordovician rocks (Faill, 1973). The Gettysburg Basin lies on the same structural trend as the Culpepper Basin to the south and the Newark Basin to the northeast. The rocks within the Gettysburg Basin are part of the Newark Supergroup, which is the sedimentary sequence found throughout the eastern North America rift basins. The Newark Supergroup includes a series of basal conglomerates and fanglomerates, arkosic and lithic arenites, gray to black siltstones and shales, and red siltstones and shales. Smoot and Olsen (pers. commun., 1994) suggest that the basal conglomerates may represent a thick regional regolith, unconformably overlain by Triassic deposits. There are also locally occurring limestones, calcretes, basaltic lavas, diabase dikes, evaporites, coal and kerogen-rich beds (Van Houten, 1962, 1964; Gore, 1988; Robbins et al., 1988; Robinson, 1988; Whittington, 1988; Birney de Wet and Hubert, 1989; Kruge et al., 1989; Mertz and Hubert, 1990).

As rifting commenced in the late Triassic (late Carnian-Norian), the Gettysburg Basin was located 14°N of the paleoequator, drifting northwards during Jurassic time (Gierlowski-Kordesch and Huber, 1995). Recent paleoclimatic models for the late Triassic-early Jurassic suggest that the Gettysburg Basin

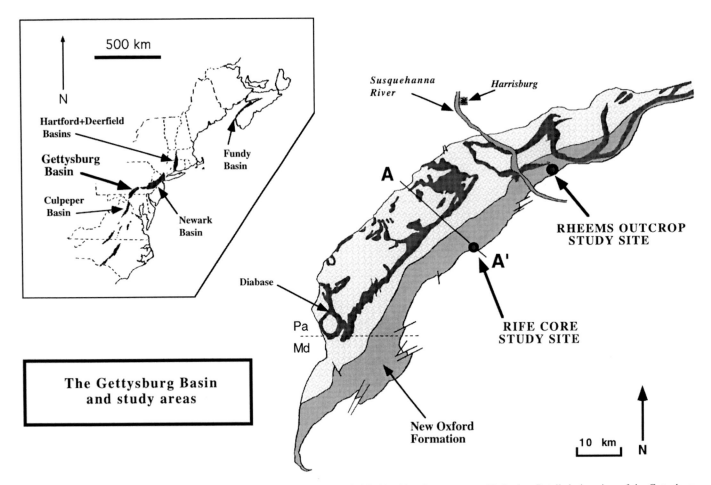

FIG. 1.—Line of onshore Triassic-Jurassic rift basins in North America (in black) with reference to specific basins. Detailed plan view of the Gettysburg Basin in southeastern Pennsylvania and western Maryland showing the New Oxford Formation (in dark gray), line of cross section shown in Figure 2, and location of the Rheems outcrop and Rife Farm core. (Adapted from Gore, 1988; Glaeser, 1966.)

would have been located in a tropical to sub-tropical belt with strong monsoonal seasonality (Parrish, 1993, Gierlowski-Kordesch and Huber, 1995).

The focus of recent research in the Newark Supergroup strata is in understanding the balance between tectonically-forced and climatically-moderated sedimentation. Border faults and syndepositional, intra-basinal faults were active throughout early Mesozoic time and had a major impact on depositional patterns (Letourneau, 1985; Schlische and Olsen, 1990; Hubert et al., 1992; Schlishe, 1993; de Wet, 1995).

Research on cyclicity within intrabasinal lacustrine strata has, however, focused more on climatic influence as the controlling factor (Van Houten, 1962; Wheeler and Textoris, 1978; Birney-de Wet and Hubert, 1989; Olsen et al., 1989; Schlische and Olsen, 1990; Smoot, 1991; Gierlowski-Kordesch and Rust, 1994). Current views suggest that generally small-scale sedimentation changes related to lacustrine water-level fluctuations are climatically controlled and may have a Milankovitch periodicity (climatic forcing at the 21–ky wavelength, e.g., Olsen, 1990), while larger-scale changes in drainage patterns are tectonically controlled (Gierlowski-Kordesch and Rust 1994).

Rocks in the Gettysburg Basin, between 5–9 km thick, constitute the New Oxford and Gettysburg Formations (Glaeser, 1966; Sumner, 1978; Schlische, 1993) (Fig. 2). The New Oxford Formation probably spans the Carnian-Norian boundary, from approximately 227–223 my, based on pollen data (Cornet and Olsen, 1985), although lowermost beds could be as old as middle Triassic (Smoot, pers commun., 1994). The formation consists of a basal, petromictic orthoconglomerate dominated by local Paleozoic quartz and quartzite, or limestone and dolomite clasts, (25–50 m thick) overlain by red to gray, micaceous sub-arkosic sandstone and red mudstone with caliche horizons (2100–2700 m thick; Root, 1988). Limestone units occur within the sandstones and overlie the basal conglomerate. The New Oxford Formation grades laterally and vertically into the younger Gettysburg Formation (6700 m thick), which is composed primarily of cyclical red and gray mudstones and arkosic sandstones (Root, 1988), interpreted as largely lacustrine in origin (Schlische, 1993). No recent facies analyses have, however, been published on Gettysburg Formation rocks, so it is unclear what relationships exist between the carbonate rocks of this study and younger lake strata.

Location and Significance of Study Sites

Two lacustrine carbonate units on the southern margin of the Gettysburg rift basin were studied (Fig. 1). One is an outcrop

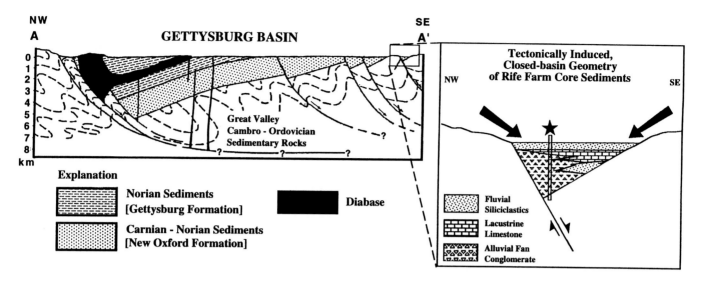

Fig. 2.—Cross section of line in Figure 1 illustrating the complex half-graben nature of the rift basin, the lower New Oxford Formation, surrounding Cambro-Ordovician rocks (predominantly limestones and dolomites) and expanded view of part of the faulted southern margin. The Rife Farm core penetrated strata shown in the faulted sub-basin (rider blocks of Schlische, 1993). Rife Farm core location noted by star. Arrows in enlarged view indicate bi-directional sediment supply (Adapted from Glaeser, 1966; Root, 1988).

located near Rheems, Pa., which contains a 6–m section of lacustrine carbonate within a 300–m section of braided-river clastic deposits. The second site is a core taken from near Thomasville, Pa., by the Thomasville Drilling and Testing Company (Cloos and Pettijohn, 1973). The core was drilled on the southern margin of the Triassic onlap, where it was expected to penetrate approximately 60–100 m of Triassic rocks. However, the core extended 263 m before encountering Cambrian basement. This depth indicated that the southern margin of the Gettysburg Rift Basin in the Thomasville area was not simple onlap but a small, fault-bounded subbasin (Fig. 2). The core contains approximately 23 m of lacustrine strata which is underlain by conglomerate and overlain by red and gray sandstone and siltstone.

Correlation between the sites is imprecise, but both are located less than 50 m from the present basin margin. There has probably been erosion of the Triassic strata and considerable loss of coarse-grained basin-margin facies, particularly at the Rheems site. Until new palynological data becomes available we have not tried to directly correlate these two lake sites. This study addresses the depositional environment of the carbonates and their diagenetic history. Field work at Rheems, Pa., the stratigraphy of the Rife Farm core, thin section (plane light and cathodoluminescence) microscopy and stable isotope geochemistry form the basis for the material presented here.

STRATIGRAPHY

Rheems Outcrop

The Rheems outcrop carbonate section lies sharply upon and is truncated abruptly by coarse-grained sub-arkosic sandstone (Fig. 3). The lacustrine beds consist of a basal, nodular siltstone, overlain by nodular and massive limestone with thin sand and shale interbeds, designated as Units 1–5 (Fig. 4). The under and overlying sandstones contain quartz, muscovite, feldspar and rock fragments, with quartz grains as large as 3 cm in diameter.

The sandstones, with beds thicker than 1 m, exhibit both large-scale cross-bedding and asymmetric ripple-marks. Channeling, with cross-sectional widths up to 50 cm, is also common.

Lacustrine lithologies: Units 1–5.—

Unit 1.—Unit 1 is a 35–cm–thick platy siltstone with light-gray carbonate nodules. The nodules increase in size and abundance upward. A silty matrix contains abundant detrital mica and clay minerals.

Unit 2.—Unit 2 is an 87–cm–thick bed of lime mudstone with green, argillaceous stringers. Thin (4–8 cm) continuous beds of terrigenous mudstone occurs 30 cm above the base of the bed and at the top of Unit 2. The unit has a variable carbonate to shale ratio both laterally and vertically. Calcite nodules contain ostracod, gastropod and bivalve shell fragments.

Unit 3.—The third unit is a 36–cm–thick brown argillaceous siltstone layer. The bed is composed of clay, silt and very fine-grained sand and contains light gray carbonate nodules up to 2 cm in diameter.

Unit 4.—Unit 4 is a 10–cm–thick green-brown shale. Although thin, it is continuous throughout the outcrop.

Unit 5.—The fifth and most massive unit of the Rheems outcrop is a blocky micritic limestone, which sharply overlies the green/brown shale unit. It is 4.5 m thick and contains ostracods, gastropods and rare charophytes. A coarse-grained sub-arkosic sandstone abruptly overlies the limestone.

Rife Farm Core

This core, first described by Cloos and Pettijohn in 1973, languished in obscurity until its rediscovery in 1993. Its 263 m record the dramatic change from very coarse-grained polymictic conglomerate 82.3 m thick to fine-grained lacustrine strata 23 m thick, overlain by cyclical sandstone-mudstone couplets 142 m thick.

The conglomerate contains Cambrian limestone, dolomite, quartzite and chert clasts in a red and gray silt matrix. The clasts

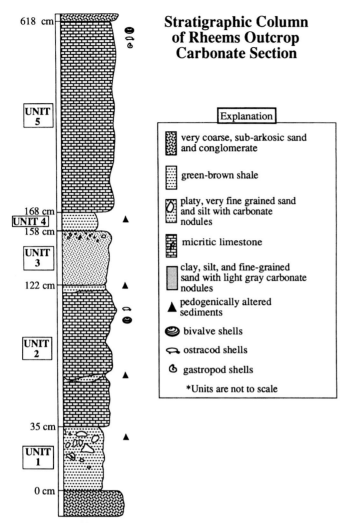

FIG. 3.—Stratigraphic column for the 6-m-thick carbonate section of the Rheems outcrop, Rheems, PA. Unit numbers are referred to in the text. Lacustrine strata containing evidence of pedogenic alteration are noted with a dark triangle.

FIG. 4.—Field photograph of the lacustrine carbonate beds at Rheems, delineated between the black lines. Numbers refer to lithologic units described in the text. The lake section is 6 m thick.

range in size from 1 cm to 1.5 m. The overlying lacustrine sequence abruptly overlies the conglomerate and is described in detail here. The division of units in the Rife Farm core lacustrine sequence have been determined by macroscopic changes in lithology and texture. For example, a change from homogeneous gray limestone to nodular limestone would be marked as a unit change. The individual units are lettered, from A to P, with A being the oldest unit to contain lacustrine carbonate and P being the youngest (Fig. 5).

Lacustrine Lithologies: Units A–P.—

The lacustrine section of the Rife Farm core contains numerous lithologies, with gradational contacts between them. Sixteen distinct units were recognized, but they have been grouped into 5 categories for brevity. The lithologies are: (1) light gray, homogeneous carbonate mudstone; (2) light gray, nodular carbonate mudstone; (3) organic-rich mudstone and wackestone; (4) dolomitic mudstone; and (5) gray and red shale with carbonate nodules. Unit letters refer to those on the stratigraphic log (Fig. 5).

1. *Light gray, homogeneous carbonate mudstone.*—The lime mudstone is composed of bioturbated, argillaceous micrite and locally contains nodules and rip-up clasts. Units A, B, C, F and J consist of this lithology.
2. *Light gray, nodular carbonate mudstone.*—Carbonate nodules, up to 33 cm in diameter, are dispersed in a micritic matrix which locally contains dark, organic-rich, argillaceous stringers (Fig. 6). The nodules are distinguished from the matrix by discrete rims or boundaries around the nodules and the presence of features, such a rhizoliths, within them. Units D, E and G consist of this lithology.
3. *Organic-rich mudstone and wackestone.*—This lithology contains mm-scale alternations of organic-rich and carbonate-rich laminae. Nodules are rare, but macroscopic plant fragments are preserved in organic-rich layers. The rocks are generally fissile due to a high clay content. Units H, I, K and L consist of this lithology.
4. *Dolomitic mudstone.*—Unit M is the only part of the core containing dolomite. Replacive sucrosic rhombs have overprinted the depositional fabric although locally patches of micrite remain unaltered. Fossils and nodules are no longer distinguishable.
5. *Gray and red shale with carbonate nodules.*—This is the uppermost lithology containing lacustrine fauna (unit P), although shale with abundant nodules also occurs in Unit N. The clay matrix is fissile, somewhat silty and contains carbonate nodules which increase in abundance up through units N and P.

The lacustrine deposits are overlain by cyclical fluvial deposits, which consist of gray and red, fine- to coarse-grained sandstone, siltstone and mudstone (Cloos and Pettijohn, 1973). Unit O consists of similar siltstone and sandstone, a thin precursor to the thick (141.7 m) succession of fluvial sediments which overlie the lake deposits. The cycles begin with a coarse-grained sandstone, usually gray, that contains small mud chips. The sandstone fines upward, grading into siltstone, overlain by mudstone. The mudstone contains numerous caliche nodules. Similar cycles have been reported from the Deerfield and Fundy

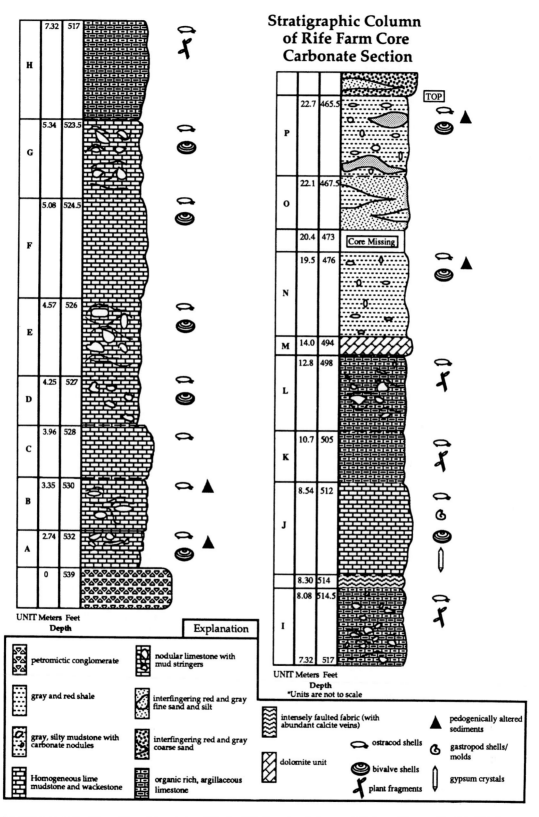

FIG. 5.—Stratigraphic column for 23 m of lacustrine strata within the Rife Farm borehole core drilled near Thomasville, PA. Unit letters are referred to in the text. Lacustrine strata containing evidence of pedogenic alteration is noted with a dark triangle.

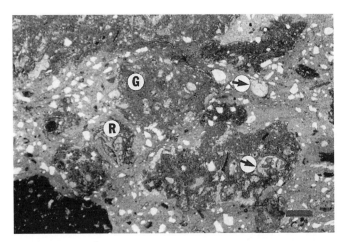

Fig. 6.—Palustrine micrite facies. Pedogenically-altered lacustrine marl showing a glaebule (G), rip-up clasts (R), abundant quartz silt grains (white grains) and partially preserved ostracod and bivalve shells (arrows). Rife Farm, unit G. Scale bar = 1 mm.

rift basins and are interpreted as braided stream deposits (Stevens and Hubert, 1983; Hubert and Forlenza, 1988).

PETROGRAPHY

Rheems Outcrop

Siliciclastic Units 1, 3 and 4.—

Unit 1 is a basal siltstone dominated by detrital quartz with subordinate muscovite, biotite, feldspar and rock fragments. Unit 3 is a compositionally immature siltstone, containing up to 2% feldspar and mica with 15% quartz and rock fragments. Unit 4 is a thin green-brown shale. Primary porosity within units 1 and 3 is filled by equant, intermediate-iron, low-Mg calcite cement. Argillaceous stringers within the siltstone units are fissile and organic-rich. All of the detrital beds contain variable quantities of carbonate, as micritic nodules.

Carbonate nodules in siliciclastic unit 1 are elliptical to subrounded in shape and range in size from 1 to 15 cm along the long axis (Fig. 7). The nodules contain up to 10% quartz grains, minor amounts of fossil fragments and numerous microstructures characteristic of calcretes. The matrix is a microcrystalline micrite with internal patches, rip-up clasts, and glaebules of darker micrite. Detrital grains and fossil debris are scattered in the micrite, producing a "floating" texture. The nodules formed *in situ* within the siltstone as incipient calcrete.

A number of different microstructures associated with calcretes are common in nodules from unit 1, although not all microstructures necessarily occur in all nodules. Rhizocretions (Fig. 8) in the nodules have an average diameter of 0.5 mm and consist of organic-rich micrite, which precipitated around plant roots from pore fluids supersaturated with respect to calcite following absorption of CO_2 by plants (Klappa, 1980; Mount and Cohen, 1984). The micrite trapped abundant organic matter and occasional silt grains as it accumulated outward from the root. The centers of the rhizocretions are filled with non-ferroan sparry calcite, which precipitated into pore space fromed by decay of the root. Also common are features that result from desiccation and rewetting, combined with expansive growth of

calcite cement: (1) crystallaria, calcite-filled cracks of varying shape; (2) circum-granular cracks and (3) non-ferroan calcite-filled coronas around quartz grains (Hubert, 1978; Allen, 1986; Wright and Tucker, 1991). Etched quartz grains may be surrounded by haloes of non-ferroan or weakly ferroan calcite cement which filled pore space around the grain margin left by dissolution of silica. Such haloes are generally 0.1 mm in width. Additional calcrete microstructures are "exploded" mica grains, formed by nucleation of calcite on the cleavage planes; and carbonate grains with micritic rims (coated grains). These 100–μm–thick micrite coats may be biogenic (Wright, 1986; Calvet and Julia, 1983) and/or abiotic in origin (Hay and Wiggins, 1980). Rip-up clasts are common and are dominantly nodule fragments that were transported and incorporated during growth of another nodule. Rip-ups range in size from .05 to 1.0 mm.

Nodules in unit 3 range in size from <1 to 2 cm along the long axis and do not exhibit the internal structure or components of unit 1 nodules (Fig. 7). Unit 3 nodules are micritic, with small cracks filled with non-ferroan or weakly ferroan calcite cement, and nodule margins are ragged or abraided.

Unit 4 contains small nodules that range in size from 0.05 to 2 mm along the long axis (Fig. 7). The carbonate occurs as microscopic rip-up clasts and diffuse boundary glaebules (orthic nodules of Wright, 1982), some of which contain calcrete textures similar to those in unit 1 nodules.

*Interpretation of siliciclastic units.—*Compositional immaturity, as indicated by the abundance of mica, feldspar and rock fragments in these units, is interpreted as local transportation in a semi-arid climate, with relatively little physical and/or chemical alteration (Hubert et al. 1978). Immature sandstones and siltstones are common within Newark Supergroup sediments and are interpreted as fluvial or shallow lacustrine deposits (Hubert, 1978; Hubert and Hyde, 1982).

Unit 1 is underlain by fluvial sandstone (Watts, 1994) and records a shallowing- and fining-upward trend. It is interpreted as an overbank deposit adjacent to a stream channel. The nodules within it contain microstructures characteristic of calcretes, indicating subaerial exposure, with infrequent flooding, on the bank margin and flood plain (Wright and Tucker, 1991). Units 3 and 4 reflect periods of increased detrital input to the lake site and a fall in lake level. The base of unit 3 and the top of unit 4 contain calcrete nodules, indicating subaerial exposure associated with lake lowstands.

*Interpretation of nodules.—*Nodules are a type of calcium carbonate accumulation associated with near surface calcrete or caliche development. Where mean annual precipitation is <50 cm/yr, a soil calcrete layer may form within a few tens of centimeters of the ground surface. With an increase in precipitation, the depth of carbonate formation increases to approximately 1 m, and the calcrete becomes nodular. All of the microstructures discussed above are typically found within calcretes, or pedogenically modified soils (Wright, 1982; Retallack, 1983; Goldstein, 1988; Goldberg, 1992), specifically in alpha calcretes as described by Wright (1990). Alpha calcretes form by pedogenic processes under arid to semi-arid conditions. They consist of microcrystalline matrices or groundmasses (crystic plasmic fabric of Brewer, 1964), with internal nodules, complex cracks and crystallaria, floating sediment grains and rhombic calcite crystals. All of these features, except isolated calcite rhombs, are

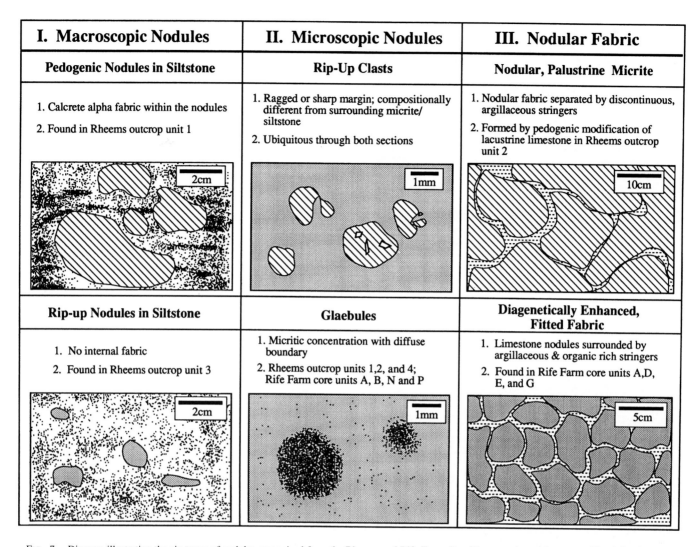

I. Macroscopic Nodules	II. Microscopic Nodules	III. Nodular Fabric
Pedogenic Nodules in Siltstone	**Rip-Up Clasts**	**Nodular, Palustrine Micrite**
1. Calcrete alpha fabric within the nodules 2. Found in Rheems outcrop unit 1	1. Ragged or sharp margin; compositionally different from surrounding micrite/siltstone 2. Ubiquitous through both sections	1. Nodular fabric separated by discontinuous, argillaceous stringers 2. Formed by pedogenic modification of lacustrine limestone in Rheems outcrop unit 2
2cm	1mm	10cm
Rip-up Nodules in Siltstone	**Glaebules**	**Diagenetically Enhanced, Fitted Fabric**
1. No internal fabric 2. Found in Rheems outcrop unit 3	1. Micritic concentration with diffuse boundary 2. Rheems outcrop units 1,2, and 4; Rife Farm core units A, B, N and P	1. Limestone nodules surrounded by argillaceous & organic rich stringers 2. Found in Rife Farm core units A,D, E, and G
2cm	1mm	5cm

FIG. 7.—Diagram illustrating the six types of nodules recognized from the Rheems and Rife Farm sites. Macroscopic nodules are readily visible in the field and core, microscopic nodules are only apparent in thin section and the nodular fabric is pervasive at both scales. The cross-hatch pattern represents micrite containing pedogenic and/or lacustrine features; stipple represents lacustrine micrite; dense, dark stipple represents micrite formed *in situ*; dashes represent argillaceous micrite. See text for further explanation.

common in the Rheems outcrop nodules. Partial dissolution of silica grains and shrinking and wetting processes create crystallaria, circum-granular, corona and intergranular porosity, which may be occluded by precipitation of calcite cement from waters that reached supersaturation with respect to calcite by partial to complete dissolution of carbonate grains and enrichment in cations from carbonate catchment areas. These waters also precipitate calcite that form exploded micas, fracture quartz grains and produce pedogenic nodules or glaebules (Brewer, 1964). Diffuse boundary nodules, termed glaebules here for clarity (Fig. 7), are the result of calcite precipitation, commonly associated with paleosols. They may be particularly well developed on previously deposited limestone (Freytet, 1973; Platt, 1989; Nahon, 1991). They are widespread in calcretes, although their origins are not well understood. Carbonate diffusion, perhaps from sites of dissolution such as circum-granular cracks, may lead to areas of preferential precipitation and displacive growth (Watts, 1978; Francis, 1986; Wright and Tucker, 1991).

All of these processes occur in soil horizons as part of calcrete formation (Buczynski and Chafetz, 1987).

Lime Mudstones: Units 2 and 5.—

These beds consist of sparsely fossiliferous, bioturbated micrite with minor terrigeneous clay and rare (<5%) quartz silt. The carbonate in unit 2 has a nodular fabric separated by discontinuous argillaceous stringers (Fig. 7). Calcrete microstructures are pervasive throughout the unit, but reach maximum concentration just below a thin shale bed at 30 cm and another thin shale at the top of unit 2. Unit 5 is weakly bedded and lacks pedogenic microstructures. Lacustrine invertebrate remains such as ostracod valves (*Darwinula* sp.), gastropods (unidentified) and bivalves (*Unio* sp.); probable conchstracans and calcispheres are scattered throughout the micritic matrix of both units (Fig. 9). Most of the shells are broken, but rare ostrocods are articulated. Rip up clasts are locally abundant. All of the carbonates are cemented with either non-ferroan or weakly fer-

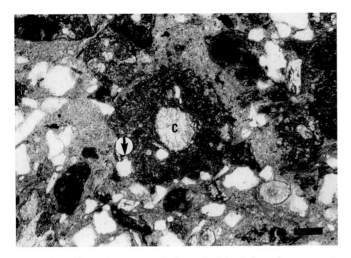

FIG. 8.—Rhizocretion composed of organic-rich micrite and sparse quartz silt grains. One of the quartz grains exhibits a partial calcite corona rim (arrow); the central root cavity (C) is filled with non-ferroan calcite spar. Rheems, unit 3. Scale bar = 1 mm.

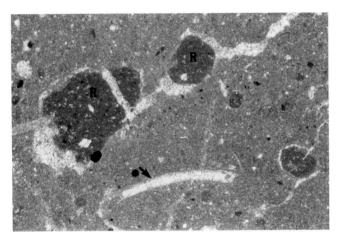

FIG. 9.—Open lacustrine facies. Fine-grained micrite containing comminuted shell fragments, rip-up clasts (R) partially surrounded by circumgranular cracks and a large bivalve fragment (arrow). The bivalve shell probably represents the genus *Unio*, a freshwater form. Rheems, unit 2. Scale bar = 1 mm.

roan (intermediate iron), equant calcite spar. Sparry, non-ferroan, low-Mg calcite cement fills small (<0.5–mm width) veins. The cement within gastropod molds is blocky, intermediate-iron, low-Mg calcite.

Interpretation.—Shallow carbonate lakes with extensive littoral zones typically contain marl, a mix of terrigeneous and carbonate mud (Jones and Wilkinson, 1978; Dean, 1981; Picard and High, 1981; Fouch and Dean, 1982). Abundant rip-up clasts within the Rheems lime mudstones suggest periodic reworking along a shallow shoreline. The assemblage of fossil invertebrates in the lime mudstones indicate a shallow, fresh, possibly alkaline lake with a muddy bottom that must have remained stable for a significant period of time (Moore et al., 1952; Majewske, 1969; Gore, 1988). The ostracod, bivalve and gastropod shells indicate well-oxygenated bottom waters (Freytet, 1973; Glass and Wilkinson, 1980). The mottled marl matrix indicates bioturbation by burrowing organisms (Murphy and Wilkinson,

1980) and the nodular fabric in unit 2 reflects pedogenic overprinting of the original lacustrine muds.

Rheems Outcrop Depositional Synthesis

Basal clastics, overlain by marl, followed by more pure carbonate, is a typical sequence interpreted as lessening of clastic input, allowing $CaCO_3$ to precipitate and remain undiluted by detrital material (Fouch and Dean, 1982). Although this is the overall trend in lacustrine strata at the Rheems site, lake deposition was not continuous there. Lacustrine sedimentation was interrupted by episodes of lake regression, resulting in pedogenic overprinting of the lake carbonate (Fig. 10). For example, the lower 30 cm of Unit 2 formed as a lacustrine carbonate. Lake-level regression resulted in pedogenic modification of the sediments. A subsequent rise in lake level deposited the upper portion of the bed, followed by another period of subaerial exposure. Shallow, vegetated lake settings which experience relatively rapid changes in water level are termed *palustrine* by Platt (1989) and *paludal* by Freytet (1973). Falling lake levels typically lead to subaerial exposure of lacustrine carbonate, overprinting it with calcrete microstructures (Platt, 1989). Platt and Wright (1992) suggest that palustrine carbonates may indicate seasonal wetlands. The Rheems strata, particularly unit 2, contain a suite of characteristics typical of such a palustrine setting. The marly matrix shows continuous input of fine-grained detrital sediment, indicating that lake levels were shallow and/or that sediment supply kept pace with the rise in lake level. The Rheems carbonate section is characteristic of an hydrologically open lake system.

The carbonate components in the Rheems site are interpreted as representative of 3 facies: (1) pedogenic calcrete facies; (2) palustrine, or pedogenically-modified lacustrine carbonate facies and (3) open lacustrine facies.

In summary, unit 1 represents the pedogenic calcrete facies associated with calcrete development in overbank silts as part of a braided river system, prior to deposition of lacustrine strata. Water ponding initiated lake development, the mechanism for water ponding is discussed later. Unit 2 represents palustrine carbonate sedimentation, interupted by a period of subaerial exposure. Unit 3, a siltstone, formed during a resurrgence of

Rheems Outcrop Depositional Unit 2

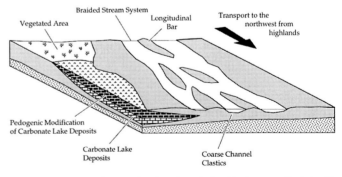

FIG. 10.—Block diagram showing the relationship between fluvial sediments and adjacent shallow lake deposits at the Rheems site. Prior to the deposition of unit 3, unit 2 lacustrine marls were subaerially exposed and calcrete nodular fabrics developed within them.

clastic input into the area. Clastics were deposited as a stream migrated back to the depositional site, essentially 'drowning out' the carbonate production of the lake. Possibly a channel migrated back into the area of deposition, thereby increasing clastic input, resulting in a decrease in carbonate content. The stream's energy was also responsible for ripping-up and redepositing some of the detrital carbonate clasts seen in this layer. Downcutting and erosion of unit 2 generated rip-up clasts that were locally reworked. Other clasts have a paleosol origin and represent calcrete precipitation which occurred during further regression. These nodules do not contain any faunal remains.

The silty shale of unit 4 represents either (a) a period of relatively deep lake or (b) a period of subaerial exposure and deposition of shoreline mud and silt. The silty matrix is very similar to that in unit 1. The presence of shallow-water clastics stratigraphically below, suggests by Walther's Law, that this layer also reflects shallow or subaerial conditions. Nodules within the shale contain glaebules and rhizoliths and are too large to have been deposited in deep, profundal regions of a lake. Many of the clasts have ragged, uneven margins, indicating abrasion and breaking involved with transport, probably when semi-lithified. The evidence suggests that the nodules were produced in an adjacent paleosol, were then ripped-up and reworked, possibly during a storm event. Together, the silt, mud and nodules indicate deposition along a shoreline or littoral setting (Renault and Owen, 1991).

Unit 5 was deposited in a shallow oligotrophic lake, the open lacustrine facies (Fig. 11). The fauna and bioturbated sediments indicate an open-water lake with oxygenated bottom water. The lack of pedogenic alteration in this unit, in contrast to older units, indicates that the sediments were never subaerially exposed. This final unit of the Rheems outcrop represents the highest lake stand. Following its deposition, a combination of lake-level fall and braided-rivers moving back into the area ended lacustrine carbonate deposition.

Rife Farm Core

All of the units designated as lacustrine contain invertebrate fossil remains indicative of a freshwater lake environment. All

Rheems Outcrop Depositional Unit 5

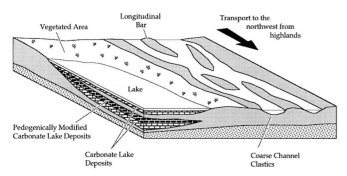

FIG. 11.—Block diagram depicting the interpreted depositional environment for unit 5, the thickest and best developed lake limestone at the Rheems outcrop site. Important features include: the close proximity of the braided river system, the deepening upward trend of the lacustrine section, and the interbedded nature of lake, palustrine and calcrete horizons.

of the fauna is the same as that at Rheems, except that there are no definitive charophyte remains. Carbonate nodules are abundant and reflect a similar range of origins as those from Rheems.

Basal Unit A.—

The underlying non-lacustrine conglomerate is *abruptly* overlain by an argillaceous micritic mudstone (unit A), containing abundant ostracod and bivalve shell fragments. The matrix is bioturbated, resulting in a mottled appearance. Carbonate nodules first occur at 2.7 m. These nodules consist of micrite cores surrounded by epitaxial bladed calcite cement (Fig. 12). There are also abundant crystallaria, glaebules and circumgranular cracks. Calcite cement precipitated in two different episodes. The first rims fractures and vugs in the micrite and is non-ferroan low-Mg calcite. The second generation of cement occludes all remaining porosity and is ferroan low-Mg calcite. The top of unit A is marked by a clay-rich horizon with calcitic nodules.

*Interpretation.—*The first unit of the carbonate section of the Rife core is a paludal or palustrine deposit. Lacustrine fauna indicate subaqueous deposition but the abundance of shrinkage cracking and glaebulization indicates subaerial exposure or vadose conditions. This unit is interpreted as an incipient lake deposit that was subaerially exposed. Displacive calcite cement growth took place in the uppermost layer of clayey matrix. Hubert (1978) suggests that such a cement fabric is the result of calcite nucleation on micritic patches, from calcite-supersaturated groundwater in soils. This fabric, therefore, provides additional evidence for subaerial exposure.

Light gray, homogeneous carbonate mudstone, Units B, C, F and J.—

These units all contain the same lacustrine fauna as Rheems, and carbonate rip-up clasts in an argillaceous micritic matrix. Shell fragments reach 6 mm in length, and articulated ostracods are common. Unit B contains glaebules and numerous fenestral cavities, many of which contain vadose silt and two generations of calcite cement: non-ferroan followed by ferroan, low-Mg calcite (Fig. 13). Rip-up clasts are so abundant in Unit F that they form a gravel layer. Unit J contains euhedral gypsum crystals randomly scattered throughout the matrix (Fig. 14).

*Interpretation.—*The homogeneous carbonate mudstone was deposited under oligotrophic conditions in a littoral to sub-littoral setting. The bioturbated matrix indicates that the sediments were deposited under oxic conditions, in the presence of burrowing organisms. Preserved, articulated, thin-walled ostracods indicate deposition below lacustrine wave base. Above wave base, shells were fragmented and mixed with micrite. Intraclasts consisting of micrite and shelly debris were eroded from near-shore areas, probably by storm events, indicating that partial lithification was an early diagenetic event around the lake margin. These clasts form gravel layers similar to the gravel described by Freytet (1973), and Renaut and Owen (1991), interpreted as beach and shoreline deposits.

The periods of subaqueous deposition were interrupted by periods of subaerial exposure and pedogenesis. Fenestral cavities, many of which contain vadose silts, indicate vadose conditions (Scholle, 1978; Wright and Tucker, 1991). These units represents a gradual shallowing of the lake, followed by a period of subaerial exposure which represents a return to palus-

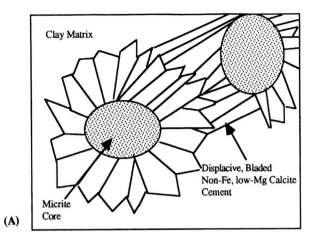

FIG. 12.—Palustrine facies. (A) Sketch of displacive, bladed calcite crystals that radiate out from micritic cores. (B) Photomicrograph of the same cement. It is always non-ferroan, low-Mg calcite and is interpreted as a calcrete precipitate. Rife Farm unit A. Scale bar = 1 mm.

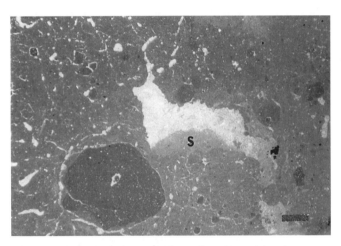

FIG. 13.—Palustrine facies. Lacustrine marl matrix containing circum-granular cracks, a fenestral cavity, partially filled with vadose silt (S) and variously sized nodules. Non-ferroan calcite cement overlays the vadose silt, followed by equant, pore-filling ferroan calcite spar. Rife Farm, unit B. Scale bar = 1 mm.

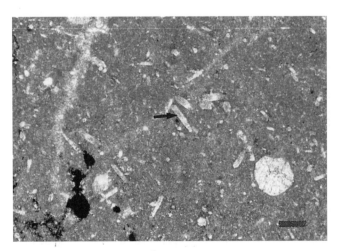

FIG. 14.—Open lacustrine facies. Euhedral gypsum crystals in a micritic matrix (arrow). Rife Farm, unit J. Scale bar = 1 mm.

trine conditions and pedogenesis. The shallow-water level of the lake is further supported by the occurrence of euhedral crystals, interpreted as gypsum based on their form, low birefringence, X-ray diffraction pattern and occurrence. They provide evidence for evaporative conditions in the Rife Farm lake site.

Light gray, nodular carbonate mudstone; Units D, E and G.—

The light gray limestone contains abundant nodules, and the mottled argillaceous micritic matrix contains ostracod, gastropod and bivalve shell fragments. Locally, it is a wackestone. Many shells are poorly preserved, occurring as broken shell hash. There are also abundant rip-up clasts, which locally form gravel packstones. The abundance of nodules, lack of vadose silts or fenestral cavities and presence of dark quartz and clay-rich mud stringers differentiates these units from the homogeneous mudstone units. Non-ferroan low-Mg calcite cement is followed by pore-occluding ferroan low-Mg calcite cement in primary intraparticle pores.

*Interpretation:.—*The types of faunal remains indicate open lacustrine conditions and the mottled bioturbated clay and micrite matrix indicates biotic activity within the sediments. The high percentage of broken shell fragments, combined with abundant micritic rip-up clasts, indicates that this site experienced extensive deposition of storm derived debris. The shells and rip-up clasts were probably transported from shallow water out to the site of deposition just below, or at, wave base. The nodular texture is the result of the interlayering of clastic laminae with carbonate clasts and micrite. Diagenetic enhancement of the texture is the result of physical and chemical compaction. Early lithification is indicated by the abundant rip-up clasts and primary pore linings of non-ferroan calcite cement.

Organic-rich mudstone and wackestone; Units H, I, K and L.—

The matrix is composed of dark gray to black, organic-rich (the Total Organic Carbon (TOC) >2% in these units), poorly laminated micritic clays with shell concentrations (Fig. 15). The shell debris is composed of thin-walled, broken ostracod shells. Macroscopic plant fragments up to 0.75 cm in length are com-

terrigeneous clay and microstructures interpreted as evidence of pedogenesis. Their differences are significant, however, and convey important information about the importance of climate versus tectonics in dictating deposition.

Differences

The Rheems site is typical of a shallow oligotrophic palustrine system where ponding occurs adjacent to a river system. It was a riparian wetland, characterized by its association with fluvial and overbank deposits and rapid and repeated transgressions and regressions of the shoreline. The lake was shallow and unstratified. Relatively little organic carbon is preserved in such a system, reflecting continual oxidizing conditions. The cementation history of the Rheems site is relatively simple: non-ferroan cements were precipitated syndepositionally and during shallow burial from oxidizing and/or iron-free pore waters. During late burial, pore waters became reducing and/or became slightly enriched in iron, and calcite cements precipitating from them contain an intermediate amount of iron, based on their response to staining. Carbon isotopic ratios range from −8 to −11.3‰ PDB; oxygen ratios exhibit a similar low spread in values, from −6.5 to −10.5‰ PDB. The greatest variability in oxygen values is related to periodic subaerial exposure in the middle stratigraphic units.

Although the Rife Farm site contains evidence for palustrine conditions during inital lacustrine formation and again at the end of lake deposition, during much of its history the lake was stratified, with dysaerobic to anoxic bottom waters where organic matter was preserved. The increase in organic matter, and preservation of laminae suggest that the lake was stratified and that the hypolimnion was oxygen-poor at least periodically. Bottom sediment in permanantly anoxic lakes, such as Lake Tanganyika and Kivu in the East African rift basin contain >10% organic matter (Stoffers and Hecky, 1978; Cohen, 1989). Storm activity, and possible turbidites, brought nearshore shell remains and pedogenic nodules into the profundal environment. The diagenetic history of the Rife Farm sediments is more complex than at Rheems. Initial non-ferroan cements are rare. Where present, they reflect early precipitation from oxygenated and/or iron-free porewaters. Subsequent burial resulted in pore waters becoming depleted in oxygen, allowing iron to become reduced to its divalent state and become incorporated into the calcite, producing ferroan cements.

Although our initial hypothesis suggested that the Rheems site was the nearshore or littoral area of the Rife Farm lake, the differences between the sites indicate that they were not connected as a single lake. Rather, the differences in underlying lithologies and hence differences in depositional setting, the differences in water depth and type of sedimentation, different diagenetic histories and different isotopic values demonstrate that two distinct lakes formed along the Gettysburg Basin's southern margin. Why two such different lakes developed in such close proximity is discussed below.

DISCUSSION AND CONCLUSIONS

When considering the sedimentology of carbonate lakes, there are a number of points that need to be addressed: (1) what was the source for the calcium carbonate, how did it precipitate and what was the chemistry of the lake water; (2) what was the nature of the basin's morphology and what were the hydrologic conditions during its existence; (3) what controlled the basin or depression in which the lake waters accumulated and (4) what was the affect of climate on the lake?

Carbonate Source and Water Chemistry

In the case of the Rheems and Rife Farm lakes, there was a ready source for $CaCO_3$ in the Cambro-Ordovician limestones and dolomites surrounding the rift basin. Chemical weathering of the adjacent highlands and solution transport provided dissolved ions which accumulated in the lake. Determining paleowater chemistry is difficult, but estimates of the Mg/Ca ratio in the water can be made using constraints discussed in Muller et al. (1972) and Hardie et al. (1978). Low-Mg calcite is the dominant carbonate mineral in both lakes and precipitates from lake water with a Mg/Ca ratio of less than 2 (Muller et al., 1972). Aragonite can form in lakes with a Mg/Ca ratio >12 (Muller et al., 1972). Aragonite is more common in saline or evaporative setting (Eardley, 1938; Hardie et al., 1978). There is no evidence that the Rheems lake was saline, or evaporative, rather the invertebrate fauna indicate fresh water. The calcite at Rheems generally precipitated as encrustations around plants, initiated by photosynthetic uptake of CO_2, termed bio-induced carbonate by Dean and Fouch (1983). Evidence for this is in the clotted, nodular fabric and light carbon isotopes of the palustrine carbonates (Platt, 1989; Wright and Tucker, 1991). Bioinduced calcite precipitation is the most common form of carbonate deposition in modern alkaline lakes (Wetzel, 1983).

The micrite deposited in the Rife Farm lake during shallow, oxygenated lake conditions is similar in origin to that of the Rheems micrite. In profundal parts of the lake, however, inorganic calcite precipitation may have formed much of the sediment. Inorganic calcite precipitation is associated with lake turnover and may be a significant source of carbonate in stratified lakes (Platt and Wright, 1991). The Rife Farm lake was meromictic with periodic overturning that may have produced calcite "whitings" of inorganic precipitation. Also, much of the calcite in the deeper parts of the core is biogenic debris, predominantly ostracod shell fragments, so biologic and bio-induced precipitation may still have been important, although the micrite is difficult to categorize because microscopic algae would have been the plants responsible for its formation (Dean and Fouch, 1983). The Rife Farm lake experienced some evaporative concentration, leading to gypsum precipitation in one horizon. Gypsum is still an early precipitation product and can form in fairly dilute waters (Hardie et al., 1978).

Basin Morphology and Hydrology

Lake-basin morphology is important in influencing facies distributions, and most lakes fall into one of two categories; either a low-gradient ramp-type or a high-gradient basin-shape. According to Platt and Wright (1991), low-gradient ramp-type margins are dominated by marginal or littoral facies, consisting largely of somewhat fossiliferous, bioturbated micrite with few siliciclastics and no evidence for local faulting. Conversely, high-gradient lake margins are associated with faulted boundaries and talus slopes or fanglomerates and exhibit better development of profundal facies, tending to indicate anoxic conditions and a higher ratio of fine-grained clastics. As documented in this paper, the Rheems site fits the description of a low-gradient, ramp morphology, and the Rife Farm site corresponds to a higher gradient, steeper margin-type basin. Lake margins may exhibit characteristics indicative of high-wave energy, such as lenticular sand bodies and/or coated grains, while low-energy margins typically consist of bioturbated micrite, perhaps with rooted vegetation (Picard and High, 1981; Platt and Wright, 1991). Pedogenesis commonly reworks the tops of regressive sequences and is preserved in low-energy settings. Both the Rheems and Rife Farm lakes are characterized by low-energy micrite-dominated shorelines.

Many of the features described from the Gettysburg Basin lake beds are similar to deposits in the East African rift basin lakes. Current research makes comparisons between the two settings (Cohen, 1990; Huc et al., 1990; Hassani and St. C. Kendall, 1994; Renaut and Tiercelin, 1994), although workers on Recent systems stress the complexity of the relationship between climate and tectonics in generating sedimentary sequences in rift basin lakes (e.g., Renaut and Tiercelin, 1994).

Although the Gettysburg Basin did not have fluvial throughput to the opening Atlantic ocean, lakes within it can be subdivided on the basis of being hydrologically open or hydrologically closed *within the basin itself* (Fisher and Brown, 1972). Being hydrologically open means that streams both enter and exit the lake basin; closed implies only stream entry into the lake, outgoing water could only be as a vapor through evaporation (Gasse et al., 1987; Gore, 1989; Bellanca et al., 1992). Hydrologically closed lakes typically form in rapidly subsiding areas where local drainage patterns are disrupted; causing catchment areas around the lake to drain towards it, rather than through it. They may contain evaporites due to frequently fluctuating shorelines (Gore, 1989), and closed basin lakes may exhibit covariance in stable isotope values (Talbot and Kelts, 1990). In an open system, a lake serves as a temporary base level, and water will move through the lake and continue downstream until it reaches another base level. In the case of basins which are closed to the sea, the next base level will likely be the lowest point in the basin, which probably would, in fact, be another lake. Gore (1989) showed that other Newark Supergroup shallow, well-oxygenated lakes, which formed on paleosols amid braided river deposits, were the result of perennial flood-plain saturation and were hydrologically linked to an adjacent river system. This type of lake deposition is most common in strata from the southern Newark Supergroup. Evidence from the New Oxford Formation shows that the Rheems lake was hydrodynamically open, being part of the overall braided stream system. The lake shows only a slight increase in isotopically heavy oxygen, as is common when water vapor preferentially incorporates $\delta^{16}O$ during evaporation, leaving the lake water enriched in $\delta^{18}O$. The Rheems lake carbonates have slightly depleted $\delta^{18}O$ values relative to the calcrete carbonate, indicating that residence time in the lake was short but that some fractionation did occur.

The Rife Farm lake deposits have characteristics more typical of hydrologically closed basins, such as the presence of evaporites (gypsum in Unit J), poorly laminated profundal sediments and preservation of organic matter indicating stratification of the water column. Such features, coupled with the greater thickness of the Rife Farm lacustrine beds, suggest that locally the basin was maintained throughout much of late Carnian to early

Norian time. The oxygen isotopic values from the Rife Farm lake beds do not exhibit clear closed basin trends because of diagenetic alteration during burial. The Rife Farm lake could represent the ultimate base level within the Gettysburg Basin during late Carnian-early Norian time, but further examples of lacustrine deposits from the basin are needed to decide conclusively. Gore (1989) notes that lacustrine deposits from northern parts of the Newark Supergroup consist of cyclic sequences, locally containing evaporites. They are interpreted as deposits of generally hydrologically-closed basins (Gore, 1989). The Gettysburg Basin is located roughly midway between the northern and southern Newark Supergroup basins, and it contains lacustrine sediments which were deposited in both hydrologically-open and hydrologically-closed systems.

The Role of Tectonics

The different characteristics of the two lakes, as outlined above, indicate that their deposition was controlled by different processes. Although the eastern margin of the Gettysburg Basin is traditionally considered to be the onlap margin, structural, geophysical and stratigraphic evidence shows that syndepositional faults locally affected deposition within the basin (Cloos and Pettijohn, 1973; Sumner, 1978; Root, 1988). The existence of two distinct lake systems only 30 km apart strongly suggests that localized faulting created different depocenters.

Evidence for faulting occurs within the Rife Farm core itself, where fractures and breccias in the lower part of the core are interpreted as a fault zone (Cloos and Pettijohn, 1973). The abrupt contact with an alluvial fan conglomerate, relative thickness of the lacustrine deposits, evidence for a high-gradient margin, episodes of evaporative concentration of lake waters, a sometimes stratified water column, poorly laminated profundal sediments and possible closed hydrologic conditions indicate that Rife Farm deposition occurred within a locally fault-bounded sub-basin (refer to Fig. 2). In contrast, the Rheems site represents slow, gradual subsidence associated with the overall Gettysburg Basin downwarp. In both cases, however, it is tectonics that creates the accomodation space for lacustrine fill. Although climate may influence rates and types of sediment deposition, at least for thick lacustrine accumulation, tectonics, specifically subsidence, must be involved to create a long-lived depression (Blair and Bilodeau, 1988).

Climatic Role in Lake Deposition

The formation of lakes within the Gettysburg rift basin could have resulted from either: (a) an increase in humidity, leading to an increase in precipitation, or (b) subsidence leading to intersection with the water table and drainage pattern changes. If the lakes formed through an increase in humidity, an increase in detrital input would have occurred due to increasing weathering and erosion rates (Leeder, 1982; Cecil, 1990). The lakes, however, are predominantly calcium carbonate, a chemical precipitate that is easily overwhelmed by high rate of clastic deposition (Dean, 1981; Cecil, 1990). Therefore, the basin development must be fundamentally the result of subsidence, albeit differential, as discussed above.

How then can carbonate lakes contribute to an understanding of paleoclimate except to indicate that clastic input was low? Low clastic input can be the result of overall low relief and sluggish detrital transport, or it may reflect fringing marshy vegetation that trapped siliciclastic debris, preventing it from reaching a lake. Both of these situations can arise during wet climate times. Very arid or very humid conditions can lead to low clastic transport, through lack of moving water, or intense vegetation and peat formation; however, both of these climates produce low-dissolved load in streams (Cecil, 1990). Triassic paleoclimatic reconstructions for eastern North America indicate a semi-arid climate (Steiner, 1983; Parrish et al., 1986). Sedimentological evidence provides a resolution of the low input versus climate problem for the Gettysburg strata.

Evidence for a semi-arid climate from the lakes and their associated deposits includes: (a) association with braided river and overbank deposits; the overbank mudstones are classified as vertisols and contain calcretes, features indicative of seasonal wetting and drying cycles (Driese et al., 1992) and semi-arid conditions (Atkinson, 1986; Wright, 1988); (b) a preponderance of alpha-type calcrete microfabrics, characteristic of arid to semi-arid climates (Wright and Tucker, 1991); and (c) the presence of the limestones themselves, demonstrating that the bed and suspended load in catchment waters was moderate to low and the dissolved load was high, as typically occurs in semi-arid settings (Cecil, 1990).

In conclusion, therefore, the study of two distinctly different carbonate lakes within the Gettysbug Basin has shown that local tectonism was responsible for generating accomodation space for lacustrine deposition, and that although the paleoclimate was semi-arid, sufficient rainfall over limestone and dolostone catchment areas led to high carbonate concentrations in the surface and groundwater, leading to significant chemical sedimentation in the lakes.

ACKNOWLEDGMENTS

D. Yocum received financial support for project research from the Franklin and Marshall College Marshall Scholars fund and Hackman Scholars fund. We thank V. Pedone for an exceptional early review of the manuscript and Christopher A. Scholz and V. Paul Wright for helpful suggestions and comments.

REFERENCES

ALLEN, J. R. L., 1986, Pedogenic calcretes in the old red sandstone facies (late Silurian-early Carboniferous) of the Anglo-Welsh area, southern Britain, *in* Wright, V. P., ed., Paleosols: Their Recognition and Interpretation: Princeton, Princeton University Press, p. 58–86.

ANDO, C. J., CZUCHRA, B. L., KLEMPERER, S. L., BROWN, L. D., CHEADLE, M. J., COOK, F. A., OLIVER, J. E., KAUFMAN, S., WALSH, T., THOMPSON, J. B., JR., LYONS, J. B., AND ROSENFELD, J. L., 1984, Crustal profile of mountain belt; COCORP deep seismic reflection in New England Appalachians and implications for architecture of convergent mountain chains: American Association of Petroleum Geologists Bulletin, v. 68, p. 819–837.

ATKINSON, C. D., 1986, Tectonic control on alluvial sedimentation as revealed by an ancient catena in the Capella Formation (Eocene) of northern Spain, *in* Wright, V. P., ed., Paleosols: Their Recognition and Interpretation: Princeton, Princeton University Press, p. 139–179.

BARRON, E. J., 1990, Climate and lacustrine petroleum source prediction, *in* Katz, B. J., ed., Lacustrine Basin Exploration—Case Studies and Modern Analogs: Tulsa, American Association of Petroleum Geologists Memoir 50, p. 1–18.

BAUM, J. S., BAUM, G. R., THOMPSON, P. R., AND HUMPHREY, J. D., 1994, Stable isotopic evidence for relative and eustatic sea level changes in Eocene to Oligocene carbonates, Baldwin County, Alabama: Geological Society of America Bulletin, v. 106, p. 824–839.

BELLANCA, A., CALVO, J. P., CENSI, P., NERI, R., AND POZO, M., 1992, Recognition of lake-level changes in Miocene lacustrine units, Madrid Basin, Spain. Evidence from facies analysis, isotope geochemistry and clay mineralogy: Sedimentary Geology, v. 76, p. 135–153.

BIRNEY DE WET, C. C., AND HUBERT, J. F., 1989, The Scots Bay Formation, Nova Scotia, Canada, a Jurassic carbonate lake with silica-rich hydrothermal springs: Sedimentology, v. 36, p. 857–873.

BLAIR, T. C. AND BILODEAU, W. L., 1988, Development of tectonic cyclothems in rift, pull-apart, and foreland basins: sedimentary response to episodic tectonism: Geology, v. 16, p. 517–520.

BREWER, R., 1964, Fabric and Mineral Analysis of Soils: New York, Wiley and Sons, 470 p.

BUCZYNSKI, C. AND CHAFETZ, H. S., 1987, Siliciclastic grain breakage and displacement due to carbonate crystal growth: an example from the Lueders Formation (Permian) of north-central Texas, USA: Sedimentology, v. 34, p. 837–843.

BURTON, R., KENDALL, C. ST. C., AND LERCHE, I., 1987, Out of our depth: on the impossibility of fathoming eustasy from the stratigraphic record: Earth-Science Reviews 24, p. 237–277.

CALVET, F. AND JULIA, R., 1983, Pisoids in the calcite profiles of Tarragona, north-east Spain, *in* Peryt, T.M., ed., Coated Grains: Berlin, Springer-Verlag, p. 456–473.

CECIL, C. B., 1990, Paleoclimate controls on stratigraphic repetition of chemical and siliciclastic rocks: Geology, v. 18, p. 533–536.

CERLING, T. E., 1984, The stable isotopic composition of modern soil carbonate and its relationship to climate: Earth and Planetary Science Letters, v. 71, p. 229–240

CERLING, T. E., 1991, Carbon dioxide in the atmosphere: evidence from Cenozoic and Mesozoic paleosols: American Journal of Science, v. 291, p. 377–400

COHEN, A. S., 1989, Facies relationships and sedimentation in large rift lakes and implications for hydrocarbon exploration: examples from Lakes Turkana and Tanganyika, *in* Talbot, M.R. and Kelts, K., eds., The Phanerozoic Record of Lacustrine Basins and Their Environmental Signals: Paleogeography, Paleoclimatology, Palaeoecology, v. 70, p. 65–80.

COHEN, A. S., 1990, Tectono-stratigraphic model for sedimentation in Lake Tanganyika, Africa, *in* Katz, B. J., ed., Lacustrine Basin Exploration: Tulsa, American Associtation of Petroleum Geologists Memoir 50, p. 137–150.

CLOOS, E. AND PETTIJOHN, F. J., 1973, Southern border of the Triassic Basin, West of York, Pennsylvania: Fault or Overlap: Geological Society of America Bulletin, v. 84, p. 523–536.

CORNET, B. AND OLSEN, P. E., 1985, A summary of the biostratigraphy of the Newark Supergroup of eastern North America, with comments on early Mesozoic provinciality, *in* Weber, R., ed., Symposio Sobre Flores del Triasico Tardio su Fitografía y Paleoecología, Memoria. Procedings III Latin-American Congress on Paleontology: Instituto de Geología Universidad Nacional Autonoma de Mexico, p. 67–81.

DANSGAARD, W., 1964, Stable isotopes in precipitation: Tellus, v. 16, p. 436–468

DEAN, W. E., 1981, Carbonate minerals and organic matter in sediments of modern north temperate hard-water lakes, *in* Ethridge, F. G. and Flores, R. M., eds., Recent and Ancient Nonmarine Depositional Environments: Models for Exploration: Tulsa, Society of Economic Paleontologists and Mineralogists Special Publication 31, p. 213–232.

DEAN, W. E. AND FOUCH, T. D., 1983, Lacustrine Environment, *in* Scholle, P. A., Bebout, D. G. and Moore, C. H., eds., Carbonate Depositional Environments: Tulsa, American Association of Petroleum Geologists Memoir 33, p. 98–130.

DE WET, C. B., 1995, Ooids and earthquakes — evidence from the western Newark Basin, (Triassic), Pennsylvania, (abs.): Northeastern Section Geological Society of America, v. 27, p. 39.

DRIESE, S. G., MORA, C. I., AND WALKER, K. R., 1992, Paleosols, Paleoweathering Surfaces, and Sequence Boundaries: Knoxville, University of Tennessee, Department of Geological Sciences, Studies in Geology 21, 111 p.

DRUMMOND, C. N. AND WILKINSON, B. H., 1993, Aperiodic accumulations of cyclic peritidal carbonates: Geology, v. 21, p. 1023–1026.

EARDLEY, A. J., 1938, Sediments of Great Salt Lake, Utah: American Association of Petroleum Geologists Bulletin 22, p. 1305–1411.

FAILL, R. T., 1973, Tectonic development of the Triassic Newark–Gettysburg Basin in Pennsylvania: Geological Society of America Bulletin, v. 84, p. 725–740.

FISHER, W. L. AND BROWN, L. F., 1972, Clastic Depositional Systems — A Genetic Approach to Facies Analysis: Annotated Outline and Bibliography:

Austin, Bureau of Economic Geology, The University of Texas at Austin, 211 p.

FLORES, R. M. AND PILLMORE, C. L., 1987, Tectonic control on alluvial paleoarchitecture of the Cretaceous and Tertiary Raton Basin, Colorado and New Mexico, in Ethridge, F. G., Flores, R. M., and Harvey, M. D., eds., Recent Developments in Fluvial Sedimentology: Tulsa, Society of Economic Paleontologists and Mineralogists Special Publication 39, p. 311–320.

FOUCH, T. AND DEAN, W., 1982, Lacustrine and associated clastic depositional environments, in Scholle, P. A. and Spearing, D., eds., Sandstone Depositional Environments: Tulsa, American Association of Petroleum Geologists Memoir 31, p. 87–114.

FRANCIS, J. E., 1986, The calcareous paleosols of the basal Purbeck Formation (upper Jurassic), southern England, in Wright, V. P., ed., Paleosols: Their Recognition and Interpretation: Princeton, Princeton University Press, p. 112–138.

FREYTET, P., 1973, Petrography and paleo-environment of continental carbonate deposits with particular reference to the upper Cretaceous and lower Eocene of Languedoc (Southern France): Sedimentary Geology, v. 10, p. 25–60.

FRIEDMAN, I. AND O'NEIL, J. R., 1977, Compilation of stable isotope fractionation factors of geochemical interest: Washington, D.C., United States Geological Survey Professional Paper 440–KK, 96 p.

GASSE, F., FONTES, J. C., PLAZIAT, J. C., CARBONEL, P., KACZMARKSA, I., DE DECKKER, P., SOULIE-MARSCHE, I., CALLOT, Y. AND DUPEUBLE, P. A., 1987, Biological remains, geochemistry, and stable isotopes for the reconstruction of environmental and hydrological changes in the Holocene Lakes from North Sahara: Palaeogeography, Palaeoclimatology, Palaeontology, v. 60, p. 1–46.

GIERLOWSKI-KORDESCH, E. AND RUST, B. R., 1994, The Jurassic East Berlin Formation, Hartford Basin, Newark Supergroup (Connecticut and Massachusetts): a saline lake-playa-alluvial plain system, in Renaut, R. W. and Last, W. M., eds., Sedimentology and Geochemistry of Modern and Ancient Saline Lakes: Tulsa, SEPM (Society for Sedimentary Geology) Special Publication 50, p. 249–265.

GIERLOWSKI-KORDESCH, E. AND HUBER, P., 1995, Lake sequences of the central Hartford Basin, Newark Supergroup, in McHone, N. W., ed., Guidebook for fieldtrips in Eastern Connecticut and the Hartford Basin: State Geological and Natural History Survey of Connecticut Guidebook No. 7, Northeast Section, Geological Society of America: p. B–1–B–39.

GLAESER, J. D., 1966, Provenance, dispersal, and depositional environments of Triassic sediments in the Newark-Gettysburg basin, (General Geology report 43): Harrisburg, Pennsylvania Geological Survey fourth series, Commonwealth of Pennsylvania, 167 p.

GLASS, S. W. AND WILKINSON, B. H., 1980, The Peterson Limestone-early Cretaceous lacustrine carbonate deposition in western Wyoming and southeastern Idaho: Sedimentary Geology, v. 27, p. 143–160.

GOLDBERG, P., 1992, Micromorphology, soils, and archaeological sites, in Holliday, V. T., ed., Soils in Archaeology, Landscape Evolution and Human Occupation: Washington, Smithsonian Institution Press, p. 145–167.

GOLDSTEIN, R. H., 1988, Paleosols of late Pennsylvanian cyclic strata, New Mexico: Sedimentology, v. 35, p. 777–803.

GORE, P. J. W., 1988, Late Triassic and early Jurassic lacustrine sedimentation in the Culpeper basin Virginia, in Manspeizer, W., ed., Triassic-Jurassic Rifting: Amsterdam, Elsevier, p. 369–400.

GORE, P. J. W., 1989, Toward a model for open- and closed-basin deposition in ancient lacustrine sequences: the Newark Supergroup (Triassic-Jurassic), Eastern North America, in Talbot, M. R. and Kelts, K., eds., The Phanerozoic Record of Lacustrine Basins and Their Environmental Signals: Paleogeography, Paleoclimatology, Palaeoecology, v. 70, p. 29–51.

HARDIE, L. A., SMOOT, J. P., AND EUGSTER, H. P., 1978, Saline lakes and their deposits: a sedimentological approach, in Matter, A. and Tucker, M. E., eds., Modern and Ancient Lake Sediments: Oxford, Blackwell Scientific Publications, p. 7–42.

HAQ, B. U., HARDENBOL, J., AND VAIL, P. R., 1987, Chronology of fluctuating sealevels since the Triassic: Science, v. 235, p. 1156–1166.

HASSANI, S. A. AND KENDALL, C. ST. C. , 1994, Sedimentology and alkaline geochemistry of rift Lake Manyara, northern Tanzania, east Africa: an analogue to lacustrine fill in some early Mesozoic lakes of south east USA: Southeastern Geology, v. 34, p. 129–137.

HAY, R. L. AND WIGGINS, B., 1980, Pellets, ooids, sepiolite and silica in three calcretes of south-western United States: Sedimentology, v. 27, p. 559–576.

HUBERT, J. F., 1978, Paleosol caliche in the New Haven Arkose, Newark Group, Connecticut: Paleogeography, Paleoclimatology, and Palaeoecology, v. 24, p. 151–168.

HUBERT, J. F., REED, A. A., DOWDALL, W. L., AND GILCHRIST, J. M., 1978, Guide to the Mesozoic Redbeds of Central Connecticut: Hartford, State Geological and Natural History Survey of Connecticut: Department of Environmental Protection, Guidebook Number 4, 129 p.

HUBERT, J. F. AND HYDE, M. G., 1982, Sheet-flow deposits of graded beds and mudstones on an alluvial sandflat-playa system, Upper Triassic Blomidon redbeds, St. Mary's Bay, Nova Scotia: Sedimentology, v. 29, p. 457–474.

HUBERT, J. F. AND FORLENZA, M. F., 1988, Sedimentology of braided-river deposits in Upper Triassic Wolfville redbeds, southern shore of Cobequid Bay, Nova Scotia, Canada, in Manspeizer, W., ed., Triassic-Jurassic Rifting: Amsterdam, Elsevier, p. 232–247.

HUBERT, J. F., FESHBACH-MERINEY, P. E., AND SMITH, M. A., 1992, The Triassic-Jurassic Hartford rift basin, Connecticut and Massachusetts: evolution, sandstone diagenesis and hydrocarbon history: American Association of Petroleum Geologists Bulletin, v. 76, p. 1710–1734.

HUC, A. Y., LEFOURNIER, J., VANDENBROUCKE, M., AND BESSEREAU, G., 1990, Northern Lake Tanganyika — an example of organic sedimentation in an anoxic rift lake, in Katz, B. J., ed., Lacustrine Basin Exploration: Tulsa, American Association of Petroleum Geologists Memoir, v. 50, p. 169–185.

JONES, F. AND WILKINSON, B., 1978, Structure and growth of lacustrine pisoliths from recent Michigan Marl Lakes: Journal of Sedimentary Petrology, v. 48, p. 1103–1110.

KLAPPA, C. F., 1980, Rhizoliths in terrestrial carbonates: classification, recognition, genesis and significance: Sedimentology, v. 27, p. 613–629.

KRUGE, M. A., HUBERT, J. F., BENSLEY, D. F., CRELLING, J. C., AKES, R. J., AND MERINEY, P. E., 1989, Organic geochemistry of a lower Jurassic synrift lacustrine sequence, Hartford Basin, Connecticut, U.S.A.: Advances in Organic Geochemistry, v. 16, p. 689–701.

LEEDER, M. R., 1982, Sedimentology: Process and Product: London, George Allen and Unwin, 344 p.

LETOURNEAU, P. M., 1985, Alluvial fan development in the Lower Jurassic Portland Formation, central Connecticut — implications for tectonics and climate: Reston, United States Geological Circular, p. 17–26.

MAJEWSKE, O. P., 1969, Recognition of Invertebrate Fossil Fragments in Rocks and Thin Sections: Leiden, International Sedimentary Petrographical Series, v. XIII, E.J. Brill, 101 p.

MANSPEIZER, W., 1988, Triassic-Jurassic rifting and opening of the Atlantic: an overview, in Manspeizer, W., ed., Triassic-Jurassic Rifting: Amsterdam, Elsevier, p. 41–79.

MANSPEIZER, W. AND HUNTOON, J., 1989, Early Mesozoic Rift Basins of Eastern North America: Origin and Evolution, in Slingerland, R. and Furlong, K., eds., Sedimentology and Thermal-Mechanical History of Basins in the Central Appalachian Orogen: Washington, D.C., 28th International Geological Congress Field Trip Guidebook T152, American Geophysical Union, p. 25–42.

McCREA, J. M., 1950, The isotopic geochemistry of carbonates and a paleotemperature scale: Journal of Chemical Physics, v. 18, p. 849–857.

MERTZ, K. A. AND HUBERT, J. F., 1990, Cycles of sand-flat sandstone and playa-lacustrine mudstone in the Triassic-Jurassic Blomidon redbeds, Fundy rift basin, Nova Scotia: implications for tectonic and climatic controls: Canadian Journal of Earth Science, v. 27, p. 442–451.

MOORE, R. C., LALICKER, C. G., AND FISCHER, A. G., 1952, Invertebrate Fossils: New York, McGraw-Hill, 760 p.

MOUNT, J. F. AND COHEN, A. S., 1984, Petrology and geochemistry of rhizoliths from Plio-Pleistocene fluvial and marginal lacustrine deposits, East Lake Turkana, Kenya: Journal of Sedimentary Petrology, v. 54, p. 263–275.

MULLER, G., IRION, G., AND FORSTNER, U., 1972, Formation and diagenesis of inorganic Ca–Mg carbonates in the lacustrine environment: Naturwissenschaften, v. 59, p. 158–164.

MURPHY, D. H. AND WILKINSON, B. H., 1980, Carbonate deposition and facies distribution in a central Michigan marl lake: Sedimentology, v. 27, p. 123–135.

NAHON, D. B., 1991, Introduction to the Petrology of Soils and Chemical Weathering: New York, John Wiley & Sons, 313 p.

OLSEN, P.E., 1990, Tectonic, climatic and biotic modulation of lacustrine ecosystems — examples from Newark Supergroup of eastern North America, in Katz, B. J., ed., Lacustrine Basin Exploration: Tulsa, American Association of Petroleum Geologists Memoir 50, p. 209–224.

OLSEN, P. E., 1993, Early Mesozoic Lacustrine Record of Cyclical Climate Change from Core and Outcrops of the Newark Basin: Tulsa, SEPM (Society for Sedimentary Geology) Field Trip Guidebook, 21 p.

OLSEN, P. E., GORE, P. J. W., AND SCHLISCHE, R. W., 1989, Tectonic, depositional, and paleoecological history of Early Mesozoic rift basins, eastern

North America: Washington, D.C., 28th International Geological Congress, Field Trip Guidebook T–351, 174 p.

PARRISH, J. T., 1993, Mesozoic climates of the Colorado Plateau, *in* Morales, M., ed., Aspects of Mesozoic Geology and Paleontology of the Colorado Plateau: Flagstaff, Museum of Northern Arizona Bulletin, v. 59, p. 1–11.

PARRISH, J. M., PARRISH, J. T., AND ZIEGLER, A. M., 1986, Permian-Triassic paleogeography and paleoclimatology and implications for Therapsid distribution, *in* Hotten, N., III, MacLean, P. D., Roth, J. J., and Roth, E. C., eds., The Ecology and Biology of Mammal-like Reptiles: Washington, D.C., Smithsonian Institution Press, p. 109–131.

PETTIJOHN, F. J., POTTER, P. E., AND SIEVER, R., eds., 1987, Sand and Sandstone: New York, Springer-Verlag, 553 p.

PICARD, M. D. AND HIGH, L. R., 1981, Physical stratigraphy of ancient lacustrine deposits, *in* Ethridge, F. G. and Flores, R. M., eds., Recent and Ancient Nonmarine Depositional Environments: Models for Exploration: Tulsa, Society of Economic Paleontologists and Mineralogists Special Publication 31, p. 233–260.

PLATT, N. H., 1989, Lacustrine carbonates and pedogenesis: sedimentology and origin of palustrine deposits from the Early Cretaceous Rupelo Formation, W. Cameros Basin, N. Spain: Sedimentology, v. 36, p. 665–684.

PLATT, N. H. AND WRIGHT, V. P., 1991, Lacustrine carbonates: facies models, facies distributions and hydrocarbon aspects, *in* Anadon, P., Cabrera, Ll., and Kelts, K., eds., Lacustrine Facies Analysis: Oxford, Blackwell Scientific Publications, p. 57–74.

PLATT, N. H. AND WRIGHT, V. P., 1992, Palustrine carbonates and the Florida Everglades: towards an exposure index for the fresh-water environment?: Journal of Sedimentary Petrology, v. 62, p. 1058–1071.

RAST, N., 1988, Variscan-Alleghenian orogen, *in* Manspeizer, W., ed., Triassic-Jurassic Rifting: Amsterdam, Elsevier, p. 1–27.

RATCLIFFE, N. M. AND BURTON, W. C., 1985, Fault reactivation studies for origin of the Newark Basin and studies related to eastern U.S. seismicity, *in* Robinson, G. R., Jr. and Froelich, A. J., eds., Proceedings of the Second U.S. Geological Survey Workshop on the Early Mesozoic Basins of the Eastern United States: United States Geological Survey Circular, v. 946, p. 36–45.

RENAUT, R. W. AND OWEN, R. B., 1991, Shore-zone sedimentation and facies in a closed rift lake: the Holocene beach deposits of Lake Bogoria, Kenya, *in* Anadon, P., Cabrera, Ll., and Kelts, K., eds., Lacustrine Facies Analysis: Oxford, Blackwell Scientific Publications, p. 175–198.

RENAUT, R. W. AND TIERCELIN, J.–J., 1994, Lake Bogoria, Kenya rift valley — a sedimentological overview, *in* Renaut, R. W. and Last, W. M., eds., Sedimentology and Geochemistry of Modern and Ancient Saline Lakes: Tulsa, SEPM (Society for Sedimentary Geology) Special Publication 50, p. 101–123.

RETALLACK, G. J., 1983, Late Eocene and Oligocene palaeosols from Badlands National Park, South Dakota: Special Paper Geological Society of America, v. 193, 82 p.

ROBBINS, E. I., WILKES, G. P., AND TEXTORIS, D. A., 1988, Coal deposits of the Newark rift system, *in* Manspeizer, W., ed., Triassic-Jurassic Rifting: Amsterdam, Elsevier, p. 649–682.

ROBINSON, G. R., 1988, Base- and precious-metal mineralization associated with igneous and thermally altered rocks in the Newark, Gettysburg, and Culpeper Early Mesozoic Basins of New Jersey, Pennsylvania, and Virginia, *in* Manspeizer W., ed., Triassic-Jurassic Rifting: Amsterdam, Elsevier, p. 621–648.

ROOT, S. I., 1988, Structure and hydrocarbon potential of the Gettysburg basin, Pennsylvania and Maryland, *in* Manspeizer, W., ed., Triassic-Jurassic Rifting: Amsterdam, Elsevier, p. 353–367.

ROSS, W. C., HALLIWELL, B. A., MAY, J. A., WATTS, D. E., AND SYVITSKI, J. P. M., 1994, Slope readjustment: A new model for the development of submarine fans and aprons, Geology, v. 22, p. 511–514.

SCHLISCHE, R. W., 1993, Anatomy and evolution of the Triassic-Jurassic continental rift system, eastern North America: Tectonics, v. 12, p. 1026–1042.

SCHLISCHE, R. W. AND OLSEN, P. E., 1990, Quantative filling model for continental exensional basins with applications to Early Mesozoic rifts of eastern North America: Journal of Geology, v. 98, p. 135–155.

SCHOLLE, P. A., 1978, A Color Illustrated Guide to Carbonate Rock Constituents, Textures, Cements, and Porosities: Tulsa, American Association of Petroleum Geologists Memoir 27, 241 p.

SMOOT, J. P., 1985, The closed-basin hypothesis and its use in facies analysis of the Newark Supergroup, U.S. Geological Survey Circular 946, p. 4–10.

SMOOT, J. P., 1991, Sedimentary facies and depositional environments of early Mesozoic Newark Supergroup basins, eastern North America: Palaeogeography, Palaeoclimatology, Palaeoecology, v. 84, p. 369–423.

SPÖTL, C. AND WRIGHT, V. P., 1992, Groundwater dolocretes from the upper Triassic of the Paris Basin, France: A case study of an arid, continental diagenetic facies: Sedimentology, v. 39, p. 1119–1137.

STEINER, M. B., 1983, Mesozoic apparent polar wandering and plate motions of North America, *in* Reynolds, M. W. and Dolly, E. D., eds., Mesozoic Paleogeography of the West-Central United States: Denver, The Rocky Mountain Section Society of Economic Paleontologists and Mineralogists, p. 1–12.

STEVENS, R. L. AND HUBERT, J. F., 1983, Alluvial fans, braided rivers, and lakes in a fault-bounded semiarid valley: Sugarloaf Arkose (late Triassic-early Jurassic), Newark Supergroup, Deerfield Basin, Massachusetts: Northeastern Geology, v. 5, p. 8–22.

STOFFERS, P. AND HECKY, R. E., 1978, Late Pleistocene-Holocene evolution of the Kivu-Tanganyika Basin, *in* Matter, A. and Tucker, M. E., eds., Modern and Ancient Lake Sediments: Oxford, Blackwell Scientific Publications, p. 43–54.

SUMNER, J. R., 1978, Geophysical investigation of the structural framework of the Newark-Gettysburg Triassic basin, Pennsylvania: Geological Society of America Bulletin, v. 88, p. 935–942.

TALBOT, M. R. AND KELTS, K., 1990, Paleolimnological signatures from Carbon and Oxygen isotopic ratios in carbonates from organic carbon-rich lacustrine sediments, *in* Katz, B. J., ed., Lacustrine Basin Exploration: Case Studies and Modern Analogs: Tulsa, American Association of Petroleum Geologists Memoir 50, p. 99–112.

VAN HOUTEN, F. B., 1962, Cyclic sedimentation and the origin of analcime-rich upper Triassic Lockatong Formation, west-central New Jersey and adjacent Pennsylvania: American Journal of Science, v. 260, p. 561–576.

VAN HOUTEN, F. B., 1964, Cyclic lacustrine sedimentation, Upper Triassic Lockatong Formation, Central New Jersey and Adjacent Pennsylvania: Kansas Geological Survey Bulletin, v. 169, p. 497–531.

WATTS, N. L., 1978, Displacive calcite: evidence form recent and ancient calcretes: Geology, v. 6, p. 699–703.

WATTS, S., 1994, Fluvial sandstones and overbank mudstones of the Gettysburg basin (Triassic), Pennsylvania: unpublished Honors Thesis, Department of Geology, Franklin and Marshall College, Lancaster, 40 p.

WETZEL, R. G., 1983, Limnology: Philadelphia, Saunders College Publishing, 767 p.

WHEELER, W. H. AND TEXTORIS, D. A., 1978, Triassic limestone and chert of playa origin in North Carolina: Journal of Sedimentary Petrology, v. 48, p. 765–776.

WHITTINGTON, D., 1988, Chemical and physical constraints on petrogenesis and emplacement of ENA Olivine Diabase Magma Types, *in* Manspeizer, W., ed., Triassic-Jurassic Rifting: Amsterdam, Elsevier, p. 557–578.

WRIGHT, V. P., 1982, Calcrete paleosols from the Lower Carboniferous Llanelly formation, South Wales: Sedimentary Geology, v. 33, p. 1–33.

WRIGHT, V. P., 1986, The role of fungal biomineralization in the formation of early Carboniferous soil fabrics: Sedimentology, v. 33, p. 831–838.

WRIGHT, V. P., 1988, Paleokarsts and paleosols as indicators of paleoclimate and porosity evolution: a case study of the carboniferous of South Wales, *in* James, N. P. and Choquette, P. W., eds., Paleokarst: New York, Springer-Verlag, p. 329–341.

WRIGHT, V. P., 1990, A micromorphological classification of fossil and recent calcic and petrocalcic microstructures, *in* Douglas, L. A., ed., Soil Micromorphology: A Basic and Applied Science: Amsterdam, Elsevier Developments in Soil Science, v. 19, p. 401–407.

WRIGHT, V. P. AND TUCKER, M. E., eds., 1991, Calcretes: an introduction, *in* Calcretes: Oxford, International Association of Sedimentologists Reprint Series 2, p. 1–22.

PART IV
EOLIAN DEPOSITS

CONTINENTAL SEQUENCE STRATIGRAPHY OF A WET EOLIAN SYSTEM: A KEY TO RELATIVE SEA-LEVEL CHANGE

MARY CARR-CRABAUGH

Dept. of Geology and Geological Engineering, Colorado School of Mines Golden, CO 80401

AND

GARY KOCUREK

Department of Geological Sciences, The University of Texas at Austin, Austin, Texas 78712

ABSTRACT: The Middle Jurassic San Rafael Group of eastern Utah is interpreted as the product of a wet eolian system, in which accumulation of a sequence results from a relatively rising water table with sediment capture within the rising capillary fringe. Sequence boundaries of bypass or erosion result from a static or falling relative water table. Four complete sequences are correlated along a 365-km transect of eastern Utah stretching from near Dinosaur National Monument to Bluff. The areal extent of these sequences suggests that the controls on relative change in the level of the water table were regional in nature and related to relative sea-level fluctuation caused by eustasy, regional and local subsidence. The rate of relative sea-level rise balanced against the rate of sediment supply controlled the depositional facies that form the accumulation. However, the rate of sediment supply was not an independent variable and appears to parallel the rate of relative sea-level rise. The probable cause of this relationship is that as the sea transgressed it delivered sediment to the coastal areas upwind of the San Rafael Group. Because both marine sequences and wet eolian sequences accumulate with a rising relative sea level and form bounding surfaces on the relative sea-level fall, they can be tied into the same sequence-stratigraphic framework. The linking of these two sedimentary systems into a common stratigraphic framework makes it possible to more accurately relate events in the basin center to sedimentary facies on the basin margin.

INTRODUCTION

Previous stratigraphic reconstructions of the Middle Jurassic Entrada Sandstone and other eolian portions of the San Rafael Group have relied on lithostratigraphic correlations. This study attempts to place the San Rafael Group in a sequence stratigraphic framework and determine the controls on its facies geometry. For some time now it has been recognized that areally extensive major bounding surfaces exist in eolian sequences (Loope, 1985; Talbot, 1985; Kocurek, 1988; Havholm et al., 1993). These surfaces, termed "super surfaces" (Kocurek, 1988), can be used as regional correlation tools in eolian units. A super surface represents the end of an eolian accumulation event and therefore may mark a change in the factors controlling accumulation of eolian sediments in the basin (Kocurek and Havholm, 1993). The basic eolian stratigraphic unit of an accumulation and the super surface that caps it can be used to reconstruct the regional eolian genetic architecture and is a sequence as defined by Mitchum et al. (1977, p. 53) as a "stratigraphic unit composed of a relatively conformable succession of genetically related strata and bounded at its top and base by unconformities or their correlative conformities".

Kocurek and Havholm (1993) also recognized three basic types of eolian depositional systems; wet, dry and stabilized. Previous work (Crabaugh and Kocurek, 1993) demonstrated that the Entrada Sandstone in the Uinta Mountains of northeastern Utah represents a wet eolian system. Wet eolian systems are characterized by a shallow water table, such that the capillary fringe is at or near the floor of the interdune flats, thereby raising the grain threshold values to the point where the surface is largely protected from deflation. Accumulation within a wet eolian system occurs because of a relative rise in the water table, and these accumulations are characterized by the presence of climbing dune and interdune-flat deposits (Fig. 1). Falling or static relative water table results in the formation of erosional or bypassing super surfaces, respectively. As predicted for a wet eolian system, the sequences identified in northeastern Utah are composed of inclined dune and interdune accumulations capped by horizontal super bounding surfaces (Crabaugh and Kocurek, 1993). Likewise, the occurrence of sedimentary fea-

FIG. 1.—Basic stratigraphic element in a wet eolian system is composed of climbing dune and interdune strata capped by an erosional or bypassing super bounding surface. Refer to text for further explanation. Modified from Crabaugh and Kocurek (1993).

tures such as corrugated surfaces (microtopography caused by differential wind deflation of damp cross-strata; Simpson and Loope, 1985), water-ripple structures, ball-and-pillow structures and other soft-sediment deformation features indicate a high water table effected accumulation of the Entrada Sandstone (Crabaugh and Kocurek, 1993). This study will illustrate that wet eolian sequences identified within the Entrada Sandstone of northeastern Utah can by traced 250 km to the south to a 100–km transect of the San Rafael Group in southeastern Utah. Linking of local sequences into a regional sequence stratigraphic framework should make possible the distinction of the effects of basin scale factors such as sea level, subsidence, and sediment supply on the facies architecture of a wet eolian system.

METHODS

To illustrate the stratigraphic architecture of the San Rafael Group of southeastern Utah (Entrada Sandstone, Carmel, and Wanakah Formations) 29 vertical sections have been used to construct an oblique-to-paleotransport transect that stretches from 24–km north of Monticello, Utah to Bluff, Utah (Figs. 2, 3). The generalized cross section (Fig. 3) has been hung on the base of the Black Steer Knoll Member of the Wanakah Formation to reflect as rationally as possible the depositional facies geometries as well as the erosional relief of the J–5 surface that caps the San Rafael Group. Sections were measured using jacob staff and hand level as well as an Elta 4 electronic tacheometer. These sections were initially measured and correlated into a

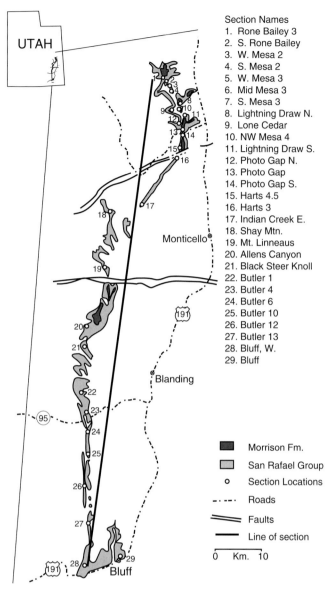

Section Names
1. Rone Bailey 3
2. S. Rone Bailey
3. W. Mesa 2
4. S. Mesa 2
5. W. Mesa 3
6. Mid Mesa 3
7. S. Mesa 3
8. Lightning Draw N.
9. Lone Cedar
10. NW Mesa 4
11. Lightning Draw S.
12. Photo Gap N.
13. Photo Gap
14. Photo Gap S.
15. Harts 4.5
16. Harts 3
17. Indian Creek E.
18. Shay Mtn.
19. Mt. Linneaus
20. Allens Canyon
21. Black Steer Knoll
22. Butler 1
23. Butler 4
24. Butler 6
25. Butler 10
26. Butler 12
27. Butler 13
28. Bluff, W.
29. Bluff

■ Morrison Fm.
▨ San Rafael Group
○ Section Locations
-·-·- Roads
═══ Faults
─── Line of section

0   Km.   10

FIG. 2.—Location map of measured sections in southeastern Utah. Line of section used for stratigraphic reconstruction in Figure 3 is oriented N8°E. Modified from O'Sullivan (1980a).

regional stratigraphic framework by O'Sullivan (1980a, b) and O'Sullivan and Pierce (1983). The nomenclature used in the stratigraphic cross section (Fig. 3) is that established by O'Sullivan (1980a, b) and O'Sullivan and Pierce (1983). This study has refined these stratigraphic correlations by tracing individual surfaces and units on foot where possible and by photomosaics where not. Foreset orientations were measured using a Brunton compass in order to reconstruct bedform geometry and migration direction.

GENERAL SETTING

Paleogeographic Setting

The San Rafael Group (Bathonian-late Callovian) in eastern Utah is bounded by the J–2 and J–5 regional unconformities

(Figs. 4, 5; Pipiringos and O'Sullivan, 1978). Figure 6 shows the general paleogeographic evolution during deposition of the San Rafael Group. The J–2 surface is overlain by the areally restricted Page Sandstone that is represented in the study area by a very restricted thin unit (Fig. 6A; Havholm, 1991). The Page Sandstone was transgressed by the Carmel Sea and interfingers with and is overlain by the marginal marine and sabkha accumulations of the Carmel Formation and its lateral equivalent, the Dewey Bridge Member of the Entrada Sandstone (Peterson, 1986).

The Entrada Sandstone makes up the bulk of the San Rafael Group in the study area. It is a complex and extensive eolian and marginal marine unit stretching from Utah to the Texas panhandle and from Arizona and New Mexico north to southern Wyoming (Blakey et al., 1988; Fig. 6B). Paleogeographic and stratigraphic reconstructions of the Entrada erg units show that it was bounded on the north, west and partly on the south by units interpreted as representing marine and sabkha environments (Imlay, 1980; Kocurek and Dott, 1983; O'Sullivan and Pierce, 1983; Blakey et al., 1988; Peterson, 1988; Fig. 6B). Stratigraphically, after initial progradation of the Entrada Sandstone to the west during Slick Rock deposition, eolian- and marine-sabkha units vertically stack and then onlap onto the Entrada eolian sand during deposition of the Moab Tongue, indicating that the Entrada Sandstone was deposited during an overall relative rise in sea level (Fig. 4; O'Sullivan and Pierce, 1983). The Entrada Sandstone was subsequently transgressed by the Curtis Sea (Fig 6C). This long-term rise is also shown in eustatic sea-level curves for the Jurassic (Haq et al., 1988).

The southern portions of the study area have been significantly impacted by events within the Blanding and Todilto basins (Figs. 6B, C). The Blanding Basin is a shallow feature with poorly-defined boundaries except on the west where it is bounded by the Monument Upwarp and to the north where it is partially bounded by the Abajo Mountains (Kelley, 1955). Kelley (1955) ascribed the formation of the Blanding Basin to Laramide deformation and later igneous activity. However, this study will show activity in the Blanding Basin during early San Rafael Group deposition. The Todilto Basin is at the site of a long-standing tectonic sag that has had only mildly episodic activity since the Early Paleozoic (Stevenson and Baars, 1977; Woodward and Callendar, 1977; Kelley, 1955). Initial Middle Jurassic subsidence and flooding of the Todilto Basin is marked by the Todilto Limestone and evaporites of the Wanakah Formation, which overlie eolian cross-strata of the Entrada Sandstone and are restricted to the central portions of the basin (McCrary, 1985; Ridgley, 1984; Vincelette and Chittum, 1981; Fig. 6C). In southeastern Utah, the first unit that is correlated to subaqueous flooding within the Todilto Basin is the Lower Member of the Wanakah Formation (Fig. 3).

Paleoclimatic Setting

Early and Middle Jurassic global circulation models of the western United States indicate prevailing wind directions fluctuated seasonally because of a strong monsoonal influence (Parrish and Curtis, 1982; Parrish and Peterson, 1988; Fig. 6). These same studies also suggest the summer winds (from NNE) were stronger than the winter winds (from NE). Between Middle and Late Jurassic time, paleogeographic changes resulted in the

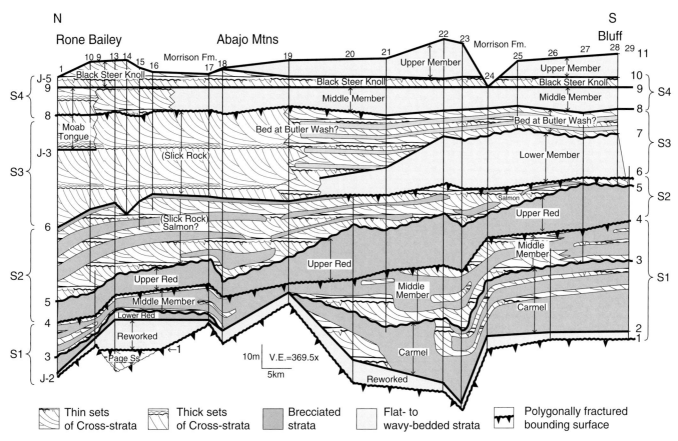

FIG. 3.—Sequence reconstruction using measured sections shown in Figure 2, transect orientation N8°E. Surfaces are numbered and sequences are indicated by brackets and lettered numbers. Four complete sequences have been identified in southeastern Utah. Stratigraphic nomenclature established by O'Sullivan (1980a, 1980b) and O'Sullivan and Pierce (1983).

breakdown of the monsoonal circulation. However, during the Middle Jurassic time, the monsoonal circulation was probably weakening and may have resulted in a degree of variability in the wind directions.

SEQUENCE DESCRIPTIONS

Wet eolian sequences, as initially described in Crabaugh and Kocurek (1993), consist of a depositional system of climbing dune and interdune strata capped by a super surface (Fig. 1). Regional study of the San Rafael Group (Fig. 3) shows this description should be modified to reflect the occurrence of a basal sabkha-dominated depositional system that is overlain by the climbing cross-stratified dune and interdune facies. Figure 3 indicates there are at least four sequences in the San Rafael Group that consist of two distinct depositional systems: a basal unit composed of sabkha-dominated facies capped by a flat or load-deformed surface and an upper unit of interbedded dune and interdune strata capped by a regional super surface that is characteristically corrugated, polygonally fractured or featureless (Fig. 3). Because the position of the relative water table controls accumulation in a wet eolian system, the regional super surfaces (Fig. 3, Surfaces 4, 6, 8, and 10) are presumed to have been relatively horizontal at the time of their formation, because they would have mimicked the static or falling relative water table in this broad, flat coastal plain. Each sequence also shows

a distinct lateral variation, with a dominance of eolian cross-strata to the north and sabkha-dominated facies to the south.

The J–2 surface shows significant paleotopography in the cross section, with the most pronounced relief in the section at Photograph Gap (Fig. 3, section 13). Figure 3 shows approximately 19 m of paleotopographic relief, which was subsequently filled by the Page Sandstone. The Page Sandstone, the basal unit in the San Rafael Group, represents the deposits of a significantly different eolian system, which has been thoroughly described by Havholm (1991). The Page Sandstone is interpreted as the deposits of a dry eolian system that blanketed much of the area west of the study area and was punctuated by super surfaces exhibiting features formed under damp or wet conditions (Havholm, 1991; Kocurek et al., 1991a; Havholm et al., 1993). The Page Sandstone will not be addressed further in this study.

Previous work in the area north of the Abajo Mountains has extended the Slick Rock Member of the Entrada Sandstone from the top of the Lower Red to the base of the Moab Tongue (O'Sullivan, 1980a, 1980b; Pierce and O'Sullivan, 1983; Fig. 3). This study has subdivided the Slick Rock by extending surfaces and stratigraphic units recognized in the south (Middle Member, Upper Red, Salmon, and Bed at Butler Wash) into the northern outcrops (Fig. 3). Recognition of these units throughout the outcrop belt has enhanced subdivision of the San Rafael Group into sequences.

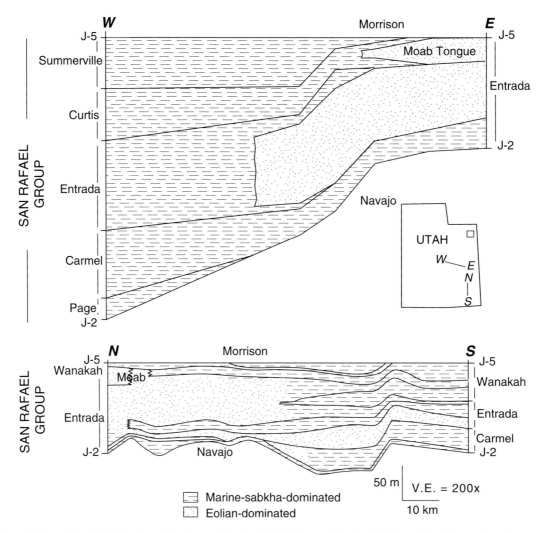

FIG. 4.—General facies relationships for the San Rafael Group of eastern Utah. Marine-sabkha-dominated deposition in the south and west grades into eolian-dominated strata in the northeast. Refer to Figure 5 for paleogeographic setting. Because eolian strata are contemporaneous with units of a marginal marine origin, and these vertically stack, then onlap onto the main body of the Entrada, accumulation of the eolian strata is believed to have occurred during an overall relative sea-level rise. Modified from O'Sullivan and Pierce (1983).

To illustrate the evolution through time of a portion of the San Rafael Group facies geometry, two diagrams have been constructed that depict the facies geometries of the individual sequences (Fig. 7). The illustrations were constructed based on the previously stated assumption that the depth of erosion for the capping super surface was controlled by the relative water-table level and that the surfaces would have been relatively flat over the study area. Therefore, these reconstructions represent snapshots of an evolving basin margin, illustrating how the facies and resulting accumulations responded through time to changing environmental conditions.

SEQUENCE 1 (S1)

Sequence Description

The basal sequence, bounded by Surfaces 1 and 4, thickens to the south varying from 22.9 to 49 m (Figs. 3, 7A). The lithostratigraphic units that make up the basal sabkha depositional system are the Reworked zone, the Lower Red Member

of the Entrada Sandstone, and the Carmel Formation. These units are dominated by a continuum of deformed strata from flat-and wavy-bedded to brecciated strata (Fig. 8A). Much of the character of these deposits results from their initial deposition on the irregular salt-ridge microtopography common to modern sabkhas (Glennie, 1970). Further penecontemporaneous and postdepositional deformation of the strata is attributed to displacive evaporite growth, desiccation, salt dissolution-collapse, fluid escape and dune loading on water-saturated sabkha flats (Fryberger et al., 1983, 1984; Hunter et al., 1992). Although the Carmel Formation is dominated by brecciated to wavy laminae, it also has isolated sets of cross-strata from tens of centimeters to 6.7 m in thickness. The cross-strata can be simple or compound sets showing at least one scale of cyclicity.

The cross-strata-dominated depositional system is represented by the Middle Member of the Entrada Sandstone, which is separated from the sabkha-dominated system by undulatory Surface 3. The undulatory nature of Surface 3 is the result of migration of bedforms across the damp water-saturated sabkha

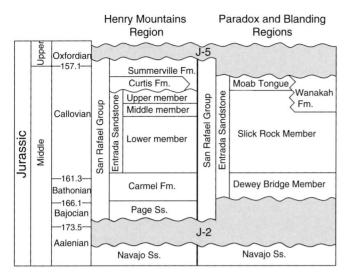

FIG. 5.—Stratigraphic units in southeastern Utah showing major unconformities, stratigraphic divisions and age assignments. Note the time scale based on Harland et al. (1989) is not linear. Modified from Peterson (1986) and Riggs and Blakey (1993).

accumulations. The Middle Member is dominated by thin cross-stratified sets interbedded with flat- to wavy-bedded strata. The cross-stratified sets vary in thickness from a few centimeters to 4 m, averaging around 1 m thick (Fig 8B). At the base of the Middle Member just south of the Abajos (Fig. 3, sections 20 and 21), there is an anomalously thick 8–m set of cross-strata. The cross-stratified sets commonly have a corrugated upper surface, a loaded base and internally are simple or compound. Surface 4, which caps S1, is corrugated and polygonally fractured over the entire length of the transect (Fig. 8C).

Interpretation of Controls

S1 illustrates the effects of relative water-table fluctuations and sediment dispersal patterns on the facies geometries of an eolian sequence (Figs. 3, 7A). Accumulation in a wet eolian system is dependent upon the relative rise of the water table. That is the water-table level must rise relative to a fixed point in the sediment column in order for passing eolian sediments to be trapped within the sedimentary record. In this low-relief coastal setting, it is postulated that the relative water-table level would be controlled by changes in relative sea level (eustatic sea level and regional and local subsidence). Local tectonic subsidence is illustrated in S1 by the abrupt thickening south of the Abajo Mountains into the Blanding Basin (Kelley, 1955; Fig. 7A). The significantly higher rates of subsidence south of the Abajo Mountains caused increased rates of relative water-table rise, and as a result the rate of accumulation was also increased. Within the Blanding Basin, S1 is composed of significantly thicker cross-stratified units than the time equivalent units north of the Abajos. It is probable that dunes were migrating across both the northern and the southern areas, but only formed significant accumulations in the south where the rate of water-table rise was greater, and thereby forming more accumulation space.

South of the Abajos there is also a distinct segregation of cross-stratified units from north to south. The cross-stratified

accumulations are concentrated in the areas immediately south of the Abajos Mountains and decrease in number and thickness to the south. This lateral segregation can most easily be explained by the manner in which sediment was transported across this area. Bedform reconstructions for the Entrada Sandstone in northeastern Utah (Crabaugh and Kocurek, 1993) indicate the bedforms were relatively large transverse bedforms migrating to the west-southwest with superimposed bedforms migrating along the lee face to the northeast or southeast. Sediment was delivered to the northern shorelines by longshore drift (Peterson, 1986) and transported to the southwest in the form of eolian bedforms under the prevailing northeasterly winds (Fig. 6; Parrish and Peterson, 1988). A similar situation exists in the modern Namib Desert of southern Africa where discreet eolian transport corridors have been identified (Corbett, 1993). Sand is picked up from cuspate portions of the Namib coastline and moved inland through relatively straight corridors in the form of barchan dunes. The facies distribution within the Blanding Basin and bedform reconstructions suggest that the bulk of sand moving into and through the sand seas was concentrated in the northern portion of the Blanding Basin and north of the Abajo Mountains with lesser amounts of sediment being spread to the south by the superimposed bedforms.

The polygonally fractured nature of Surface 4 suggests that the relative water table was static for some period during the formation of the super surface. As discussed previously, supersurface formation in a wet eolian system occurs when the relative water table is falling or stable and thereby not allowing for accumulation of eolian sediments. Polygonal fractures are caused by repeated fracturing of an evaporite-cemented sandstone, suggesting the capillary fringe of the relative water table was at or near the sediment surface for an extended period of time (Kocurek and Hunter, 1986).

SEQUENCE 2 (S2)

Sequence Description

Unlike S1, the second sequence shows a thickening to the north varying from 38.5 in the north to 17.8 m thick in the south (Figs. 3, 7B). S2, bounded by Surfaces 4 and 6, shows a distinct division of depositional systems both vertically and laterally. The basal sabkha depositional system, represented by the Upper Red Member, thickens to the south (Fig. 3) and is characterized by red, brecciated, silty sandstone, and red wavy-laminated sandstone (Fig. 9A). There are a few thin discontinuous cross-stratified sets that are capped by polygonally fractured surfaces. The sabkha depositional system is separated from the dune-interdune depositional system by Surface 5, which is inclined and loaded.

The dune-interdune depositional system, represented by the Salmon Sandstone and the lower portion of the Slick Rock Member, shows a pronounced thinning to the south, from 29.5 to 2.1 m (Figs. 3, 7B). These units are dominated by compound cross-stratified sets separated by thin flat, wavy and brecciated units (Fig. 9B). The compound cross-stratified sets and interbedded flat- to wavy-bedded units probably represent the accumulations of climbing dunes and interdune areas. Because the outcrops are generally oriented in a perpendicular to paleoflow direction, it was not possible to document bedform climb in the Salmon Sandstone. However, the facies characteristics

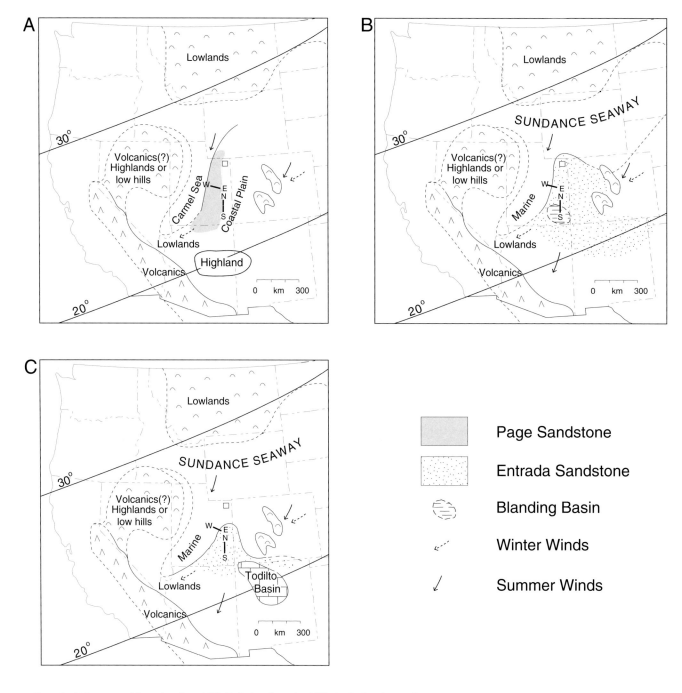

Fig. 6.—Paleogeographic setting for middle Bajocian through middle Oxfordian San Rafael Group. (A) Indicates general paleogeographic setting during Page deposition. (B) Paleogeographic setting during Entrada Slick Rock deposition, when the Entrada erg was most developed. (C) Upper San Rafael Group depositional setting as Curtis Sea transgressed over the northern portions of the Entrada erg. Study area of Crabaugh and Kocurek (1993) indicated by box in northeastern Utah. Section lines for Figure 4 shown by W–E and N–S dark lines. Modified from Peterson (1988) and Havholm (1991).

are very similar to those described by Crabaugh and Kocurek (1993) as resulting from the accumulation of climbing dunes and interdune areas. The cross-stratified sets show two scales of cyclicity and vary in thickness from 0.10 to 3.3 m. These sets are heavily deformed by loading at their bases and polygonal fracturing of their upper surfaces. The internal bounding surfaces are commonly corrugated, that is they show microtopography caused by differential wind deflation of damp cross-

strata. Surface 6 caps the Salmon Sandstone and varies along its length from a polygonally fractured, deformed surface in the south to a smooth erosional surface in the north.

Interpretation of Controls

The facies geometries of S2 illustrate the time transgressive nature of the surface that separates the sabkha and cross-stratified depositional systems. The sabkha accumulations of the

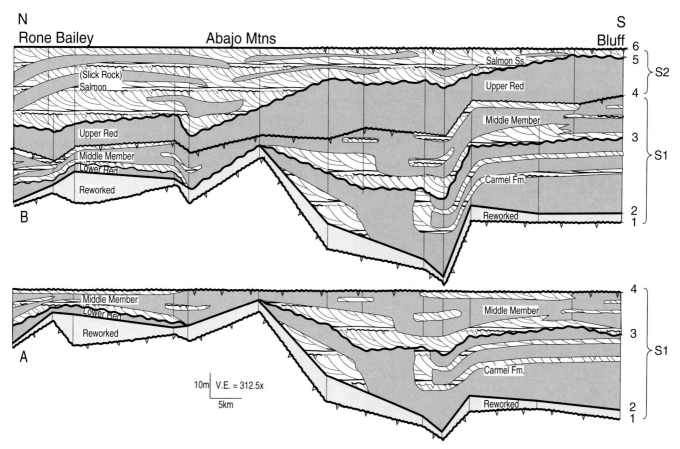

FIG. 7.—Reconstructions of depositional facies geometries for Sequences 1–2. Symbols are the same as Figure 3. Cross sections are hung from the capping super surface of each genetic package. Super surfaces are assumed to have been relatively horizontal at their time of formation, because they would have mimiced the relative water table. (A) Represents facies geometries of S1 illustrating increased accumulation south of Abajos Mountains because of increased subsidence in the Proto-Blanding basin. (B) During S2 deposition, the greatest accumulation was north of the Abajos Mountains indicating cessation of differential subsidence in the Proto-Blanding basin and increased rates north of the Abajos Mountains.

Upper Red Member are separated from the Salmon Sandstone by a loaded surface (Surface 5) that rises to the south. The rising nature of Surface 5 suggests it is time transgressive in nature. This would correspond to the suggestion that the dune fields of the San Rafael Group were better developed in the north and expanded to the south with time over the sabkha depositional system. As the initial bedforms that resulted in the Salmon Sandstone migrated across the area, they sank into the damp sabkha environment of the Upper Red Member resulting in the loading observed along Surface 5.

S2 shows an overall thickening to the north in contrast to S1, which thickens to the south into the Blanding Basin (Kelley, 1955, 1960; Fig. 7B). This change in facies geometry suggests the Blanding Basin south of the Abajos Mountains was no longer actively subsiding. The northward thickening of S2 suggests increased subsidence north of the Abajo Mountains probably as a result of renewed activity along late Paleozoic faults associated with the Nequoia-Abajo Lineament and related salt flowage within the Paradox Basin (Stevenson and Baars, 1977).

Surface 6 marks the cessation of the S2 accumulation event. The flat erosional portions of Surface 6 overlie facies that show numerous features consistent with deposition under the influence of a high water table. For these facies to be eroded, the water table must have dropped, but because Surface 6 is not polygonally fractured, erosion did not apparently reach the water table. In contrast, in the south, Surface 6 is corrugated suggesting the capillary fringe of the water table was stable at the surface for an extended period of time.

SEQUENCE 3 (S3)

Sequence Description

S3, bounded by Surfaces 6 and 8, varies in thickness along the transect from 46.7 m in the north to 28.7 m in the south. S3 is divisible into two depositional systems, however, the distribution of the basal sabkha depositional system, represented by the Lower Member of the Wanakah, is limited to the southern portions of the study area (Fig. 3). In the northern portions of the study area, the sequence is dominated by the dune cross-stratified depositional system of the Upper Slick Rock Member and Lower Moab Tongue, which interfinger to the south with the cross-stratified Bed at Butler Wash.

The Lower Member is a poorly exposed red silty sandstone (Fig. 10A). It has thin resistant sandstone lenses that are composed of cross-bedded to wavy laminae and commonly have a polygonally fractured upper surface. The Lower Member is equivalent to the basal portions of the Beclabito Member of the Wanakah Formation in the Todilto Basin of northwestern New Mexico and eastern Arizona as described by Condon and Huffman (1988). The Beclabito Member immediately overlies the Todilto Limestone, which results from the initial flooding of the Todilto Basin. The basal surface of the Todilto Limestone in the Todilto Basin is stratigraphically equivalent to Surface 6 and overlies dune cross-strata of the Entrada Sandstone (Condon, 1989).

The Lower Member, at section 29 (Figs. 3, 4), has three sharp truncation surfaces (Fig. 10A). The truncation surfaces are horizontal; however, the underlying truncated sandstones and shales are distinctly undulatory, showing large-scale soft sediment deformation. Section 29 is the only area where this deformation is evident, probably because of the poor quality of exposures immediately north of section 29 and perhaps the lack of deformation farther to the north. Surface 7, which separates the sabkha depositional system from the dune cross-stratified system, is a poorly-exposed rising surface that varies from loaded to featureless.

The dune cross-stratified depositional system changes facies laterally from thick sets of cross-strata in the northern Slick

A

B

C

A

B

FIG. 8.—Features of S1: (A) Brecciated to wavy laminae of the Lower Red Member at Rone Bailey Mesa. Lens cap for scale. (B) Compound cross-strata of the Middle Member of the Entrada Sandstone at Butler Wash 1. Person for scale. (C) Polygonally fractured surface 4, which caps S1. Polygons are approximately 2 m in diameter. Person for scale.

FIG. 9.—Features of S2: (A) Brecciated to wavy strata of the Upper Red Member at section 20. 20-cm stick for scale. (B) Interbedded compound cross-stratified sets and wavy-bedded to brecciated strata of the Salmon Sandstone at Rone Bailey Mesa. Trough in center of photo is 2.8 m thick.

Rock Member and Lower Moab Tongue to thin sets of cross-strata interbedded with wavy-bedded units in the southern Bed at Butler Wash. In the north, the doubly compound sets of cross-strata vary from 1 to 8 m in thickness and are interbedded with flat- to wavy-bedded units, from 1 to 4 m thick (Fig. 10B). The cross-stratified sets show two scales of internal cyclicity. The most prominent is a broad scoop or scallop, which results from the migration of superimposed bedforms along the lee face of the main bedform (Fig. 10). Bounding surface orientations suggest the main bedforms were migrating to the west. The smaller scale cyclicity averages about 50 cm thick and is characterized by an erosional basal contact overlain by wind-ripple laminae, which are in turn overlain by grainflow strata that are truncated by the erosional base of the next successive cycle. The foreset orientations of the smaller scale cyclicity indicate the superimposed bedforms were migrating to the northwest and southeast. These cycles represent fluctuations in paleoflow direction,

FIG. 10.—Features of S3 and S4: (A) Several units exposed at the Bluff section are indicated by letters: A = Lower Member of the Wanakah Formation, B = Bed at Butler Wash, C = Middle Member of the Wanakah Formation, D = Bed at Black Steer Knoll, E = Upper Member of the Wanakah Formation. Arrows indicate three truncation surfaces in the Lower Member of the Wanakha Formation. (B) Thick set of cross-strata in the Upper Slick Rock Member at section 12. Person for scale.

probably on an annual basis. These cycles are also recognized in outcrops of the Entrada Sandstone in northeastern Utah and are discussed in greater detail in Crabaugh and Kocurek (1993). Surface 8, which caps S3, is corrugated in the north and featureless in the south.

Interpretation of Controls

Thickening of S3 to the north suggests that subsidence rates north of the Abajos Mountains continued to be greater than areas to the south. These increased rates of subsidence may be related to reactivation of Paleozoic faults and salt flowage. However, subsidence may have been enhanced by loading and compaction of the underlying sediments by the sand sea itself.

In the south, the sandy sabkha accumulations of the Lower Member mark the initial influences of a rising water table in the Todilto Basin. However, three sharp erosional surfaces exposed near Bluff apparently represent short-term drops in the relative water-table level resulting in eolian deflation of the wet sabkha accumulations. Because these surfaces cannot be traced to the north because of poor outcrop quality, their regional significance is uncertain.

The thin sands and shales of the Lower Member interfinger and are prograded over by the cross-bedded strata of the Slick Rock Member of the Entrada Sandstone (Fig. 3). Once again, the cross-bedded strata are concentrated in the northern portions of the study area. Initial dune development occurred in the north as evidenced by the cross-stratified sets that immediately overlie Surface 6. While the main body of the Slick Rock erg migrated west, it also gradually expanded and prograded southward over the Lower Member as the Bed at Butler Wash. Therefore, Surface 7 represents another time transgressive surface similar to Surface 5.

The surface separating the Slick Rock Member from the Moab Tongue is correlated to the J–3 surface, which marks the base of the Curtis Formation (Fig. 5; Imlay, 1980; O'Sullivan and Pierce, 1983). J–3, therefore, corresponds to the flooding and reworking of the Entrada Sandstone in the northern portions of Utah by the Curtis Seaway. Previously the Curtis transgression has also been correlated to flooding of the Todilto Basin (Fig. 6C; Harshbarger et al., 1957; Ridgley, 1984; McCrary, 1985). The facies reconstruction for the study area (Fig. 3) suggests that flooding of the Todilto Basin at the base of S3 and the J–3 surface were not coincident. However, because marine incursions are time-transgressive, these events can be related to one another within the same sequence. Flooding of the Todilto Basin marks the inital rise of sea level and the J–3 surface marks the erg interior response to maximum transgression during the accumulation of S3.

Migration of the Curtis shoreline to the south may have implications for increased sediment supply to the area during southward progradation of the Moab Tongue and its equivalent the Bed at Butler Wash. The Curtis transgression was a fairly high-energy system as indicated by the mass-wasting of Entrada Sandstone dunes in northeastern Utah by a high-energy tide-dominated sea (Eschner and Kocurek, 1986). This southward-transgressing high-energy system would have brought significant amounts of sediment to the shorelines north and east of the transect, from where, under the influence of a northeastern wind, these sediments would have served as an ample sand source for the Moab Tongue.

SEQUENCE 4 (S4)

Sequence Description

Extending from Surfaces 8 to 10, this sequence is the thinnest, ranging from 21 to 13.5 m from north to south. The basal sabkha depositional system in S4 is restricted to the central and southern portions of the transect and is represented by the Middle Member of the Wanakah Formation. The Middle Member is dominated by wavy-bedded to brecciated strata, but wind-ripple and water-ripple laminae occur as well (Fig. 10A). The Middle Member is prograded over by the cross-stratified facies of the Upper Moab Tongue and the Bed at Black Steer Knoll. The lateral transition from wavy-bedded strata of the Middle Member to the thick compound sets of the Moab Tongue occurs over roughly 6 km. Approximately 50% of the Bed at Black Steer Knoll is composed of thin discontinuous simple sets of grainflow and wind-ripple laminae. The sets are separated by wavy-bedded laminae and some brecciated horizons. The surface capping the S4 is featureless to polygonally fractured in the southern areas and has been eroded and overprinted by the J–5 unconformity in the north.

Capping the Bed at Black Steer Knoll in the area is the Upper Member of the Wanakah Formation. This unit is dominated by wavy and deformed strata and probably represents the basal portion of a fifth sequence (Fig. 10A). This fifth sequence either never developed fully or was removed by erosion associated with the formation of the J–5 surface.

Interpretation of Controls

The final complete sequence in the San Rafael Group (S4) shows the same characteristic initial development of the cross-stratified depositional system in the northern portions of the transect and its progradation over the southern sabkha depositional system (Fig. 3). The sequence maintains a relatively constant thickness across the study area suggesting there was not significant differential tectonism during S4 time.

REGIONAL CORRELATION OF SEQUENCES

Previous work, by Crabaugh and Kocurek (1993) on the Entrada Sandstone near Dinosaur National Monument, has demonstrated the existence and viability of the use of super surfaces in reconstructing the facies response of wet eolian systems to basinal controls such as relative sea level and tectonism. Four of the bounding surfaces associated with the sequences in the Dinosaur National Monument area have been correlated 265 km to southeastern Utah (Fig. 11). Correlation of the J–2, J–3 and surface A at Dinosaur National Monument to Surface 3 at Rone Bailey Mesa are based on regional studies (Fig. 11; Pipiringos and O'Sullivan, 1978; O'Sullivan and Pierce, 1983; Peterson, 1986). Two other surfaces are correlated over this 265–km area based on their surface characteristics and related facies changes. The capping super surface of S1 (Surface G) in Dinosaur National Monument appears to correspond to the surface capping S1 (Surface 4) in southeastern Utah (Fig. 11). Both surfaces cap eolian cross-strata, are polygonally fractured and are immediately overlain by a red brecciated facies. The eolian cross-strata of S1 appear better developed in the Dinosaur National Monument area, perhaps because the sediment supply was once again greater to the north.

The second surface identified by this study caps the second sequence in both the northern and southern study areas (Surface K and Surface 6). These surfaces are correlated based on their stratigraphic location, the nature of the surfaces, as well as the accumulations immediately below and above. In the Dinosaur National Monument outcrops of the Entrada Sandstone, Surface K is the only surface that shows erosional relief; likewise in the south at Rone Bailey Mesa Surface 6 is also an irregular erosional surface. At both Rone Bailey Mesa and Dinosaur National Monument, the surface marks an abrupt facies change from facies heavily influenced by a high water table (breccias, soft-sediment deformation, corrugated surfaces) to thick cross-stratified sets. The thick cross-stratified sets on top of Surface 6 at Rone Bailey Mesa and Dinosaur National Monument are very similar, consisting of compound sets with two scales of cyclicity (Crabaugh and Kocurek, 1993). By contrast, in the southern most outcrops, Surface 6 becomes polygonally fractured, and unlike the northern sections it marks a transition to wetter depositional facies, corresponding to flooding of the Todilto Basin.

The upper portion of the Entrada Sandstone in Dinosaur National Monument, between Surfaces K and J–3, has been divided into two complete sequences and a third incomplete package that was truncated and subsequently onlapped by the Curtis transgression (Crabaugh and Kocurek, 1993). These individual sequences are not distinguishable in the southern outcrops and are correlated to the lower portion of S3 bounded by Surfaces 6 and J–3.

DISCUSSION

It has long been recognized that the fundamental controls of sedimentation on continental margins are eustasy, subsidence and sediment supply (Grabau, 1924; Curray, 1964; Jervey, 1988; Swift and Thorne, 1991). In particular, Schlager (1993) demonstrated that marine sequences and their systems tracts are controlled by the balance between the rate of change of accommodation space (eustasy and subsidence) and the rate of sediment supply. With the recognition of wet eolian sequences in coastal settings, use of this paradigm can now be extended to continental systems where relative sea-level fluctuations control the formation of accumulation space and less directly the type of depositional system that fills the space.

Accommodation space as defined for marine systems is the space available for potential sediment accumulation above which non-deposition or even erosion will occur. Therefore accommodation space in marine systems represents both accumulation space (as used in this paper) and preservation space because sea level also marks the baseline of erosion. In eolian systems, accumulation space and preservation space do not necessarily coincide (Kocurek and Havholm, 1993). For example, many Saharan ergs have built significant accumulations that rest on the stable Saharan basement and under present conditions will not be incorporated into the stratigraphic record (e.g., Kocurek et al., 1991b).

The repetitive nature expressed by the occurrence of five eolian accumulations and their capping super surfaces suggests that during San Rafael Group deposition the available accumulation space along the margin of this basin fluctuated through time. Sedimentary features in both the sabkha- and dune-dom-

Sierra Grande uplift, 123–130
Sinemurian, 73
sinuous channel fills, 22, 24, 58, 88, 102, 112, 130
Slick Rock Member, 214, 215, 217, 218, 219, 220–221
slope onlap surface, 49
Snorre field, 73, 74, 75, 77, 78, 79
soil, 5, 19, 23, 25, 32, 35, 36, 37, 38, 39, 40, 41, 43, 44, 72, 73, 75, 78, 90,
    91, 165, 196, 197, 202, 203
South Dakota, 133–145
South Saskatchewan River, 100, 104
Statfjord field, 73, 74, 75, 77, 78, 79, 105
Statfjord Formation, 73–79, 105
storeys, 78, 96, 97, 98, 100, 101, 102, 104
Straight Cliffs Formation, 65
stratigraphic base level, 86, 102, 104, 105, 144
stream equilibrium profile, 65–73, 74, 75, 78, 79
stream profile, 18, 19, 66, 67, 68, 70, 72
strike slip fault, 118, 141
subsidence, 17, 22, 32, 46, 51, 60, 61, 66, 71, 73, 88, 90, 102, 104, 109, 118,
    134, 139, 142, 143, 144, 186, 206, 213, 214, 219, 221, 222, 223, 224,
    225, 226
subsidence rates, 22, 49–62, 73, 102, 105, 187, 221
sulphur, 155, 156, 159, 166
Summerville Formation, 93, 216, 217
Sundance Formation, 135, 137, 140, 141, 142, 143, 145
super surface, 213, 215, 216, 219, 222, 226
suspended load, 8, 85, 87, 90
Sverdrup Basin, 21
swamp, 57, 164
Sydney Basin, 152–161, 167
systems tract, 23–25, 31, 44, 65, 72, 83, 86–88, 90, 118, 119, 151, 159, 161,
    162, 163, 167, 169, 182, 185, 187, 226

T

Taiwan, 12
Tana River, 100
tectonic subsidence, 102, 217
tectonics (also "tectonism"), 11–13, 14, 17, 18, 20–22, 23, 26, 68, 69, 73, 78,
    90, 91, 102, 104, 111, 123, 124–130, 133–145, 186, 214, 222
temperature, 4, 8, 10, 37, 44, 179, 202, 203
Teredo, 127
Tertiary, 32, 34, 46, 58, 61, 109, 110, 118, 123, 152, 161–166, 203
Texas, 19, 24, 25, 26, 31–46, 214
Thalassinoides, 115, 116, 124, 127
thermal sag, 73
Tidwell Member (also just "Tidwell"), 93, 95, 102, 103, 104
Todilto Basin, 214, 220, 221, 222
Todilto Limestone, 214, 220
topogenous mire, 161
topset deposits, 102, 112, 115, 178, 180, 181, 182, 184, 185, 187
trace fossil, 52, 126, 127, 138
Transcontinental Arch, 133, 134, 138, 139, 141, 142, 143, 144, 145

transgression, 17, 32, 34, 37, 39, 43, 44, 45, 46, 52, 59, 69, 70, 72, 74, 78, 79,
    117, 119, 124, 151–152, 153, 157, 159, 161–162, 164, 179, 186, 205, 221,
    222, 223, 225
transgressive sands, 52, 77, 93, 119
transgressive surface, 49, 52–55, 56, 117, 118, 119, 124, 151, 159, 160, 167,
    221
transgressive systems tract, 23, 50, 65, 72, 118, 119, 151, 159, 161, 162, 167,
    182, 185, 187, 226
tree stumps, 115, 164, 165
Triassic, 73, 75, 76, 77, 83, 142, 191–206
trough cross-bedding, 84, 97, 99, 102, 138
turbidite, 111, 116, 205
turbidity currents, 182, 201

U

ultimate base level, 65, 66, 79
unconformity (and unconformities), 22, 31, 35, 38, 42, 44, 49, 51–52, 55, 57,
    61, 62, 66, 70, 83, 109, 133, 136, 137, 139, 140, 143, 145, 153–154, 159,
    167, 214
United Kingdom, 73–79, 99
uplift, 11–13, 14, 17, 19, 20, 22, 25, 55, 61, 66, 73, 93, 109, 116, 118, 119,
    123–130, 133, 139, 144, 177, 187
Upper Atanikerdluk Formation, 109
Upper Cretaceous, 25, 32, 34, 83–91, 93, 101
Upper Mannville Group (include "Upper Mannville"), 51, 52, 55–62
upper-flow regime, 102
uranium, 95, 96, 140
Utah, 65, 93–105, 177–187, 213–227
Uzen field, 105

V

valley, 17, 18, 19, 20, 23, 33, 34, 35, 44, 52, 58, 109–119, 125, 151
valley fill, 17, 18, 19, 21, 23–25, 26, 33, 40–42, 43, 44, 52, 57, 58, 111–115,
    116, 117, 119, 125
Viking graben, 73, 74, 75, 76
vitrinite, 155, 156, 157, 158, 159, 161, 170
volcanics, 111, 116, 117, 118, 119, 135, 160, 161, 162

W

Wabash River, 100
Wanakah formation, 213, 214, 217, 219–220, 222
water table, 8, 59, 65, 67, 87, 123, 201, 206, 213, 215, 216, 217, 219, 221,
    222, 223, 224, 225, 226
Western Interior Seaway, 93, 142, 151
Westwater Canyon Member, 104
wet eolian system, 213–227
Wind River Basin, 134, 135, 137, 140, 141
Wittingham Coal Measures, 152, 167
wood, 126, 127, 137, 142, 164, 165
Wynn Seam, 152–161, 168, 169, 170
Wyoming, 116, 133–145, 214